SCANNING PROBE MICROSCOPES

Applications in Science and Technology

K.S. Birdi

CRC PRESS

Boca Raton London New York Washington, D.C.

Library of Congress Cataloging-in-Publication Data

Birdi, K. S., 1934–
Scanning probe microscopes : applications in science and technology / K.S. Birdi.
p. cm.
Includes bibliographical references and index.
ISBN 0-8493-0930-1
1. Scanning probe microscopy. 2. Science. 3. Technology. I. Title.

QH212.S33 B57 2003
502′.8′2—dc21 2002038821

Direct all inquiries to CRC Press LLC, 2000 N.W. Corporate Blvd., Boca Raton, Florida 33431.

International Standard Book Number 0-8493-0930-1
Library of Congress Card Number 2002038821
Printed in the United States of America 1 2 3 4 5 6 7 8 9 0
Printed on acid-free paper

Dedication

to

Lilian, Leon, and Esma

Preface

Mankind has always been keen about being able to see through the microscope, in order to understand all kinds of natural phenomena. The degree of resolution of microscopes has indeed increased over the decades from a few hundred to almost a million times. The latter techniques like the electron microscope allowed one to be able to see with molecular resolution. However, only two decades ago a new technique was invented, the scanning probe microscope (SPM), which revolutionized the whole microscopy application area. Actually the first type was based on scanning tunneling microscopy (STM), which resulted in the award of Nobel prize. Some years later atomic force microscope (AFM) was added to these SPMs. SPMs thus allowed one to see and analyze molecules under ambient laboratory conditions. Later, one could even investigate under fluids (later under high pressure or vacuum). The three-dimensional (digital) images could be analyzed by digital procedures, thus enhancing the analyses. Under dynamic conditions one can see live molecular details, such as gas adsorption.

The aim of this book is to guide the reader through the vast developments of SPMs which have taken place over the last decades. STM and AFM both have been found to provide useful information at molecular level of all kinds of molecules (small molecules [inorganic or organic compounds or lipid-like substances]; macromolecules [biopolymers; cells; viruses]). Besides STM and AFM, recently new dimensions have been added. This latter development has introduced a new term in the industrial revolution, the so-called *nanotechnology.* Nanotechnology, is called the science and technology of precisely controlling the structure of matter at the molecular level. Without any doubt, this is widely viewed as the most significant technological frontier currently being explored. Materials and devices at the nanoscale (a nanometer is one billionth of one meter: or roughly a thousandth of the thickness of this sheet of paper in the book) hold vast promise for innovation in virtually every industry and public endeavor including health, electronics, transportation, the environment and national security, and have been heralded as "the next industrial revolution."

The book starts with an introduction to the development of SPMs. The basics of STM and AFM are described. The other SPMs (friction force microscope (FFM), SNOM) are also described. The different apparatus are described and the method of calibration is delineated besides other parameters, along with extensive references and experimental details.

The rest of the book is divided into chapters related to different kinds of molecular species and systems from real life. The lipid-like molecules are described under a separate chapter. This includes the vast area of research which is going on about the self-assembly monolayers (SAMs). The contribution of SPMs to the understanding of SAMs has been immense, since this has provided information in three-dimensions, for the first time in literature. The subject of SAMs is becoming a vast application area in both industry (micro-electronics; computer chips) and biological (vesicles; sensors) applications.

The SPM's application to the understanding of macromolecules is described under a separate chapter. The very first image of DNA was actually obtained by using SPM. Later, other biological molecules were investigated (both by STM and AFM), and are described in detail. The experimental procedures are extensively described. SPMs also provide dynamics of various systems, such as the rates of reactions or adsorption on surfaces. SPMs also are found to provide information of reactions which take place in nanoreactors.

The nanotechnoloigical developments are very extensive and are described accordingly in much detail. These subjects include nanolithography and other applications. The developments around friction force microscopy (FFM) are described in detail.

The main theme in the book has been to provide systematic and in-depth experimental details covering the various aspects of SPM applications in science and technology (nanotechnology). Nanotechnology has been given high priority support from all the national science foundations worldwide. The reader can thus easily determine the experimental conditions and thus follow these with a vast number of pertinent references. This is an attempt to help the reader in designing his own experiments to almost all kinds of applications of these SPMs.

The aim of this book thus has been to provide basic and advanced information, hitherto not easily available in one volume. The chapters are arranged in such a way that it can be used as a textbook about SPMs. Further, the extensive information provided is also useful to advanced researchers in this field. The book is even more useful for a beginner, since the detailed data and description of various applications will guide one through the extensive literature covered.

Additionally, there are presented data in the form of two- and three-dimensional figures, wherever these are pertinent. This is intended to provide an impressive image gallery to the reader, which should present a clear view of the application posibilities of SPMs to science and technology.

I would like to thank all the past students and research associates (especially D.T. Vu, Danish Technical University, Lyngby) who have contributed in many ways in the research on SPMs over the past decades as reported in this book. It is a pleasure to thank DME, Herlev, Denmark, for assisting in the operation of SPMs. I want to thank the staff at CRC who have helped in different ways to make the publication as smooth as possible.

The Author

Professor K.S. Birdi received his undergraduate education (B.Sc. Hons. Chem.) from Delhi University; Delhi, India, in 1952. He also majored in chemistry at the University of California at Berkeley. After graduation in 1957, he joined Standard Oil of California, Richmond.

In 1959, Dr. Birdi became chief chemist at Lever Brothers in Denmark. He became interested in surface and colloid chemistry and joined the Institute of Physical Chemistry as an assistant professor. He initially did research on surface science aspects (e.g., thermodynamics of surfaces, detergents, micelle formation, adsorption, Langmuir monolayers, biophysics). During the early exploration and discovery stages of oil and gas in the North Sea, Dr. Birdi became involved in Danish Research Science Foundation programs, with other research institutes around Copenhagen, in the oil recovery phenomena and surface science. Later, research grants on the same subject were awarded from European Union projects. These projects involved extensive visits to other universities and an exchange of guests from all over the world. Professor Birdi was appointed research professor in 1985 (Nordic Science Foundation), and was then appointed, in 1990, to the School of Pharmacy, Copenhagen, as professor in physical chemistry. Since 1999, Professor Birdi has been actively engaged in consultancy to both industrial and university projects.

Professor Birdi is a consultant to various national and international industries. He is and has been a member of various chemical societies, and a member of organizing committees of national and international meetings related to surface science. He has been a member of selection committees for assistant professor and professor, and was an advisory member (1985 to 1987) of the ACS journal *Langmuir.*

Professor Birdi has been an advisor for some 90 advanced student projects and various Ph.D. projects. He is the author of some 100 papers and articles (and a few hundred citations).

To describe these research observations and data he realized that it was essential to write books on the subject of surface and colloid chemistry. His first book on surface science was published in 1984: *Adsorption and the Gibbs Surface Excess,* with Chattorraj, D.K., Plenum Press, New York. This book remains the only one of its kind in the present decade. Further publications include *Lipid and Biopolymer Monolayers at Liquid Interfaces,* Plenum Press, New York, 1989; *Fractals — In Chemistry, Geochemistry and Biophysics,* Plenum Press, New York, 1994; *Handbook of Surface and Colloid Chemistry,* CRC Press, Boca Raton, FL, 1997 (CD-ROM, 1999), 2nd edition, 2002; and *Self-Assembly Monolayer,* Plenum Press, New York, 1999. Surface and colloid chemistry has remained his major research interest throughout these years.

List of Abbreviations

AES Auger electron spectroscopy
AFM Atomic force microscope/microscopy
ASF Atomic sensitivity factor
ATR Attenuated total relection
BLM Bilipid membranes
BM Bridging model
CD Corona discharge
CE Contact electrification
CFM Chemical force microscopy
CP AFM Conducting probe atomic force microscopy
CV Cyclic voltammogram
DFM Dipping force microscope
DLVO Derjaguin–Laudau–Verwey–Overbeek
DSIMS Dynamic secondary ion mass spectrometry
EFTEM Energy-filtered analytical transmission electron microscopy
EM Electron microscope/microscopy
EQCM Electrochemical quartz crystal microbalance
FEG SEM Field-emission gun scanning electron microscopy
FESEM Field-emission scanning electron microscopy
FFM Friction force microscope
FFT Fast Fourier transform
FM Frequency modulation
FTIR Fourier transform infrared spectroscopy
HOPG Highly oriented pyrolytic graphite
HREM High-resolution electron microscopy
HRTEM High-resolution transmission electron microscopy
ICFM Inverted chemical force microscopy
IEP Isoelectric point
JKR Johnson–Kendall–Robert
LB Langmuir–Blodgett
LEED Low-energy electron diffraction
LFM Lateral force microscope
LRDS Laser reflection detection system
MFM Magnetic force microscope
MPD Mean patch diameters
MS Mass spectrometry
NEMS Nanoelectromechanical systems
NMR Nuclear magnetic resonance
OTS Octadecyl trichlorosilane

PBC Probe beam deflection
PCM Patch charge model
PCS Photon correlation spectroscopy
PMMA Poly(methylmethacrylate)
PSD Position sensitive detector
PV Phase–volume (adj.)
PZC Point-of-zero charge
QCM Quartz crystal microbalance
QELS Quasielectric light scattering
RHEED Reverse high energy electron diffraction
RIE Reactive ion etching
SAM Self-assembly monolayer
SCM Scanning capacitance microscopy
SEM Scanning electron microscope
SEMPA Scanning electron microscope with polarization
SEPM Scanning electric potential microscopy
SFA Surface force apparatus
SFM Surface force microscope
ShFM Shear force microscope
SIMS Secondary ion mass spectrometry
SNOM Scanning near-field optical microscopy
SNOM–AFM Scanning near-field optical microscopy and atomic force microscopy
SPM Scanning probe microscope
SPR Surface plasmon resonance spectroscopy
STM Scanning tunneling microscope
STNR Signal-to-noise ratio
STXM Scanning transmission x-ray microscopy
SUV Small unilamellar vesicles
TDFM Transverse dynamic force microscopy
TEM Transmission electron microscope
TM AFM Tapping-mode atomic force microscopy
UHV Ultrahigh vacuum
XPS X-ray photoelectron spectroscopy

Contents

1 Introduction

1.1 BACKGROUND

At the end of the 20th century, we found a big surge in the development of important techniques available for science (nanoscience) and technology [self-assembly structures (vesicles); biomolecules; biosensors; surface and colloid chemistry; nanotechnology]. In fact, these developments indicated that there was no end to this trend, as regards the vast expansion in the sensitivity and level of information; therefore, in this chapter, we would like to explain some of these exciting new developments to the reader.

Typical of all humans, seeing is believing, so the microscope attracted much interest for many decades. All of these inventions, of course, were basically initiated on the principles laid out by the telescope (as invented by Galileo) and the light-optical microscope (as invented by Hooke).[1] Over the years, the magnification and the resolution of microscopes improved. However, for the man to understand the nature, the main aim of mankind has been to be able to see atoms or molecules. This goal has been achieved, and the subject as described here will explain the latest developments which were invented only approximately 15 years ago.

The ultimate aim of scientists has always been to be able to see molecules while active. In order to achieve this goal, the microscope should be able to operate under ambient conditions. Further, all kinds of molecular interactions between a solid and its environment (gas or liquid or solid), initially, can take place only through the surface molecules of the interface. It is obvious that when a solid or liquid interacts with another phase, knowledge of the molecular structures at these interfaces is of interest. The term "surface" is generally used in the context of gas–liquid or gas–solid phase boundaries, while the term "interface" is used for liquid–liquid or liquid–solid phases. Furthermore, many fundamental properties of surfaces are characterized by morphology scales of the order of 1 to 20 nm [1 nm = 10^{-9} m = 10 Å (Å = Angstrom = 10^{-8} cm).

Generally, the basic issues that should be addressed for these different interfaces are as follows:

1. What do the molecules of a solid surface look like, and how are the characteristics of these different than the bulk molecules? In the case of crystals, what about the kinks and dislocations?
2. Adsorption on solid surfaces requires the same information about the structure of the adsorbates and the adsorption site and configurations.

3. Solid–adsorbate interaction energy is also required, as known from the Hamaker theory.[2,3]
4. Molecular recognition in biological systems (active sites on the surfaces of macromolecule, antibody and antigen) and biological sensors (enzyme activity, biosensors) should be addressed.
5. Self-assembly structures should be found at interfaces.
6. Semiconductors should be addressed.

Depending on the sensitivity and experimental conditions, the methods of molecular microscopy are many and varied.[4] The applications of these microscopes is also varied and extensive.

For example, information about crystal structures and three-dimensional configurations of macromolecules have been obtained.

The most popular application of microscopy is the study of molecules at surfaces. Generally, the study of surfaces is dependent on understanding not only the reactivity of the surface but also the underlying structures that determine reactivity. Understanding the effects of different morphologies may lead to a process for enhancement of a given morphology and, hence, to improved reaction selectivities and product yields.

Atoms or molecules at the surface of a solid have fewer neighbors as compared with atoms in the bulk phase, which is analogous to the liquid surface; therefore, surface atoms are characterized by an unsaturated, bond-forming capability and, accordingly, are quite reactive. Until a decade ago, electron microscopy and some other similarly sensitive methods provided some information about the interfaces. Although, there were always limitations inherent in all of these techniques, which prompted needed improvements.

A decade ago, the best electron microscope images of globular proteins were virtually little more than shapeless blobs. However, these days, due to relentless technical advances, electron crystallography is capable of producing images at resolutions close to those attained by x-ray crystallography or multidimensional nuclear magnetic resonance (NMR) imaging. In order to improve upon some of the limitations of the electron microscope, newer methods were needed. A decade ago, a new procedure for molecular microscopy was invented and will be delineated herein. The new scanning probe microscopes not only provide new kinds of information than that known from x-ray diffraction, for example, but also open a new area of research (for example, nanoscience and nanotechnology).

The basic method of these scanning probe microscopes (SPMs) was essentially to be able to move a tip over the substrate surface with a sensor (probe) with molecular sensitivity (nm) in the longitudinal and height directions (Figure 1.1). This may be compared with the act of sensing with a finger over a surface, or more akin to the old-fashioned record player with a metallic needle (a probe for converting mechanical vibrations to musical sound) on a vinyl record.

The sensor movement was controlled under a highly sensitive feedback system, which when coupled to a variety of signals, could provide atomic surface details.

This gave almost a new dimension "in the nanometer (nm = 10^{-9} m = 10 Å) range" to surface chemical research, which was much needed at this stage. Due to

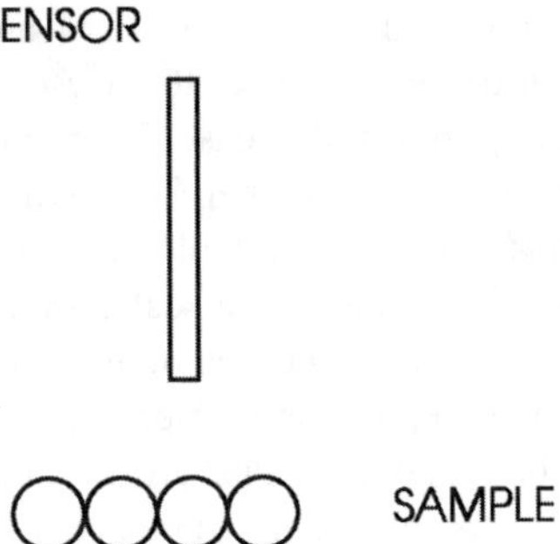

FIGURE 1.1 A schematic drawing of the sensor (tip, cantilever, optical, magnetic device) movement over a substrate in *x*, *y*, *z* direction with nanometer sensitivity (controlled by piezomotor) (at solid–gas or solid–liquid interface).

these developments, the last decade has experienced a new aspect of science, the so-called nanotechnology. Originally, two types of microscopes were invented: scanning tunneling microscopes (STM) and atomic force microscopes (AFM). The AFM later developed into a large variety of so-called scanning force microscopes. Additionally, other types of microscopes are being developed, such as the friction force microscope (FFM), etc. AFM thus provides much more useful information than the two-dimensional images of a surface produced using an SEM. AFM can be used for thin films, polymer coatings, and single-crystal substrates. The aim of this book is to describe all of these developments extensively. This approach is one of a kind in the current literature.

The scanning tunneling microscope was invented by Binnig and Rohrer in 1982 (for which they were awarded the Nobel prize in 1986).[5–9] Later, the atomic force microscope was developed based on the principles of the STM but with resolving surface structures for nonconducting and conducting materials. The advent of these scanning microscopes revolutionized this field. In recent years, other modifications of original STM and AFM have been designed, which are also described herein.

STM and AFM proved to be powerful tools for obtaining information on the packing order of molecular adsorption on a surface.[10–13] Data obtained from STM images can be useful in providing information on the relative importance of molecule–molecule and molecule–substrate interactions, as well as the types of forces responsible for the packing order at the surface. This is useful in such applications as epitaxial growth of thin films, chromatography, lubrication, and microelectronics fabrication, each of which involves interactions between molecules on a surface and can be investigated by using these procedures. The complementary properties of STM and AFM make it attractive to apply these methods to organic surfaces and interfaces. Some typical examples can be as follows:

- Ultra-thin adsorbed layers
- Charge transfer complexes
- Macromolecules and polymers
- Biological matter

High resolution is an important feature of STM and AFM, but the ability to provide different, original information make them useful also at lower magnifications. A fast scan range is required in this case. Instruments able to scan 200 μm × 200 μm × 12 μm are now becoming commercially available.

As regards biological material analysis [biological cells; biopolymers (DNA, proteins, enzymes, lipid assemblies)] of molecular scale, prior to STM and AFM methods, specimens were generally analyzed by the scanning electron microscope (SEM).[14,15]

The main objectives of the preparation methods have been based upon stabilizing the specimen, preventing shrinkage and other artifacts during dehydration, and rendering the sample electrically conductive. In some cases, water was replaced by other fluids, such as glycerol or triethylene glycol, with the possibility of artifacts. These requirements were not necessary in the case of STM and AFM methods. The samples could be analyzed under ambient conditions without additional drying or treatment. Radiation damage is the main problem that prevents determination of the structure of a single biological macromolecule at atomic resolution using any kind of microscopy. This is true whether neutrons, electrons, or x-rays are used. Furthermore, the electron microscope was first applied to the study of monomolecular films, and it was shown that monolayers could be investigated after depositing the films on glass slides (as Langmuir–Blodgett films) and shadowing with evaporated metallic films. The techniques have been refined and extended considerably by later investigators.[3] In order to examine a sample in the electron microscope, it must be supported in some appropriate way, and a thin evaporated carbon or plastic film is used. A thin layer of organic matter on such films gives insufficient contrast for its detection. In order to achieve contrast, heavy atoms may be incorporated in the sample, as a stain or by coating the sample with them to give a silhouette for the specimen. The staining of the biological specimens, for example, is typically carried out by treating them with OsO_4 or $KMNO_4$ or uranyl salt solutions. More detailed information about the organization of the monomolecular film layers has been obtained by the shadowing method. In this procedure, the monolayer is deposited on a suitable solid support, and a film of Cr or Pt metal is evaporated in a vacuum chamber. It is these shadows that give the monolayer structure. Whereas the technique for examining the monolayer structures by these methods seems to be straightforward for the experienced investigator, the interpretation requires extreme caution.[3] The main criticism that may be raised is from the fact that the monolayers originally present on the surface of water are deposited on a metal solid surface. This obviously is not the same state as compared to the liquid (water) surface; therefore, on a molecular level, there may or may not be interactions that would give rise to different results. However, the sensitivities of the STM and AFM are high enough to resolve these problems, as described in the following.

At this stage in the literature, it can be determined that STM and AFM can operate under fluids,[16] which is technically impossible by electron microscope. This means that, for the first time in history, molecular dimensional analyses of surfaces and molecules situated at surfaces can be carried out in a liquid. The latter invention, the most important discovery in surface science instrumentation, allows one to see molecules in fluids, hitherto impossible with any electron microscope. Furthermore, based on these basic principles, a variety of scanning probe microscopes are now being developed.

TABLE 1.1
Areas of Application for the SPMs (STM and AFM)[3]

Areas of Application
Lipid monolayers (as Langmuir–Blodgett films)
Different layered substances on solids
Self-assembly structures at interfaces
Solid surfaces
Langmuir–Blodgett films
Thin-film technology
Interactions at surfaces of ion beams and laser damage
Nanoetching and lithography
Nanotechnology (diverse applications: nanomachining)
Semiconductors
Atomic switches
Mineral surface morphology
Metal surfaces (roughness)
Microfabrication techniques
Optical and compact discs
Ceramic surface structures
Catalyses
Single-molecule studies
Surface adsorption (metals, minerals)
Surface manipulation by STM or AFM
Synthetic polymers
Biopolymers (peptides, proteins, DNA, cells, virus)
Vaccines
Biosensors

This is the most dramatic development, because surface features from interatomic spacing to fractions of a millimeter can be studied with the same instrument. The various systems analyzed by STM and AFM are many and varied (Table 1.1). Additionally, these so-called surface force microscopes allow the possibility of measuring the interfacial forces (at nm distances). In later years, other variations of STM and AFM were developed, where other detection probes have been utilized [such as light (UV, VIS, IR, fluorescence), current, pressure, etc.]. This development also vividly indicates that there will be expected intensive inventions in the future about STM and AFM. In addition, by utilizing the nanometer (nm) manipulation, one has applied STM and AFM for manufacturing nano-sized objects or surfaces. The future developments will, of course, mean that STM or AFM will be combined with other sensors, and such combinations will add much more to these nanomicroscopic developments. The application of SPMs to nanoscale microengineering is one of the most recent new inventions.

1.2 HISTORY OF MICROSCOPY

Based on the concept of "seeing is believing," humans have always been keenly interested in the science of microscopy. Before describing the details about SPMs

(STM and AFM), it is useful to describe the historical background of the microscope. The development of microscopy and human interest in it, needs a remark about its history. Probably no other scientific techniques have contributed so much to scientific development in biology, medicine, and material science, as the different microscopy techniques. Even if they might seem totally different in construction and performance, the different microscopy techniques have one thing in common — they magnify our ability to "see" small features that we otherwise would not be able to see with our naked eyes. To this, some microscopic techniques can also contribute other information, such as spectroscopic data and crystallographic properties of material. In this book, one of the latest developed microscopy techniques, scanning force microscopy, is described. Only less than a decade after its invention, it is hard to evaluate the scientific importance of this technique, as it is still undergoing rapid progress. Refinement of microscopy techniques has always been an interplay between scientists' urging for better visualization possibilities and scientific engineers. Scanning force microscopy development is no exception. In the following, a short description about the different microscopy techniques and a brief historical review of microscopy will be presented.

1.2.1 Optical Microscopy

The first optical microscope that consisted of more than one lens, a so-called compound microscope, was most likely built in Italy around 1590 by Giambattista della Porta and in Holland around 1595 by the eyeglass maker Hans Jensen.[17] The true scientific developer of the optical microscope (as well as the telescope) around 1610 was Galilei Galileo, who, in turn, called it "a Dutch invention" and Johannes Kepler, who further developed the technique shortly after that time. The problems of chromatic and spherical aberration were discovered early, and in 1637, Rene Descartes explained how to avoid spherical aberration by using hyperbolic lenses, which were impossible to manufacture in that time, and in 1671, Isaac Newton explained the reason for chromatic aberration. The technique to assemble convex and concave lenses in order to reduce chromatic aberration was invented by Chester Moor Hall in 1733 and was commercialized by John Dollond 25 years later.

Today, modern optical microscopes have been optimized to near perfection, so the fundamental physical laws set the limits of the resolution in these microscopes. The diffraction limit for maximal resolution is set by the wavelength of the light to be around 100 nm in extreme UV light but is normally not better than 250 nm (in immersion oil). Frits Zemike was awarded the Nobel Prize in Physics in 1953, "for his demonstration of the phase-contrast method, especially for his invention of the phase-contrast microscope."

An optical microscope can be described as follows:

Optical microscope
- Light source
- Optics
- Magnification
- (Maximum 1000×)

1.2.2 Electron Microscopy

The first electron microscope was built in 1931.[17] As electrons are particles that can be accelerated, focused, and detected, these particles have given a shorter "wavelength" than visible light and illuminate the specimen, giving higher resolutions than in the optical microscope. The transmission electron microscope (TEM) is conceptually similar to the transmission optical microscope, in the arrangement of specimen, "light source," and image plane, with the difference being that here, electrons instead of light are used, and electrostatic and magnetic lenses replace glass lenses.[17] The detection here is done with a fluorescent screen. Another important type of electron microscope is the scanning electron microscope (SEM),[18] which can almost be seen as a link between the TEM and the SPM. Here, the electron beam is focused onto a small spot and scanned in parallel linear scans (raster) over the surface of the specimen, and the electrons that are going back out from the surface are detected simultaneously with an electron detector. The principles of the SEM technique were first explained in 1938. Today, the resolution of the SEM is set by the quality of the lenses, which also involves the diffraction limit, as the opening angle must be kept small, and the interaction process, when electrons bit the specimen.[18] Ernst Ruska was awarded the Nobel Prize in Physics in 1986 "for his fundamental work in electron optics."

Until two decades ago, the electron microscope method and some other similarly sensitive methods provided some of this information about the interfaces, although there were always some limitations inherent in all of these techniques, which needed improvements.[17]

However, these days, due to relentless technical advance, electron crystallography is capable of producing images at resolutions close to those attained by x-ray crystallography or multidimensional NMR. A little more than a decade ago, a new procedure for molecular microscopy was invented, and that will be delineated herein.

The basic method was merely, in principle, to be able to move a sharp tip over a solid surface with a *sensor* with molecular scale sensitivity in the longitudinal and height directions. The idea to measure the surface topography using a sharp conducting tip was already described in 1972.[19] This gave almost a new dimension [in the nanometer (nm = 10^{-9} m = 0.1 Å) range] to surface chemical research, which was much needed at this stage. All of these procedures can be performed at ambient temperature and pressure, unlike with electron microscopy. All of the images (digital) thus obtained are presented three-dimensionally, and in some SPMs, one can even manipulate molecules.

The characteristics of the nanosensor have been found to be of a variety of kinds. In fact, due to these developments, the last decade has experienced a new aspect of science, so-called *nanotechnology*.[3,20] Originally, two types of microscopes were invented: the scanning tunneling microscope (STM) and later, the atomic force microscope (AFM).

The areas of applications of SPMs are vast and expanding rapidly, as will be described in this book extensively (Table 1.1). These developments clearly indicate how the SPMs are contributing to the scientific developments, where the main aim is to be able to see at a molecular level (in three dimensions) and to be able to manipulate.

2 Scanning Probe Microscopes (SPMs)

There are basically two kinds of SPMs which were originally invented, and these will be described here in detail. However, recently, various other kinds of SPMs of related scanning microscopes have been developed, as also described in the following. The reason for this is that STM and AFM have developed to such an extensive state that the literature reports are immense. On the other hand, the new kinds of SPMs are of recent origin and are currently under development. The reader is given ample literature references with discussions to guide through these SPMs. In the following, STM and AFM apparatus will be described in detail, with applications and extensive evaluation of the literature results. However, in the rest of the book, the more specific details will be delineated wherever necessary, for STM and AFM as well as for other SPMs. The aim of this book is to explain the function and applications of SPMs, in such a way that the reader can easily use these microscopes at all levels of application.

2.1 SCANNING TUNNELING MICROSCOPE (STM)

2.1.1 STM Apparatus

STM and other atomic resolution microscopes, based on SPM principles, are practical realizations of what was formerly considered thought experiments, which is to be able to image on the scale of the atomic dimension (10^{-8} cm = Å = 10^{-9} m) under ambient conditions and in three dimensions. Furthermore, these new microscopes could operate under a variety of experimental conditions as close to natural phenomenon as desired. STM showed up about 1981 in the laboratory, and it has since distinguished itself as a very important instrument for microscopy of surfaces.

STM can image the surface of conducting materials with atomic-scale detail. As with other microscopes, to use the STM to extend microscopy, we conventionally record an image that is a map of the trajectory of a probe tip over a surface, where the height of the probe tip is constantly adjusted to maintain a constant tunneling current between the tip and the surface.

It has proved to be of high value in research on solid states of matter, and scientists can use it to detect individual surface atoms and even manipulate the atomic landscapes of conductive hard surfaces. These latter procedures are just beginning to materialize and will be expected to lead to a whole new area of science based solely on nanoscience and nanotechnology (nanomachines; nanotools; nanosensors).

In literature, scientists have debated the theoretical definition of resolution and the practical ways of measuring its predicted values for centuries.[10] The object points that one sees with such microscopes as STM and AFM are in the range of atomic dimensions, i.e., nm = 10^{-9} m = 10 Å. The magnification is around 10 million times. Therefore, what one actually can see will depend largely on the smallest and closest objects that can be seen in the image. In simple terms, the resolution of an image formed in any optical instrument is defined by the smallest distinguishable distance between two closely spaced features in the sample. In the case of any microscope, the resolution of two objects at a distance equal to the centers of their diffraction centers[10] will depend on the degree of overlap of the intensity distribution peaks.

The scanning tunneling microscope (STM) was invented by Binnig and Rohrer in 1982 (for which they were awarded the Nobel prize in 1986).[4–8] Later, the atomic force microscope (AFM) was developed, based on the principles of the STM but with resolving surface structures for nonconducting and conducting materials. These microscopes have since become real revolution in the field of research on surface and colloid science.

STM and AFM have proved to be powerful tools for obtaining information on the packing order of molecular adsorption on a surface.[9–12,21]

Data obtained from STM images can be useful in providing information on the relative importance of molecule–molecule and molecule–substrate interactions as well as the types of forces responsible for the packing order in the surface. This is useful in such applications as epitaxial growth of thin films, chromatography, lubrication, and microelectronics fabrication, each of which involves interactions between molecules on a surface, which can be investigated in this procedure. In STM, the tunneling current is strongly dependent on the chemical and mechanical properties of the tip. As regards to the biological material analyses, prior to STM and AFM methods, the specimens were generally analyzed by scanning electron microscope (SEM).[15] While STM has become a widely used tool for obtaining atomically resolved images of surfaces under a variety of experimental conditions, there is still a great deal of debate regarding the mechanisms. In particular, under debate is the determination of the nature of the image contrast observed for electrically insulating adsorbates, such as substituted long-chain hydrocarbons imaged at the liquid–solid interface.

The main objectives of the preparation methods have been based upon stabilizing the specimen, preventing shrinkage and other artifacts during the dehydration, and rendering the sample electrically conductive. In some cases, water was replaced by other fluids, such as glycerol or triethylene glycol, with the possibility of artifacts. These requirements were not necessary in the case of STM and AFM methods. The samples could be analyzed under ambient conditions without additional drying or treatment.

This means that it is for the first time in history that molecular dimensional analyses of surfaces and molecules situated at surfaces can be carried out under liquid. The latter invention is the most important discovery in surface science instrumentation, where one can see images of molecules under fluids, hitherto impossible

TABLE 2.1
Different Materials Investigated by STM Under Varying Conditions

Air	Vacuum	Under Fluids
Gold	Silicon	Platinum
Platinum	GaAs	Nickel
Silver	HgCdTe	Graphite
Rhodium	Aluminium	Gold
Graphite	Copper	Stainless steels
Nickel		
Iron		
PbS		
$PbO_2Co_3O_4$		

with any electron microscope. Furthermore, based on these basic principles, now a variety of scanning probe microscopes is being developed.

Actually, while the monolayer technique as invented by Langmuir provided information on the molecular interactions at the air–water interface, STM and AFM provided the molecular orientation information on solid surfaces.[20] The combination of these two techniques is the most important useful source of information, as currently evident from a vast number of publications in relevant scientific journals.

This is the most dramatic development, whereby surface features can be studied from interatomic spacing to a fraction of a millimeter with the same instrument. The various systems analyzed by STM and AFM are many and varied (Table 2.1). Additionally, these so-called surface force microscopes (SFM) allow the possibility of measuring the interfacial forces (at nm distances). In Table 2.1 are various materials that can be investigated under varying conditions by STM.

2.1.2 Description of STM

The principle of STM, in simple terms, can be described as follows. Two conducting electrodes separated by some isolator (air or liquid) form a barrier for the electrons inside the electrodes. If the barrier is thin enough, electrons can pass through it by a quantum-mechanical process called *tunneling*. In an STM, the barrier is the gap of 1 nm. The operation of STM is based on the principle that a probe (made of W or Pt-Ir alloy) scans the surface of a sample with the help of a piezoelectric device (Figure 2.1) at a distance of 5 to 10 Å (0.5 to 1 nm).[3] At very small distances of separation, the wave functions of molecules at the surfaces of the sample and the tip overlap. As described below, the gap may be air (gas) or fluid. In fact, recently, it was mentioned that STM could lead to new materials being made by arranging atoms.[22] The absolute conductance of an STM junction, with a given geometry and tip–sample separation, depends strongly on the potential felt by a tunneling electron. A major contribution to this potential depends on the extent to which the tunnel junction is polarized by the electron's field. The great increase in developments

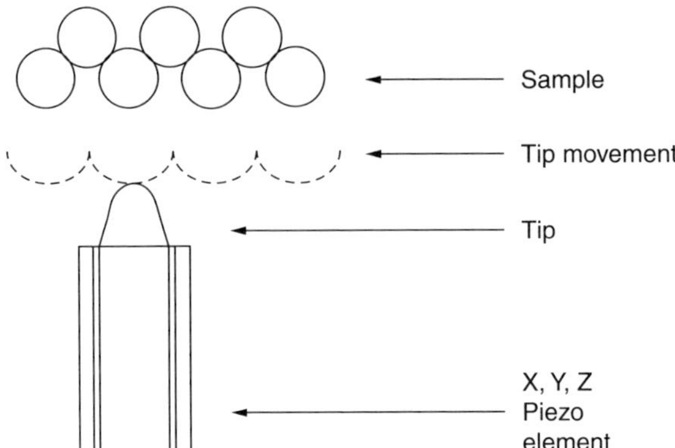

FIGURE 2.1 A typical STM apparatus: STM tip movement and setup in a commercially available instrument of piezomotor control over a solid surface (gas–solid or liquid–solid).

around the basic STM to quantum-mechanical phenomena is suggestive of many future applications to come from these scanning probe instruments.

If a bias voltage is applied to the sample, an electron tunneling current flows between the two phases. The *tunneling current* between the conductive sample and the tip is measured as a function of the distance traveled in the X and Y directions. The direction of the electron flow depends on the sign of the bias applied to the sample. Measurement of the current is sensitive of the order of magnitude necessary to resolve atomic or molecular corrugations on the surface of the sample. If electrons tunnel from a sharp metallic tip to a flat metallic electrode or similar substrate, two different situations are observed.[23] In the case of large tip and substrate separation [much greater than the de Broglie wavelength, $\lambda_B = h/(2 \text{ m e U})^{1/2} = 1.2/(U)^{1/2}$, in nm, where h is the Plancks constant, and U is the applied voltage (volts)], electrons can only penetrate the gap when high electric voltage is applied, which is the case in field emission. On the other hand, when the distance between the tip and substrate is approximately λ_B, as is the case in STM, electrons can tunnel directly across the gap when a small bias voltage is applied. This allows one to obtain information about the surface at the atomic resolution. The interaction between the probe tip and the sample varies exponentially with the distance between the sample and the most extreme tip on the probe. Thus, in the case of STM, the atom at the tip of the probe allows the tunneling current to pass through. In a recent study, the energy exchange processes occurring in tunneling microscopy were described.[23]

The resolution in the image, which is only a surface or subsurface image, is defined in the vertical direction and laterally on the plane of the sample. The magnitude of resolution can be achieved in the range of fraction of 1 Å (10^{-10} m = 0.1 nm). Obviously, the size and shape of the probe tip will determine the lateral resolution. The tip moves at ca. 10 Å/V applied as to a translator. Typically, a lead zirconate–lead titanate (PZT) polycrystalline ceramic is used. This is a polycrystalline ceramic material that has strong piezoelectric properties. It is manufactured with various compositions of the ingredients that give the materials different characteristics. The sample is mounted on a scanner that enables movement in all three dimensions, x, y, and z, with a resolution better than 1 Å and with a maximum scan length of the order of a few hundred μm. The movements are controlled by voltage applied over the different segments of the piezoelectric tube. The PZT exhibits some degree of nonlinearity, hysteresis, and creep. Some of these defects can be corrected by controlling the charge applied to the electrode. It can, however, be almost completely corrected by software that uses calibration samples, as will be described later.

STM is based on the principle that a probe (made of W or Pt-Ir alloy) scans the surface of a sample with the help of a piezoelectric device[17] (Figure 2.2) at a distance of 5 to 10 Å (0.5 to 1 nm). At small distances of separation, the wave functions of molecules at the surface of the sample and the tip overlap. If a bias voltage is applied to the sample, an electron tunneling current flows between the two phases. The tunneling current between the conductive sample and the tip is measured as a function of the distance traveled in the x and y directions. The direction of the electron flow depends on the sign of the bias applied to the sample. The absolute conductance of an STM junction, with a given geometry and tip–sample separation, depends strongly on the potential felt by a tunneling electron. Measurement of the current is sensitive to the order of magnitude to resolve atomic or molecular corrugations on the surface of the sample. In a recent study, field emissions in STM resonances have been investigated.[24]

A typical STM image of graphite (highly oriented pyrolytic graphite, HOPG) is given Figure 2.2a.[3] First, observe the molecular resolution achieved by STM. This is achieved under ordinary laboratory conditions and without any treatment of the substrate. Second, the three-dimensional capability is the most useful aspect of this method (Figure 2.2b). The digital data (three-dimensional) can be processed by software available from all the commercial STM or AFM microscopes. The images can be depicted in all ranges of color contrasts. The morphology of objects in three dimensions provides much more useful information than that previously available from electron microscopes. Other software image processing features, such as filters, z-magnification, Fourier transform, and others will be described later. Furthermore,

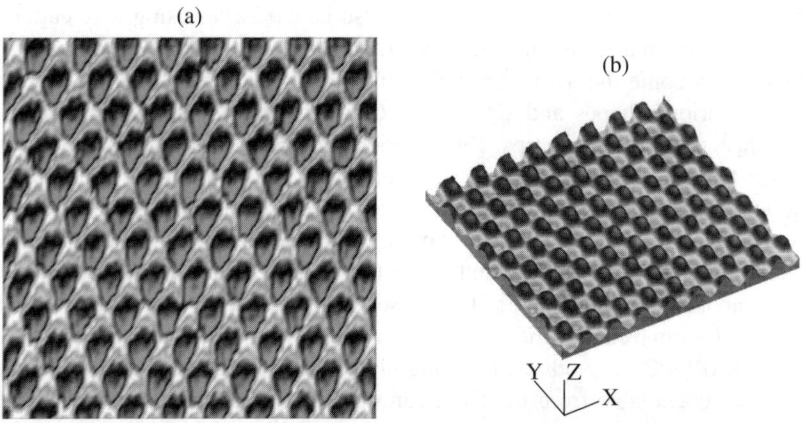

FIGURE 2.2 Image of HOPG (highly oriented pyrolytic graphite) by STM (23.2 Å × 23.2 Å × 2.19 Å): (a) two-dimensional image; (b) three-dimensional image.

with the help of commercially available image manipulation software, images can be smoothed and sharpened, missing data can be added, and many other sophisticated image treatments can be utilized. Some of these image analyses procedures will be presented in this book.

It was found that the apparent corrugation is dependent mainly upon the tunneling resistance defined by the ratio of bias voltage to tunneling current, and which is expected to have a maximum of about 0.3 nm on a freshly cleaved surface. The surface of graphite has a layer structure in which each layer is composed of carbon atoms in a honeycomb arrangement. The distances between carbon atoms was 0.246 nm (2.46 Å) and 0.142 nm (1.42 Å). Adjacent layers are separated by 0.335 nm.[25] High-resolution images of HOPG were studied by STM. These images clearly showed a trigonal lattice forming the so-called large hexagons characteristic of the graphite. This is used regularly for control and calibration of STM and AFM. It is worth mentioning that HOPG images are highly reproducible, which affirms the sensitivity and reproducibility of the scanning microscopes. All of the literature images of HOPG are in agreement, where investigations were carried out on different kinds of SPMs.

In STM, the apparatus in general consists of the following parts:

- Sample holder
- Tip setup
- Electronic controller
- Computer driver for the electronic controller
- Software for image processing

The sample holder is generally made as a simple clip to keep the sample fixed. However, more complex setups as may be necessary have been reported. The tip moves from above or from below the microscope. The tip is usually commercially available, which secures high reproducibility. Tungsten wire (0.5 mm diameter) and Pt or Pt-Ir wire have been used for tips. The etching of the tip is carried out in a solution of KOH or HCl. These wires can also be used after using wire cutters. On the microscopic scale, one finds that asperities as small as single atom clusters will potentially become the active tip during tunneling. All sorts of custom-made tips, made in various shapes and sizes, can be bought commercially. However, some researchers make their own tips. The electric controller is, in most cases, just a black box for the researcher in surface science. The software used for the controls is different for various commercially available microscopes. The software has continued to develop throughout the history of the STM. However, in general, the software capabilities are similar. In the author's STM and AFM, the image is acquired in IMG or another suitable type of file. The software provided by the STM can convert this to TIFF format easily. The TIFF files can be easily converted to other file formats, such as BMP, PCX, or others, by using other image analyses programs. However, the software available from the STM can only convert two-dimensional images to three-dimensional images. Further, there are available software programs that can treat and manipulate these image files, which can also rotate and manipulate images in various parameters. For example, one may wish to manipulate missing data (by applying linear or other extrapolation methods). These various image procedures will be described in detail here with appropriate image examples.

2.2 ELECTRON TUNNELING

Electron tunneling originates from the overlap of wave functions between the molecules at the tip and surface atoms of the substrate. The tunneling current between two metals is given by the following:[26]

$$I = I_o \cdot \exp(-A\sqrt{t}(\phi z)) \tag{2.1}$$

where $A = 1.025$ eV$^{-1/2}$ Å^{-1}, ϕ is the barrier height, and z is the distance between the electrodes. For typical barrier heights of 4 eV, the tunneling current decays one order of magnitude when the distance z is increased 1 Å. Therefore, the tunneling tip is at close proximity to the sample, typically at distances of 5 to 10 Å. In order to calculate the sensitivity of tunneling, a small modulation Δz is applied. The relative variations of the tunneling current are then given by:[26]

$$\Delta I/\mathrm{I} = l - \exp(-A\sqrt{t})\phi z)] \sim A\sqrt{t}(\phi z) \sim \sqrt{t}(\phi z) \tag{2.2}$$

where Δz is in A and ϕ in eV. A distance modulation of 0.1 Å causes a current variation of 20%, which shows that tunneling is a sensitive method allowing distances as small as 0.01 Å to be measured. However, typical tunneling currents are of the order of nanoampere (Na), requiring high amplifications that limit the bandwidth to a few kHz. In the dynamic mode, the lever oscillates at its resonance frequency, which is typically 30 to 100 kHz for microfabricated cantilevers. Therefore, most of the applications are performed in the static mode. Another handicap of the tunneling detection method is its sensitivity to contaminants such as oxides or hydrocarbons. The presence of contaminants causes the tip to press and the lever to oscillate until the nominal tunneling current is achieved. Forces as high as 10^{-6} N have been reported under these conditions.[27] For ideal tunneling between metals, the forces are reduced to 10^{-9} N (= nN). Empirically, it is found that freshly evaporated gold films and Pt-Ir tips are necessary to achieve stable tunneling conditions at ambient pressure.[28] For standard applications in air, the tunneling method is too delicate, and optical methods are more reliable and easier to operate. For UHV and low-temperature experiments, the low power of Nw and the possibility of minimizing the dimensions seem to be advantageous. The first successful low-temperature AFM experiments in ultra-high vacuum and UHV[27] have clearly demonstrated that the tunneling detector is well-suited to these conditions.

The role of the *tip* in STM is being increasingly recognized and discussed, as the STM cannot be expected to provide a completely technically clean molecular image of a free surface, unless the role of the tip is absent or negligible.[29] Additionally, theoretical calculations based on the effect of tip size on any observed image are being exhaustively investigated, as described later. Two conducting electrodes are separated by some isolator that forms a barrier for the electrons inside the electrodes (Figure 2.1). If the barrier is thin enough, electrons can pass through it by a quantum-mechanical process called tunneling. In a scanning tunneling microscope, the barrier is a gap of 1 nm.

The resolution in the image, which is only a surface or subsurface image, is defined in the vertical direction and laterally on the plane of the sample. The magnitude of resolution can be achieved in the range of a fraction of 1 Å (10^{-10} m = 0.1 nm). Obviously, the size and shape of the probe tip will determine the lateral resolution. The tip moves at ca. 10 Å/V applied to a translator. Typically, a lead zirconate–lead titanate (PZT) polycrystalline ceramic is used.

A typical STM image of graphite (highly oriented pyrolytic graphite, HOPG) is given in Figure 2.2. A detailed surface analysis of HOPG was given in a recent study.[18] The molecular resolution achieved by STM and the three-dimensional capability are the most useful aspects of this method. The morphology of objects in three dimensions provides much more useful information than previously available by electron microscopes. HOPG is generally used as a substrate for studying other molecules placed on it by STM and other SPMs.

STM studies under ultrahigh vacuum (UHV/STM) have been carried out in many recent investigations.[19]

Because the STM tip measures current as a function of distance of separation, most software as supplied by STM microscopes also allows for the measurement of hysteresis curves of current (I) versus volt (V), while the tip is kept stationary at a fixed spot on the sample. Hysteresis curves of V versus I plots are obtained, which are related to the capacitance of the substrate, besides thickness and other parameters.[30]

2.3 ATOMIC FORCE MICROSCOPE (AFM)

Soon after STM was invented, it was evident that a similar kind of scanning microscope was needed that could be used for nonconductor surfaces. Historically, there were some inventions that operated on principles which could be useful as a basis for such instrumentation. The general principle of operation is almost similar in STM and AFM. The first scanning force microscope (SFM) or atomic force microscope (AFM) was developed 5 years after the introduction of the STM.[3,7,20] Based on the same principles as for STM, the atomic force microscope (AFM) was later developed for nonconducting solid surfaces.[7]

Thus, AFM can operate on conducting and nonconducting surfaces. A cantilever scans the solid surface (while keeping the force constant between the tip and the substrate), and the deflection of the cantilever is detected by the reflection of a laser beam (Figure 2.3).

Accordingly, the AFM was mainly developed for nonconducting solid surfaces.[10–14] Thus, AFM can operate on conducting and nonconducting surfaces.[3] A cantilever scans the solid surface, and the deflection of the cantilever is detected by the reflection of a laser beam (Figure 2.4). As it is easily recognized, one can modify the cantilever by various means and obtain other information about surfaces than that merely related to forces. These modified force microscopes will be described throughout the text.

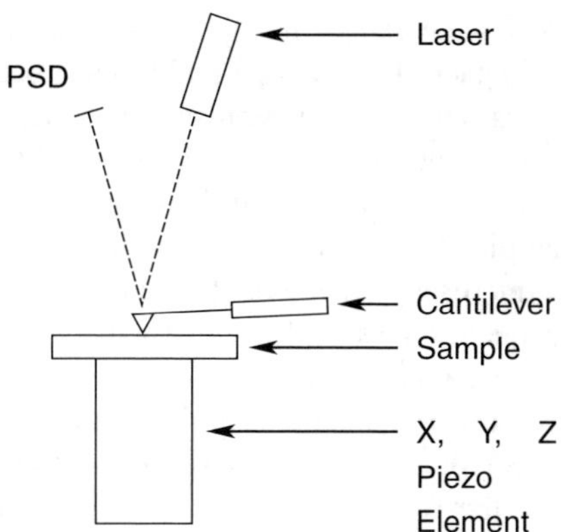

FIGURE 2.3 AFM apparatus principle. Cantilever movement on a substrate under controlled constant force or other parameters.

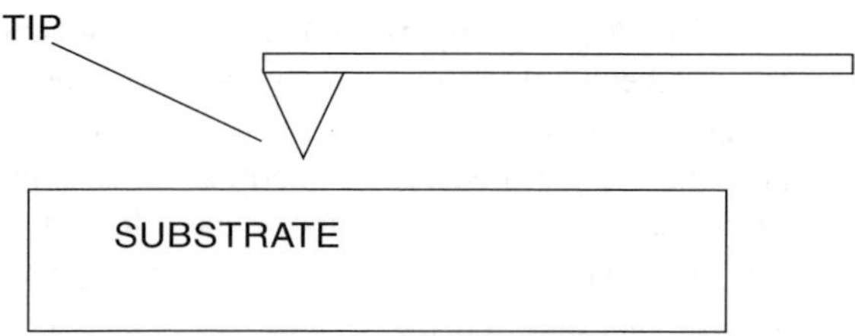

FIGURE 2.4 AFM cantilever (schematic) movement over the solid surface (beam deflection principle).

2.3.1 Basic Principles of AFM

Scanning force microscopy is based on the existence of a separation-dependency force between any two bodies. The first AFM was described 5 years after STM.[7] In the case of SFM or AFM, it is the force between the tip and the substrate that is present at close separations. Typically, pyramidal silicon nitride tips are used, which have a radius of curvature on the order of 100 Å. These are made by an etching process that removes silicon from the substrate, leaving an etched or sharpened tip behind. The force is detected by placing the tip on a flexible cantilever that deflects proportionally to the exerted force. The deflection is then measured by some convenient procedure, such as laser reflection or some other device. Actually, the main innovation may be seen as being a copy of the principle behind the record player of some decades ago, where a sharp metallic needle moved over a vinyl record to reproduce sound.

A Si_3N_4 cantilever of an approximate 200 μm length and a spring constant of 0.06 N/m or 0.12 N/m is generally used. However, other materials have also been used for tips, and these will be described in detail below. Recently, much research has been carried out on improving the shape and other characteristics of the cantilever, as delineated below in further detail. AFM has thus become a versatile modern technique of immense applications in such areas as surface and colloid science.

Because the latter has a wide range of applications in industrial processes and biological systems, the impact of SPMs is found to have been immense.

The interaction forces, attractive or repulsive, as small as few nano-Newtons (nN = 10^{-9} N) between the cantilever and the sample can be measured. The biggest advantage of AFM is that most samples can be investigated in their natural state, including biological samples (even in an aqueous environment), which is otherwise impossible by electron microscopy methods. In the case of a repulsive-force procedure, the tip physically touches the sample (with a sensitivity of nN force) and traces across the sample without damaging the morphology. In AFM, the sample moves under the cantilever, which is the main reason that the resolution by AFM is somewhat less than that for STM. However, this being a general statement, it will be extensively described in the following. The concept of using a force to image a surface is general and can be applied to magnetic and electrostatic forces, as well as to the interatomic interaction between the tip and the sample.

The AFM images of HOPG are almost of the same resolution as those by STM; however, the resolution is highly dependent on the characteristics of the sample (i.e., surface texture and even the degree of hydration), as well as on sample preparation. This latter point has not been settled completely, and one should be careful in reaching any conclusions at this stage. These considerations will be described later, wherever appropriate. The sample preparation is critical in all kinds of SPMs, and this aspect needs further investigation. Furthermore, the effect of vibration is more profound in the case of AFM than STM. It is generally believed that the main difference arises from the fact that in STM, it is the tip that moves, while in AFM, it is the sample. In any case, at this stage, these points are not completely resolved in the literature. Another parameter of importance is the scan speed limit.[31] The scan speed was calculated from the spring characteristics. The scan speed is determined by the spring constant, k_{sc}, of the cantilever, its effective mass, m_{mass}, the damping constant, D_{ca}, of the cantilever in the surrounding medium, and the stiffness of the sample. Procedures to measure k_{sc}, m_{mass}, and D_{ca} were delineated. From these calculations, the scan speeds were estimated as follows:

- In vacuum: 0.1 μm/s
- In water: 2 μm/s

The increase in water arises from the damping effect of water on the cantilever, as expected. The density of water is approximately 1000 times greater than that of air.

In general, the AFM is operated under a constant-force mode that incorporates optical beam deflection for sensing the cantilever motion. Usually, the forces exerted are in the range of 0.5 to 0.1 nN (force constant of the cantilever used is 0.06 N/m). The AFM is calibrated by using graphite and grating samples in the x–y direction. The height axis has been calibrated by using such collapsed lipid monolayers as the Langmuir–Blodgett (LB) film.[20]

The AFM apparatus operates based upon the knowledge of the magnitude of the spring constant of the cantilever. The spring constant of the cantilever must be measured directly,[21] because the calculated values may not be too reliable. However, commercially obtained cantilevers are supplied with this information. The spring constant, k_{spring}, for a beam (with Young's modulus E_Y) loaded on the end with rectangular cross section is given as follows:[21]

$$k_{spring} = E_Y w_{width}/4(t_{thickness}/l_{length})^3 \tag{2.3}$$

where w_{width}, $t_{thickness}$, and l_{length} are the beam width, thickness, and length, respectively. The calculation of k_{spring} becomes difficult when cantilevers of a thickness <1 mm are used, due to the above relation. A simple method for measuring the spring constant is based on using a procedure where the static deflection of the cantilever under the force of a known mass is used.

The cantilever with the tip senses surface forces arising from various interactions (van der Waals, electrostatic, hydration) between the tip and the sample. When the force curve is initiated at a point where the tip and the sample are far apart, the magnitude of surface forces (acting between the tip and sample) are quite negligible, and thus, there is no deflection of the cantilever. As the sample is raised toward the tip, a variety of attractive and repulsive forces deflect the cantilever. This region is called the *noncontact region,* as described below. The effect of force on molecules by the tip has been studied and is described elsewhere herein. The electronics used in AFM have many variations, and literature studies are reviewed.[26]

The principle in SFM is based on the existence of a separation-dependent force between the tip and the surface of the substrate. This force can be attractive or repulsive, which depends on the separation distance, as well known from physicochemical laws of nature.

In general, the AFM or SFM is operated under constant-force mode, which incorporates optical beam deflection for sensing cantilever motion. Usually, the forces exerted are in the range of 0.5 to 0.1 nN (force constant of the cantilever used is 0.06 N/m). AFM is calibrated by using graphite and grating samples (in the *X–Y* direction). The height axis has been calibrated[3,20] by using collapsed lipid monolayers as LB film. There are two regimes of force (repulsive and attraction) that can be measured by the tip. If the distance between the tip and the sample, r_d, is small, the force, F_r, is expressed as follows:[3,20]

$$F_r = 12B/r_d^{13} - 6D/r_d^7 \tag{2.4}$$

where B and D are constant depending upon the substrate and the tip. For small separation distances, the first term (repulsive) will dominate. The resolution is best when the tip is in contact with the substrate, due to the rapid decay of the repulsive force term. This is also due to the fact that an image is being made by the interaction between more than one atom of the tip. It may also be considered that the tip is behaving as an average of a number of atoms at the apex. This is consistent with the fact that different tips give exactly the same image resolution (for example, of HOPG). AFM thus provides information about the surface morphology and about the surface forces when two bodies are at nm separations.

In a recent study, it was shown that there may be a possible source of artifacts in force measurements that is due to digitalization.[32] Normally, the sample is moved in discrete steps, due to the fact that the piezomotor is controlled by digital signals. The step-wise approach of the sample can cause oscillations of the cantilever, which might give rise to a hysteresis in the noncontact region. Cantilever oscillations can also lead to erroneous measurements of attractive forces.

2.3.2 Imaging in AFM and Tip Effects

As is well established in all kinds of physical measurement procedures, there exist artifacts if experiments are not conducted or performed under well-controlled conditions. The artifacts are generally encountered only in those cases where appropriate controls are not conducted. Image analysis, in general, is a complicated process involving the image produced by the microscope and the interpretation presented by the investigator. An image obtained using a microscope needs to be calibrated by using suitable samples with known grids, etc. However, when dealing with such microscopes as electron microscopes or STM or AFM, grids with molecular dimension scale are needed in order to eliminate any artifacts. The safest procedure for avoiding artifacts is to design experiments in which one expects to see differences in images, which would eliminate such artifacts. Some of these procedures, which have been reported in the literature, will be described in detail below. The most simple control is for two researchers using two different instruments to confirm images of the same system. Because, at this stage, these aspects are not completely described in the literature, we will mention them briefly here. And, systematic studies carried out on these aspects will be detailed. The image obtained is determined by the movement and sensitivity of the tip movement and controls. Review of current literature reveals that artifacts reported are not well analyzed. The use of the term "artifacts" is not quite satisfactory as applied here, because if the image of HOPG gives correct data, then the AFM is functioning correctly. If, on the other hand, control experiments are not used, then the AFM is not to blame. Furthermore, the instrument can be designed to consist of software programs to avoid any artifacts related to such a possibility in the controls. For example, most microscopes include the possibility of reversal of the scanning direction. This presents a fair amount of control on any artifacts, for example, any debris being attached to the tip. A few detailed studies have been reported in the literature, but most workers mention any such problems in their reports. It is obvious that when two bodies approach, different kinds of forces are interacting at such short distances. Furthermore, the morphological information desired from such experiments will involve some artifacts. This may be compared with the old record player, where a metallic needle, while moving in the groove of vinyl record, reproduced sound through mechanical vibrations and amplification. This led to wear and tear of the needle and the vinyl. There is no diffraction limit in the interaction between the tip and the specimen in AFM or SFM, which is obvious, as the wavelengths of the interaction are often far larger than the tip–surface separation.[21]

Actually, the limiting factor of image quality is known to be a combination of the resolution limit due to tip shape and the separation dependence of the force, specimen deformation, scanner imperfections, thermal drift, mechanical vibrations, and electronic noise and stability. Artifacts due to specimen deformation and scanner imperfections will be described later.

A force curve starts at a point where the different forces between the tip and sample are negligible, and the cantilever is not deflected.[32] As it touches the sample, this region is called the *noncontact* region. Finally, the sample and the tip contact each other, and the deflection of the cantilever follows the movement of the sample. This is the *contact region.* A possible source of artifacts in force measurements is due to normalization, as discussed in detail elsewhere. Normally, the sample is moved in discrete steps. In the *noncontact mode*, the net force detected is the attractive force between the tip and the substrate. On the other hand, in the *contact mode*, the net force measured is the sum of the attractive and repulsive forces. Thus, these two modes are different as follows:

- Contact mode: attractive forces + repulsive forces
- Noncontact mode: attractive forces

The tip–substrate can be depicted as follows:

CANTILEVER (TIP) ....
AIR OR FLUID:::::::::::::::::::::
SUBSTRATE.........................

In this case, the tip is oscillating (also called AC mode or tapping mode) in the attractive force region, sensing the force without touching the substrate. Computer simulation models have been developed in some cases to determine theoretical models of these factors in SFM.[33]

2.3.3 Analyses of Tip Effects

In all systems where a probe is used in close proximity (nm scale) of a substrate, some effects on the images may be expected. These have recently been well studied, and the various results will be described throughout the text. It is obvious that tip effects will be difficult to analyze, if such an effect was present. However, it must be stressed that because reproducibility of images by using different tips (and different researchers) is high (for example, the HOPG images have been reported with high sensitivity), the conjecture that tip effects are negligible, in general, is supported. Considering that AFM is based on the moving, in close proximity, of a tip on a substrate (Figure 2.5), the atomic image quality will be dependent on tip geometry and quality, besides other factors, as depicted below.

SHAPE OF TIP..
SHAPE OF SUBSTRATE MOLECULES....
....
IMAGE OF SUBSTRATE MOLECULE......

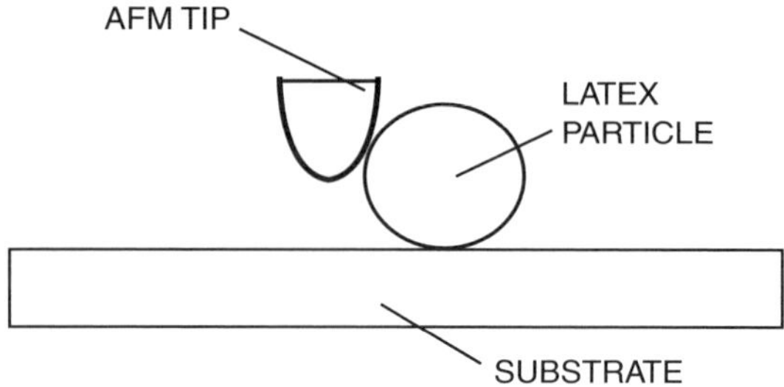

FIGURE 2.5 A schematic drawing of the AFM tip and a spherical latex particle. The AFM image will thus be determined by the radius of curvature of both the tip and the substrate molecule. The size of the tip (ca. 10,000 Å) is obviously much larger (more than 100 times) than the substrate molecule (ca. 50 Å).

Before describing these tips in detail, it is expected that a very sharp tip will be used. Generally, this has been achieved by merely cutting a thin wire with scissors. The outermost atom will exhibit the tunneling carrier. There are ample experimental data that show validity of this postulate. The various tip requirements can be classified as follows. First, it is important that the nonsignificant force contribution be maintained at a minimum. This can be achieved by having a small cone angle and small mean tip radius. In this way, the long-range attractive interactions that do not contribute to the contrast can be kept small. These attractions can otherwise lead to a large deformation of the tip and the surface in contact angle measurements. Another effect is that a cone angle that is too large will lead to a snap-in far from the surface and cause unstable regulation in attractive mode measurements. The radius of the tip at the end and the cone angle will mainly determine the image quality. However, this factor has become considerably more complex, as various kinds of tip designs have become commercially available. This factor will be described throughout the text.

The shape of the AFM tip is intimately related to the lateral enlargement phenomenon because of simple, hard-surface geometric considerations and because sources of interactive forces are distributed over the tip surface as well as the sample surface. Furthermore, some effects of the asymmetry of the tip on images have been reported.[34] In simple terms, one can show the possibility of an image of a square being scanned. This may be considered the most common artifact. The shape of the tip and the sample are what might be seen in the final image. The width of the sample will be expected to be broader than it is, unless the tip compresses the sample. However, it must be stressed here that these considerations are not well described in the current literature. The reason for this is that the calibration grids are made of metallic surfaces that provide sharp and accurate data. This means that when measuring soft materials, the images may or may not be distorted depending on many factors. As will be described later, many parameters in SPMs are not completely understood at this stage. Therefore, many reports can be found that cannot be easily explained at this stage.

Second, the tip should exhibit as a single and well-defined apex. This is especially important for specimens with a pronounced topography, where there is a risk of

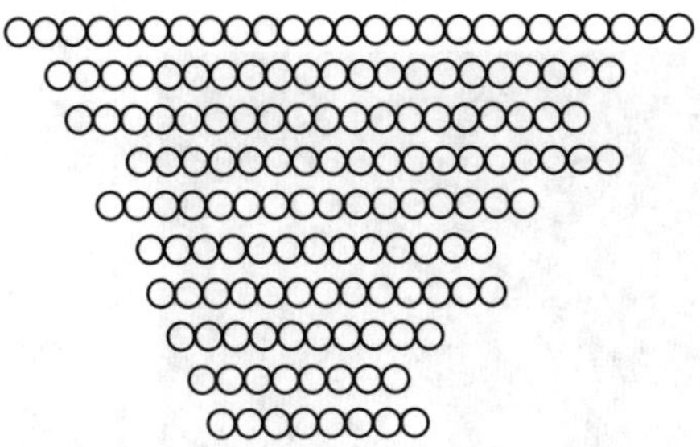

TIP ENLARGED VIEW

FIGURE 2.6 A cantilever (or tip of STM) and its (schematic) exploded view at atomic scale.

getting more than one apex close to some parts of the surface. The magnitude of cone angle has been suggested to be related to the degree of resolution. Actually, this demand is fairly satisfied when using commercially manufactured cantilevers. As depicted in Figure 2.6, it can be imagined that it is the outermost atom that gives rise to the images. However, these aspects are not well investigated at this stage (neither experimentally nor theoretically). In general, one finds that the commercially available tips are highly reproducible in obtaining images of test substrates. This means one less worry for this aspect of the SPMs.

Third, the tip should be mechanically rigid and hard, in order not to deform on the microscope scale when subjected to surface forces or having unwanted lateral compliance, which can be contrary to the need of small cone angle that will be expected for small lateral spring constants. Further, in order to provide high reproducibility, the "same" tip may be used for all comparative investigations.

The image of the tip can be obtained by various ways. One method is by using the electron microscope. Another more useful way is to obtain images of special specimens with narrow columns. Further, more useful procedures can be developed by taking images of well-defined substrates and varying the tip shape and quality. Another more realistic procedure is to be able to obtain an image of the scratch of the needle on a soft substrate (see Figure 2.7). The scratch is 500 Å wide at the top and 45 Å deep. The end of the tip is very sharp (less than 10 Å).

Common tip artifacts have been described in the current literature. However, one must be careful in accepting these preliminary conclusions. The shape of the tip and the morphology of the sample are combined in the final image. The least offensive consequence of this effect is an exaggerated sample width. However, these tip-related observations will be described in detail in the following.

Scanning tips are commercially available in various shapes and sizes. This is a critical parameter, because the image generated by the AFM is actually the sample. In many cases, as with large field scans of living cells, this problem is negligible. However, as one scans tall, steep-walled structures, or structures with dimensions approaching those of the end of the tip, the tip convolution problem becomes pronounced. In some cases, the image is heavily dominated by the tip shape, and the true shape of the sample is unrecognizable. This phenomena has led to a number

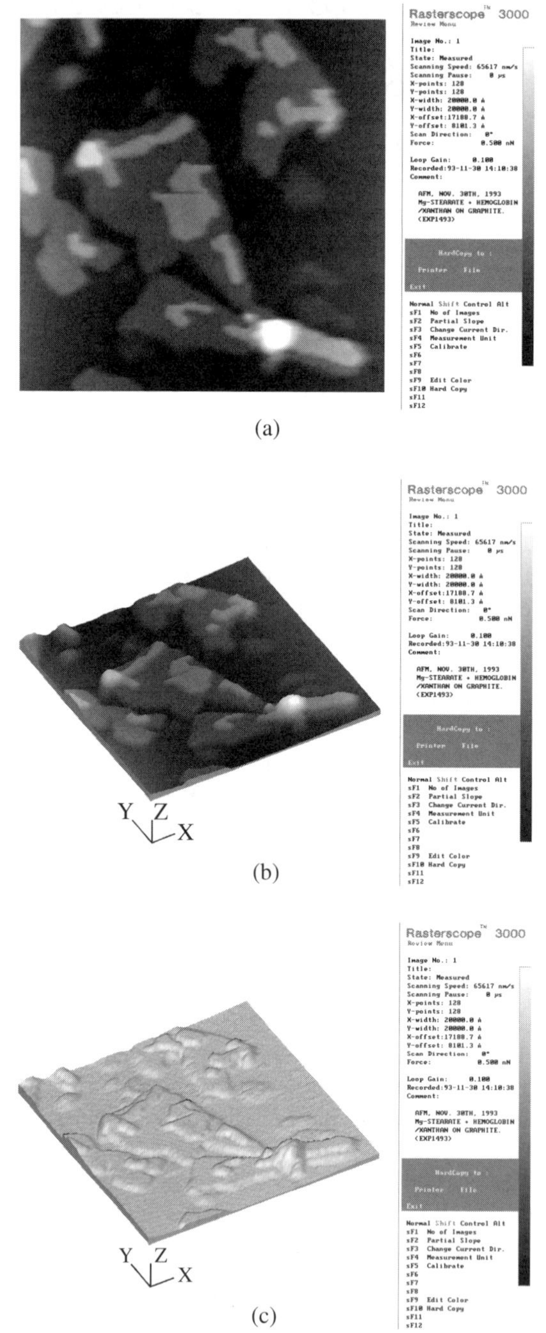

FIGURE 2.7 Scratch made by the tip and the shape of the tip (collapsed mixed-lipid monolayers as LB film on HOPG): (a) two-dimensional, (b) three-dimensional (smooth), (c) light image, (d) one-line image, (e) two-line image (20,000 Å × 20,000 Å).

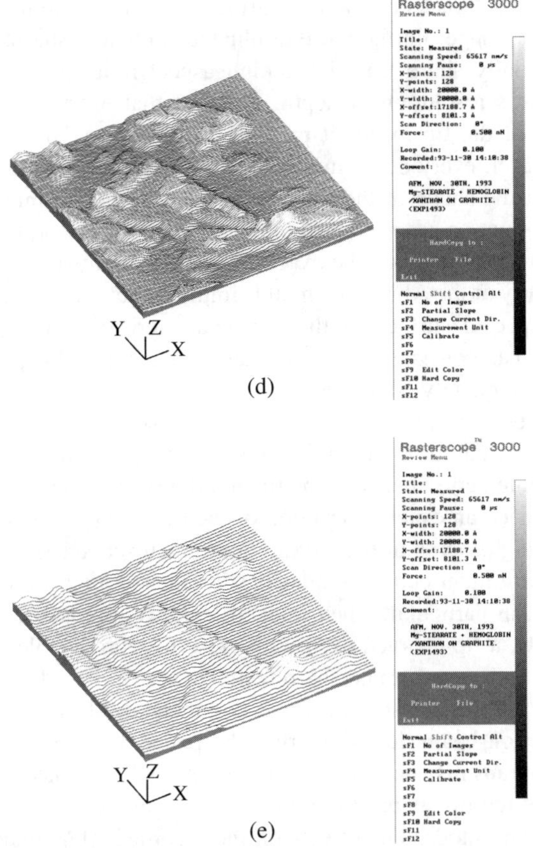

FIGURE 2.7 *Continued.*

of reports in which a particular image feature was attributed to the sample. However, as in many other analytical procedures, one must adhere to some specific standard procedures to avoid any artifacts. This subject is described in detail elsewhere. It is the author's experience that in spite of these considerations, the commercially available tips are highly sensitive with satisfactory reproducibility.

Reproducible artifacts in an AFM image can be recognized and understood by studying the shape of the tip apex at the atomic level (<30 nm). The imaging mechanism for intermittent contact mode AFM over corrugated surfaces is discussed based on the knowledge of the exact shape of the tip and the sample surface. Adsorbed contaminants on the probing tip have previously been considered to nullify the imaging process only. It has been shown that tip adsorbents and other artifacts may sometimes influence the AFM imaging in a positive way. The fact that tip-adsorbed debris is proven to act as the sensing tip in certain cases may explain why even blunt tips, compared to the features of the sample surface, can produce highly resolved images. This confirms that the probing site may actually be formed continuously during the scanning by adhesion of some surface material or by spallation, which in turn, tracks the sample surface through a material exchange process and provides the image. Furthermore, a blunted tip that is sharpened *in situ*, compared to an initially sharp tip with a high aspect ratio, gives reasonable lateral resolution but less resolution in depth. The fact that every AFM image contains information about the tip makes it possible to use the AFM to determine the tip geometry. However, every detail in the AFM images must be carefully interpreted, considering that the shape of the tip may change during scanning, e.g., due to tip wear, tip disruption, or adsorption of surface contaminants on the tip. This would make any attempts to calculate the exact tip shape from the recorded images superfluous. The quality of the AFM tips should simply be determined (*in situ*) by judging whether the expected surface of the "tip characterizer" was reproduced or not. However, if one can consistently obtain images in a series of experiments, then this effect of tip is completely controlled.

Based on these considerations, one finds in current literature extensive studies that have been carried out to determine the role of tip shape and geometry. Several models for the interaction between the tip and the surface have been suggested, but generally, they need an exact description of the geometry of the tip, the surface, or both. The imaging process of the atomic force microscope (AFM) in contact, non-contact, and intermittent contact modes is still debated after more than a decade of widespread use, in particular, when imaged features approach atomic dimensions. In a recent study, a tip characterized with close to reproducible geometry, exactly known angles of all surfaces, and sharp features was described with close to atomic dimension.[35] It has been tested on three commercial AFM probes and a laboratory-made electron-beam-deposited tip, sharpened by oxygen plasma etching. High-resolution transmission electron microscopy (HREM) has been used to unambiguously verify the tip shapes to atomic dimensions, before and after imaging in intermittent contact mode. The effects on the recorded AFM images are shown of tip shape, tip wear, spallation, and accumulation on the tip of amorphous and crystalline debris. The imaging is shown to be a dynamic event, with a continuously changing tip and occasional catastrophic events that give abrupt changes in imaging conditions. The tips are severely worn after scanning only a few centimeters, but accumulated amorphous material may still give it imaging capabilities in the nanometer range, even with a tip radius exceeding 130 nm. Accumulated amorphous material seems to be more important than previously believed. Procedures for tip *in situ* characterization and reliable imaging are suggested.

The information that can be obtained by scanning probe microscopy (SPM) depends strongly on the interaction between the probe tip and the surface. The "shape" of the probe tip must be carefully controlled, and its dimensions must be accurately known if the true profile of the surface is to be determined. However, commercially available tips are reliable and conform to specifications. Because real tips used on corrugated sample surfaces cannot be regarded as infinitely sharp point probes over a flat specimen, tip effects must be taken into account when interpreting the data acquired. The tip effects occur due to interaction of different parts on the probe tip with the sample, depending on the geometry of the surface, leading to a (nonlinear) convolution of the sample features with the tip shape. The fact that every AFM image contains information about the tip makes it possible to use the AFM image to determine tip geometry. If the object is steeper or sharper than the tip, the approach is known as inverse AFM or self-imaging. This is analogous to (direct) AFM imaging and gives a reconstruction of the tip, or rather of each point of the tip that was in contact with the sample during the image acquisition.[36]

It is commonly assumed (as one should expect from simple laws of physics) that the AFM imaging performance can be substantially improved by the use of very sharp tips (which are commercially available). The sharpness of a tip is measured in terms of macroscale, while the image obtained is in atomic scale. It can be seen that these two criteria are at least by a magnitude of 1000 times or more scale different. For a tip to be considered "sharp," that is, for it to be capable of creating images without significant convolution, it must have an end radius less than the smallest radius of curvature of interest in the sample (approximately 1/10 nm), and it must have an opening angle less than the angle of the steepest feature of the sample. However, it has been shown that even blunt tips can be used to achieve high-resolution images over flat surfaces, which suggests that a blunt tip may have one, or several, small protrusions, which act as the actual sensing tip.[37,38] A different tip effect may arise if the shape of the tip changes during scanning, e.g., tip wear, tip disruption, or adsorption of surface contaminants on the tip.[39–41] This can be avoided by measuring on a test substrate (such as clean HOPG or other suitable sample) between different series of measurements. When scanning relatively soft materials, such as biological or polymer materials, one can assume that the tip shape is maintained (unless the surface material has been adsorbed onto it). On the other hand, if the sample consists of hard solid material and exposes sharp features, the tip shape is likely to be deformed, either abruptly or gradually, during the scanning. Because such effects are difficult to anticipate, they are not easily corrected for, and consequently, the resulting images are difficult to interpret. One can minimize the risk of erroneous image interpretation by regularly checking the tip *in situ*, by means of inverse AFM using a well-defined sample surface, a "tip characterizer." The ideal tip characterizer would be an infinitely narrow spike, yielding a perfect image of the tip. Something close to this has been achieved, where unique structural features of the sample surface, i.e., needles or columns, were inversely imaged.[42,43] The region around the tip apex is, however, not imaged correctly, because the columns have a finite thickness. Using narrower columns would be a simple solution, but these columns would bend or break during AFM scanning. Characterizers consisting of lithographically patterned arrays of square pillars have been used.[44,45] The corners of these pillars could generate an image of the tip, but the limited sharpness of these corners and the size uncertainty of the pillars makes them a poor choice. Spheres of latex or gold may be useful materials, because the knowledge of the exact shape of the tip characterizer is crucial for the success of inverse AFM.[43–49] These are commercially available with diameters ranging from 5 to 500 nm with a narrow size distribution. The verification of the actual tip shape is, however, not straightforward.[48–51]

Latex spheres offer a wide size range but are not readily available in the interesting 1 to 10 nm range, where also charging could cause problems.[51] It is likely that the smallest gold particles (<5 nm) cannot be considered spherical, but instead exhibit a polyhedral structure that originates in the fcc metal structure.[52] They might also be covered with a hydrocarbon layer, which prevents an accurate size determination.[48] Lateral movement of the weakly adsorbed particles by the tip and bending of the tip axis are other aspects, which can complicate the imaging process even more.[53,54,55] Morphological tip reconstruction techniques using mathematical descriptions of tip–sample geometry are usually based on noncompressible probes and samples, obtaining a single contact point with hard-surface interaction.[56,57] That is, the surface and the tip are assumed to behave like continuous surfaces of infinitive hardness, with contact that is determined by steric hindrance. However, it has been shown that the compressibility (i.e., elastic modulus) of the tip and sample surface[57,58] and the tip wear and contamination[58] are important factors that may give theoretical modeling a difficult task. In the case of conducting tips, a method for *in situ* characterization of the tip shape using electrostatic interaction has been described.[59] This noncontact procedure avoids the risk of damaging the tip during scanning of the reference sample. A plausible procedure is using single-crystal films of $SrTiO_3$ (perovskite), which exhibit on high index surfaces a stable sawtooth surface structure with an atomically defined morphology. In this case, the consequences of lateral movement or bending of the surface features are avoided effectively.[60] The quality of the AFM tips was determined simply by visually judging whether the expected surfaces were reproduced or not. It is an indirect technique for a quick routine check of the quality of the tips before use, without going into the depth of the actual shape of the tip. For thorough understanding of the tip–sample interaction, however, a direct imaging of the tips is required. Micrometer-scale artifacts can be relatively easily recognized due to their characteristic repeated appearance in the images, often as pyramids or triangular-shaped patterns.[61] Supplementary SEM or TEM measurements have previously been performed to verify the AFM topographies and to determine the tip shape.[61,62] However, in case one wants to investigate details at the mesoscopic level (5 to 100 nm) or less, comparison with SEM is often inappropriate, because these images do not provide sufficient resolution. An exact determination of the shape of the probing site is essential in order to evaluate AFM images and to recognize artifacts unambiguously. High-resolution transmission electron microscopy (HRTEM) in combination with a powerful tip characterizer would greatly widen the understanding of the imaging mechanism at the atomic level, where forces other than true contact-point interactions may come into decisive play, e.g., attractive electrostatic or adhesive forces. Adhesive forces between tip and sample may increase the contact diameter and seriously deteriorate the imaging resolution for the sharpest tips.[62,63,64] Much work using computer simulations based on molecular dynamics has been performed to provide insights into the physics of the tip–sample system on the atomic level.[65,66] However, a practical understanding of the imaging mechanism, also including forces on cantilevers, has not yet been satisfactorily attained. As mentioned above, the tip quality of a cantilever in AFM determines the image resolution. At least, one may use this as a general argument, even though no extensive theoretical analyses as yet has been forwarded to explain this. Thin films of CeO_2 can be grown and tailored by physical vapor deposition to exhibit unique features at the atomic scale. These films have previously been used to study various properties, such as structure-sensitive catalysis.[67–69] In this study, a well-characterized thin film of CeO_2 grown on R-cut Al_2O_3 ("sapphire") was utilized to evaluate tip quality.[70] The corrugated surface exposes atomically sharp ridges elongated in two different directions, which are perpendicular to each other. Four different types of probes for intermittent contact mode (i.e., tapping mode) AFM from three different manufacturers have been characterized using HRTEM, before and after scanning in tapping mode. Tip artifacts and the imaging mechanism are discussed based on the knowledge of the exact shape of the tip and the sample surface. The results presented here show that the CeO_2 thin film can be used with the advantage of characterizing the quality of the probing tip. However, adsorbed contaminants on the tip apex and the elasticity of the probe and sample surface are important issues that must be carefully considered when performing measurements with the atomic force microscope. The CeO_2 thin films were deposited on an Al_2O_3 (sapphire) substrate in a magnetron sputtering system. These fragments of the CeO_2 films on sapphire substrates were also prepared by using the small-angle cleaving technique.[71] The fragments were mounted sideways on a single-hole copper grid, so that the top surface could be studied. Samples were analyzed using an HRTEM.

The characterizations of the structure and surface of the thin film have been described in the literature.[70] The tips have been characterized before and after AFM scanning using a TEM. In these studies, the cantilevers were cut under water using a diamond scriber in order to mount the tips in the sample holder. The probe fragments were tilted and glued on a copper grid. The electron-beam-deposited (EBD) tip was grown in an electron microscope.[62] There were two pointed peaks that represent close to atomically sharp ridges that are parallel with the electron beam, while another one on the left was a profile view of a similar ridge. The AFM measurements were performed at ambient room conditions using a commercial multimode SPM. The instrument was operated in tapping mode, i.e., intermittent contact mode. In these investigations, four different silicon probes from three different manufacturers were evaluated:

- Nano-probe etched silicon probe (TESP, high frequency)
- Point-probe silicon probe (high frequency)
- Super cone probe
- EBD tip (length ~630 nm), produced by deposition of an amorphous carbon column on a silicon probe (low frequency)

The high-resolution electron micrograph showed that the deposited CeO_2 film, grown on an oriented sapphire substrate, was strongly corrugated over the whole surface. The ridges were found to be elongated in the directions of the CeO_2 lattice, i.e., exactly perpendicular to each other. The dimensions of the ridges were in the order of 30 nm × 30 nm, with a maximum height of 35 nm. The top angle of the ridge was estimated to be 70.5, which agreed with the angle between planes in the fluorite structure of CeO_2. This means that in order to image the ridges, and not the tip, the opening angle of the tip apex must be less than 70.5. Further evidence of the exact shape derived from HRTEM tilt experiments and detailed discussions on factors governing the as-deposited film have been given elsewhere.[70] The ridges were found to be almost atomically sharp, with one to three atom columns missing at the top only. TEM micrographs showed that the dimensions and angles of the probe given by the manufacturer corresponded acceptably to the measurements (tip apex = 40 nm). Because the corners at the tip apex were still relatively sharp, one can assume that this tip may produce good images. From TEM images of the tips, the front and back angles of the tip were estimated to be 17° and 25°, respectively. The opening angle was evidently larger at the tip apex (+17%). The diameter of the plateau at the apex measured approximately 80 nm. Images showed preferential orientation in one direction, and the dimensions of the ridges were approximately 30 × 30–150 nm. The height scale analyses showed that a vertical cross-section roughness of the surfaces was low (R_{ms} = 4.8 nm).

In these investigations, the TEM analyses of a tip used for 2 h, corresponded to approximately 30 m^2 scanned area. It was found that the tapping and scanning had given rise to some to the tip, that is, the end radius and opening angle exceeded 150 nm and 120°, respectively. Even so, the damaged tip could still produce an image with fairly good quality. Furthermore, when the tip was exerted to high loads, it was found from SEM to have severe distortion at the place of surface contact. The tip radius at the probing site was estimated to be greater than 150 nm. The opening angle was over 120°. One could also notice lattice strain contrast on the right-hand side, which was absent in the unused tip. This could be most likely introduced by the impact when the tip "taps" the surface. The etched silicon probes of type B displayed a poor overall quality, where approximately 40% of the probes were rejected for imaging and discarded.

The TEM images at different magnifications of an unused etched silicon tip of type B were also obtained. The tip dimensions were similar to the A-type tips but exhibited a much blunter tip apex. In general, to avoid steric hindrance when scanning rough surfaces, the cantilever is usually slightly tilted forward during the scanning, which displaces the contacting tip apex to within the thinner right side of the tip. This particular tip exhibits a crack perpendicular to the cantilever direction. By using electron diffraction and HREM imaging, the crack direction was determined to be along the {111} plane of the cubic silicon structure. This is known to be a preferred cleavage direction in Si. This has occurred on an unused tip, but it seems likely that silicon probes are easily damaged or disrupted in a similar way during tapping, especially on hard corrugated samples. In this report, the image of a used tip of type B was obtained, which indicated some degree of severe tip degradation. The tip radius was found to be ca. 120 nm. One of the last frames to be scanned with this B-type tip showed a decrease in image contrast (equivalent to height span in height mode), compared to the AFM image recorded by the A-type tip, which also indicated an apparent decrease in measured roughness. The mean roughness value, R_{ms}, had decreased by 33%, from 4.8 to 3.2 nm. In the TEM profile image, an amorphous ramp structure was observed at the tip apex with a surface slope of approximately 15, with respect to the vertical tip axis. The adsorbed material could be amorphized silicon from the tip or amorphous carbon contamination originating from the tip or the sample surface. This flat slope in the amorphous material may reflect the fact that the cantilever is mounted slightly tilted forward in the free-standing midpoint position.

The comparison of the images of the tip and sample at an accurately scaled size had shown that it was impossible to obtain any useful image, considering the huge size of the tip compared to the surface features. A TEM micrograph of the apex of a used tip (A-type tip) was used for comparison. The silicon tips of type C fulfill their specifications reasonably well. The TEM images clearly showed a sharp, high-aspect-ratio probe. The tip radius is estimated to be below 4 nm with a cone angle of only 12°. The aspect ratio [i.e., the quotient of tip length (10 μm) and base diameter (4.5 μm)] of these tips was estimated to deviate considerably (30%). It was found that this tip reveals more details and gives a sharper image appearance with a slightly higher apparent roughness (5.0 nm), compared to the images that were recorded with the much blunter tips of types A and B. The intermittent contact mode AFM image (1 μm × 1 μm) of the CeO_2 surface was obtained by the C-type tip (scan rate 1.5 Hz). The crystalline ridges and the valleys were clearly seen in the images. The dimensions of the ridges agree well with the HREM measurements. In these studies, it was found that high image quality was observed for a short period of time. This may be related to the fact that after some time, the tip either broke or became contaminated. It was found that the image suddenly changed and gave rise to a double-tip effect.[61] However, the tip breakage is a handicap in general, although only a serious problem in some systems. For example, on very clean HOPG surfaces, one may continue to use the same tip for hundreds of images without any damage to the tip. It must be mentioned that tip damage can be a difficult problem on some substrates, probably due to excessive roughness. However, this observation is not fully understood at this stage in the literature. The topography, as obtained from the image (500 nm × 500 nm), showed that the same sample surface was monitored only two frames later. It was found that all high features of the surface had been imaged twice. This gave topographies in which each of the ordered ridges contained a substructure, which was located approximately 40 nm away with a slightly lower apparent height (~3 nm). The C-type probe that was used to record this particular topography was immediately withdrawn from the surface and prepared for TEM analysis. The tip apex was found to have been contaminated with amorphous debris, which might have caused the double-tip effect. No signs of truncation of the crystalline part of the tip apex could be detected. This was also observed for the used B-type tip. The width of the debris corresponded to the distance between the apparent double ridges. Furthermore, the height difference between the end and the lower right edge of the debris coincided reasonably well with the apparent height difference in the AFM image. This suggests that the adsorbed amorphous debris acts as an integrating part of the probing tip, and not only of the silicon tip. AFM images in intermittent contact mode (500 × 500 nm) were obtained of the CeO_2 surface. The image was recorded by the same C-type tip only a few scans after the image, but an abrupt change was seen at the tip apex. All of the largest features of the surface were imaged twice, giving the well-known double-tip effect; thus, the twin substructure was located 35 nm apart with a lower apparent height (~3 nm). An artificial preferential direction was found to be induced in the micrograph. The double-tip effect was suggestive of contamination. A correlation was found between the width of the debris and the distance between the apparent double ridges in the AFM image. In another study, the silicon tip was sharpened by depositing an amorphous carbon column at the tip apex by a focused electron beam in a scanning electron microscope.[62] In these studies, the probe was etched in oxygen plasma for a few seconds in order to increase the sharpness of the carbon column. The tip shape can be expected to be deformed through adsorption or incorporation of surface debris, and thus, it may give rise to poor image quality. This is often observed when controls are performed on clean HOPG surfaces. In the author's laboratory, all new series of experiments are initiated only after HOPG images show satisfactory resolution. TEM image was obtained of the same EBD tip, but the image was obtained after usage over the CeO_2 surface. The effect of these rather harsh scanning conditions showed no effect on the EBD tip shape. Although, a CeO_2 flake (as found from EDS) seemed to have affected the tip apex. These microscopic analyses showed that the shape of the CeO_2 flake was disk shaped and asymmetrically positioned around the tip apex. The resulting AFM image was recorded just before the tip was viewed in the TEM. It was concluded that the diffuse protrusions in the AFM topography were most likely due to the protruding CeO_2 ridges. These were probably the inverse images of the tip apex. The diameter of these diffuse protrusions corresponds with the diameter of the adsorbed CeO_2 flake (~110 nm). Even the small height difference between the flake and the end of the EBD tip (~8 nm) can be observed in the AFM image as lighter dots, or as bumps in the cross-sectional profile. Images showed lighter dots that were correlated to the displacement of the CeO_2 flake directly. The coarse contamination did not, however, result in a doubling effect. TEM micrographs of an etched silicon tip of type D were taken, onto which was deposited an amorphous carbon column (EBD tip). The radius of the amorphous carbon tip was estimated to be 16 nm with an opening angle of only 8°. TEM images of the same EBD tip (D) were taken, but they were taken after usage over the CeO_2 surface. A CeO_2 flake (analyzed by EDS), which originates from the sample surface, was found to have seriously altered the shape of the tip apex. However, the EBD tip was found to be unaffected by the harsh scanning conditions. The diameter of the flake was approximately 120 nm and was positioned 10 nm below the tip apex. The CeO_2 flake was found to have blunted the tip, which gave rise to poor spatial resolution in the image (with the apparent roughness of the surfaces being 3.3 nm).

In another study,[70] it was shown that adsorbed material on the tip may give the image information in the AFM imaging process, and not the tip. It is well-documented that tip defects, e.g., bluntness and double tips, are inversely imaged on rough surfaces. In the author's laboratory, this problem was avoided by taking routine images of standard samples, such as HOPG. This procedure avoids any artifacts related to tip defects. The most reliable procedure used has been to first take an image of the clean substrate. Second, image analyses of the sample plus substrate were conducted. However, contaminants, which are inevitably adsorbed on the tip, have previously been considered to nullify the imaging process only.[62] The idea to image structures, e.g., biomolecules, that are deliberately attached on the tip by means of inverse AFM have been proposed, but never experimentally performed, to our knowledge. However, there is always the risk one would destroy, or detach, the soft molecules if the probe pressure became too large. As described in the following, even when the tip pressure is almost zero, the attractive adhesive forces have been shown to induce damage to the sample constituents.[63,64] It was argued that if the tip is sharp, adhesive forces may increase the contact radius by elastic compression or may even dull the tip, which will result in poor imaging resolution. However, it is not easy to quantify the work of adhesion, because it is dependent on environment, surface roughness, rate of pull-off, diameter of the tip contact, and the nature of the involved materials.[65–78] As is well known, most solid surfaces absorb moisture from air, even if it corresponds to a single monolayer or multilayers of water molecules. The relative humidity of ambient air in most laboratories is typically 30 to 50%, which is high enough to induce a thick water film (~9 nm) to adsorb on the sample surface. During contact mode imaging, the presence of a liquid film is probably of little concern for large-scale topographic imaging. However, for high-resolution imaging, the presence of physico- or chemisorbed liquid molecules and the resulting increase in force due to meniscus formation could drastically affect the imaging mechanism. If the film is not homogeneously distributed on the sample, then changes in surface energy and, hence, surface forces will be convoluted with the actual topography in the collected image.[79] It is thus found that AFM scanning should preferably be performed in a controlled environment, e.g., in ultrahigh vacuum (UHV), at elevated temperatures, and immersed in a liquid of suitable polarization in order to render the van der Waals forces negligible. Although the reliability and reproducibility of AFM tips with a high aspect ratio and improved sharpness have increased during the past years, direct verification of the *in situ* AFM tip shape will remain important.[80] In general, removing the probe for inspection in a TEM is impractical. Furthermore, if the probe is asymmetrical, several tilt experiments have to be performed in the electron microscope in order to obtain the true tip shape. However, when the tip is as sharp as the C-type tip, the possibility exists to evaluate the three-dimensional shape simply by studying the change in transmission contrast or the thickness fringes of the silicon. Derivations of the exact probing shape and size of the tip apex may be performed, because the exact geometry of the ridges of certain CeO_2 films on sapphire are known. Cross sections of the AFM image illustrate the apparent shape of the ridges as recorded by the tip. By recording images with different scan directions and subtracting the crystallographic shape of the ridge, reconstruction of the tip apex, with precision to a few nanometers, may be conducted. However, when the tip is shown to wear and assume slightly different shapes during scanning, the results of a tip reconstruction might be acceptable for one particular experiment only. It has been suggested that a good estimation of the tip shape may be obtained by evaluating the quality of the AFM image. This procedure was recommended by leading probe manufacturers.[58] A tip evaluation kit based on a Ti film, which exhibits extremely sharp features, is commercially available. However, the dimensions and angles of the features of this sample vary, which makes an exact tip reconstruction impossible.

One may expect that the tip may become blunted while scanning the reference sample, thereby revealing a duller appearance than that obtained while scanning the actual sample. The reference sample, or the tip characterizer, should preferably have a modulus similar to or slightly lower than the tip. Furthermore, to prolong the AFM tip life, the pressure applied by the cantilever must be decreased. This can be achieved by reducing the drive amplitude, which will result in moving closer to the surface with a lower set point when scanning. Operating on the low-frequency side of the resonance peak may also extend the useful lifetime of the tip. Preferably, AFM tips should be made of a durable, rigid material, which is also hard and chemically inert. The tips should be long enough to reach into narrow trenches and tight corners, but excessively long and thin tips are known to be less durable and possess poor vibration stability. The material of choice has therefore been limited to Si, Si_3N_4, or EBD tips, which are all available on the market.[81] Tungsten (400 GPa) and diamond-coated silicon tips may show increased durability but are more expensive and often exhibit a greater tip radius and enhance the risk for sample wear. The sharpness and the durability of the probe will be the determining factor when choosing the AFM probe. It was found that stability was acceptable, as determined from the indication that exactly the same ridges were found again by HREM after being stored mainly in ambient air for 2 years, with a loss of only one atomic column at the sharpest ridge apex.[70] For AFM, no apparent change of the CeO_2 surface could be observed after this time. The wear of the surfaces was negligible, concluded after scanning a small area several times, followed by expanding the scanning area, including the previously scanned area. No reduction in contrast, i.e., roughness, due to wear of the surface by the tip was detected. Elastic deformations of the sample features, induced by the AFM, have been included in previous investigations, but the elastic deformation of the tip is usually neglected.[82,83] If the modulus of the tip is lower than the sample modulus, tip deformation should be predominant over sample deformation. It was found that under these conditions, the prerequisite conditions reported to give true atomic resolution imaging with AFM are present.[40] However, the geometrical factors, i.e., the radius and length of the tip, are dominant.[44] Any attempt to determine the elasticity of the CeO_2 film would be misleading, because the properties of the sapphire substrate strongly influence measurements on the thin CeO_2 film. However, the value of sintered CeO_2 powder has been reported to have slightly higher modulus (~170 GPa) than the silicon probe material, which would contribute to tip wear and disruption during scanning.[84]

It has been reported that images can be obtained with lateral atomic resolution when imaging close to atomically flat surfaces.[85,86] Under these conditions, it is the protruding end of the tip that is active. Calculations of elastic indentation make it possible to estimate the size of the contact area in a tip–sample interaction.[82] In SPMs, the size of the tip is found to be several orders of magnitude greater than the size of an atom or molecule, as expected (Figure 2.5).

These experimental facts contradict the idea that only one atom of a probe contacts the surface during imaging. In fact, one can obtain atomic resolution by using a wire merely cut by scissors in STM. The participation of several atoms in interaction with the surface is more probable. Several suggestions and theoretical models have been proposed to explain how atomic resolution can be achieved for atomically flat surfaces, but it has really never been understood how this can be obtained for more corrugated surfaces.[87]

However, experiments show that the furthermost atom on the tip is the image of the tip. What imaging mechanism is responsible for the highly resolved, but not atomically resolved, when considering the tremendous size difference between the tip and sample features? If it was accepted that TEM measurements give realistic indications of the dimensions of the CeO_2 ridges and the tip, then how can a tip, which is more than 300 nm in diameter, track and resolve features less than 40 nm wide and measure depths deeper than 30 nm? It has been shown that close to atomically sharp tips do not necessarily produce sharper images, and that they are unstable, being easily blunted, spallated, or contaminated. A simplistic idea is, therefore, that during the long scanning time with the relatively hard CeO_2 film, the broad tip eventually breaks, producing nanometric sharp and long protrusions, which then can map the surface features. However, it is highly improbable that all types of tips would break in a way that produces better images, and from our results, it does not seem relevant for EBD tips, which become contaminated instead of blunted. An alternative explanation is that the tip is actually formed during the scanning of the surface by adhesion of some surface material on it. Highly resolved images are then produced by this adhered material, and not by the tip. Depending on the applied force, this adsorbed, relatively soft debris, may be slightly deformed at one or every tip contact oscillation, and may then be molded according to the surface topography. The apparent height will consequently be recorded as the point where the voids at the tip–sample contact area have been filled with the debris of the tip, fully or partly, depending on the applied force. This may be one of the reasons why, sometimes, an increased surface roughness is observed when the cantilever force is increased. A comparison with *contact mode* AFM images suggests that the tip–sample interactions, which dominate contact and intermittent contact mode image formations, are different. The contact (repulsive) mode involves mainly a true hard-contact convolution mechanism, while the intermittent contact mode is more dependent on a tip–sample material exchange and is strongly influenced by adhesive forces.[88] In contact mode of AFM, however, prolonged scanning of the tip has been shown to give better images, provided that the scanning is maintained with identical conditions, avoiding abrupt topographic variations.[41] However, these findings were not extensively investigated and should be considered as preliminary conclusions. The adhesion of gold particles on a contact AFM tip has been shown to nullify the imaging process.[39]

But, this effect has never been observed in intermittent contact mode AFM. Considering that the contact time between the tip and sample is very short (<10 s), and the shear forces between the tip and sample surface are significantly reduced, the probability for intermittent contact mode of such relatively large material interchange is very low. Therefore, only a defined material interaction with the tip apex can be maintained, which might, as was shown, actually enhance the imaging resolution.

It has been argued that broader tips with a large opening angle induce greater lateral enlargement of features, not only due to convolution but also due to the increased friction and adhesion forces on the surface. Further, these tips are found to be less prone to fracturing and to bending and flexing around the vertical axes.[89] It has also been argued that the possibility of achieving high resolution in depth on steep and corrugated surfaces may be limited. And, fragile, super-sharp crystalline tips may mean weaker attractive forces but may minimize the possibility of attaining adsorbed material, which might enhance the imaging resolution in a positive way. This suggests that if no sufficiently sharp and durable tips can be fabricated, somewhat broader AFM tips should be preferred, compared to the smallest features of the sample surface.

TABLE 2.2
A Typical Probe Specification

Spring constant	(N/m) 20–100
Resonant frequency	(kHz) 200–400
Cantilever length	(μm) 125 (or larger)
Nominal tip radius of curvature	(nm) 5–10
Tip angles	17 side, 25 front, 10 back

At this stage in the current literature, one may conclude that based on various investigations, the sharpness of the tip is the most important parameter. This may be especially concluded from a study where four different types of probes for intermittent contact mode AFM from different manufacturers were compared before and after scanning. The results of these tips showed that well-defined corrugated CeO_2 thin-film surfaces can be used to characterize the quality of the probing tip. A large difference in shape and quality between different kinds of etched silicon probes exists, but this has a minor effect on the quality of the resulting image. The sharpness of the tip will also be vulnerable to damage. Electron-beam-deposited tips have been found to exhibit rather high durability. However, partial embedding of larger, harder fragments into the tip can easily contaminate them. On the other hand, the author's experience has been that cantilever tips generally remain unperturbed for many images. In fact, the tip is broken much too often, and the other parameters seldom occur. Summarizing, one must initiate a series of experiments with images of a test sample (such as HOPG), and one must also measure the test sample in between the series. This is the most rigorous procedure to use to avoid artifacts being measured due to the tip shape, etc.

Data of a typical tip are given in Table 2.2.

2.3.4 Effects Related to Thermal Drift

Temperature variations of surroundings can give rise to various kinds of effects.[26] However, if the images obtained can be reproduced repeatedly, then these effects may be considered to be absent or negligible. Further, because the scanning is fast, in most cases, the thermal drift is also not much of a consequence. Throughout the text, the role of thermal drift will be mentioned. However, as in all physical systems, one must perform experiments under controlled temperature conditions. The images of HOPG were repeatedly found to be consistent, which allows for the conclusion that under normal laboratory conditions, thermal drift effects are negligible.

2.3.5 Effect of Mechanical Vibrations

In all kinds of microscopes, the effects of vibration have to be kept to a minimum. It is obvious that in any sensitive instrument, such as a SPM, the effects of mechanical vibration will be important when the operation requires any mechanical movement. The vibration effects are generally avoided by special procedures as used for the suspension of the instrument, or other methods, such as the use of vibration-free tables or suspension tables. On the other hand, if vibrations become a problem, stripes are generally seen in images. The simplest procedure used is to suspend the whole device in air using rubber bands. One can also use a tripod, as used by photographers. In the author's laboratory, this has been found to give high vibration-free setup. Both STM and AFM provided sharp images of HOPG, as described later. In some cases, the use of rubber tubing has been effective.

2.4 MODES OF OPERATION OF AFM

Originally, most of the AFM operated by keeping a constant force between the tip and the substrate and by moving the substrate, which provided useful images. This was also found to be successful in most cases. However, recent investigations have shown that there are many other modes by which sharper images can be obtained, as described below. At this stage in the current literature, there is not enough data to clearly describe whether any one mode is preferable to another.

The contact mode is generally used in most AFM procedures. In direct contact, "dc," the method is almost similar to the analogy of a needle touching the record player with constant force. The SFM or AFM operates with considerably lower loads and higher resolution. The feedback signal keeps the specimen moving (i.e., the distance between the speciman and the tip) and thus keeps constant force. The tip of the "dc" mode force microscope is mounted on a cantilever that has a spring constant in the range of 0.01 to 10 N/m. The tip is in the repulsion region, as described in detail in the following.

Contact mode can also be operated under fluids with similar high resolution to that in air. The AC or (tapping) mode is achieved by oscillating the tip close to its resonance frequency (approximately 1000 Å). This has provided much improvement in images in some systems.

In tapping-mode AFM, the probe encounters the attractive and the repulsive force fields of the sample during a period of vibration of the cantilever.[90]

In noncontact mode, the tip is never in contact with the substrate, which differs from the contact mode. This mode is found to be most suitable for soft materials.

Since the introduction of AFM, a large variety of the new AFM techniques have been invented and revised.[5–9,26] At this stage in the literature, there are four main types of AFM, as described later. The first is contact-mode AFM.[9,26] This is certainly the most well known and the most resolving mode.

Influence of the repulsive force field presents the vibration amplitude variation of the amplitude of the tip when moving near the sample surface. This is related to the influence of the attractive force field. In order to observe the attractive force influence, experiments were carried out by setting a smaller vibration amplitude (~120 nm). In that phase, bistable behavior for frequencies smaller than the resonant frequency can be observed but not for frequencies larger than the resonant. The study of the amplitude of vibration of a tungsten tip (amplitude of several tens of nm, stiffness constant of the cantilever 30 N/m) in interaction with a solid surface revealed that *tapping-mode* AFM is sensitive to attractive and repulsive force fields of the sample. Indeed, the vibration being large, the tip meets with these two types of force during a period of vibration. Several experiments described the respective influence of the attractive and repulsive force fields on tip vibration.

2.5 SIMULTANEOUS AFM AND SCANNING NEAR-FIELD FLUORESCENCE (SNOM AND SNOM–AFM)

In more recent literature, STM and AFM have been combined with other scanning probe parameters. This novel technique is being pursued in different directions and most likely will be the most exciting development in future SPM applications. Scanning near-field optical microscopes (SNOM) permit optical imaging on a sub-100 nm scale by scanning a nanometer-sized optical probe in close proximity (5 to 10 nm) over the surface under investigation. The principle of SNOM can be seen from the following:

TIP (OPTICAL SENSOR)......v
SAMPLE ON____________
SUBSTRATE........................

SNOM is becoming one of the most useful microscopes of these types. The advantage of SNOM is inherent spectral information, which allows for chemical imaging of surface adsorbates (under static and dynamic conditions). The confinement of light by subwavelength scatterers is known to play a dominant role on SNOM imaging and to make resolution far beyond the diffraction limit possible. It has only been within the last half century that these concepts of superresolution microscopy in the near field have been vigorously pursued and experimentally demonstrated.[91] In this, a new family of scanning near-field optical microscopes (SNOM–AFM) have been developed. A SNOM–AFM apparatus with feedback signal from AFM in the noncontact mode was described. The liquid cell consisted of glass plate placed in contact with the microscope window. The probe was immersed in the liquid. SNOM offers the potential for imaging surfaces spectroscopically at even higher resolutions.

The SNOM–AFM microscope should be expected to be very useful for biological systems. This method will be much more useful for studying cellular structures in living cells at resolutions and contrasts unobtainable by conventional optical microscopy.

In the most common configuration of *SNOM*, the sample is illuminated with an aperture and the transmitted light is examined.[92] These applications of the floures-cence method to AFM open a vast area of research, especially of biological interest. An interesting alternative setup that was proposed uses an uncoated fiber tip for the illumination and collection of the reflected signal from the sample.[93] Not requiring a coated fiber tip and being self-aligned are two main advantages that motivated several groups to explore this configuration, experimentally and theoretically. The optical resolution of such an arrangement was investigated experimentally. It was concluded that the apparent sub-100 nm resolution that is obtained in the optical signal is topography induced. Further, it can be concluded that the true optical resolution is only of the order of λ/2, and diffraction is limited.[92]

Single-molecule spectroscopy at nanometric scale studies at ambient environment can also be studied using SNOM.[94] The tetrahedral tip[95] is used as a light-emitting probe for SNOM and, simultaneously, as an STM tip.[96] In mixed-metal films, silver grains could be easily distinguished from gold grains at a lateral resolution in the 10 to 1 nm range by their specific near-field contrast. Plasmons excited on the faces as well as on the edges of the metal-coated probe are thought to be the cause of light compression to the nanoscopic dimensions of the tip, responsible for the high lateral resolution. SNOM at a molecular resolution exploiting local plasmon excitation of the probe for contrast enhancement is a challenging perspective of SNOM with a tetrahedral tip.

2.6 FRICTION FORCE MICROSCOPY (FFM)

It is obvious that when AFM studies are based upon near contact distances (nm) between two bodies (tip and the substrate), these data will also be related to the characteristics of the *friction forces*. These friction forces have been measured by designing a FFM.[3] This application of AFM is perhaps the most important with regard to various industries, where friction and lubrication are important matters (building and tunnel, car engines, tire industry).

The interaction forces, attractive or repulsive, as small as a few nano-Newtons (nN = 10^{-9} N) between the cantilever and the sample can be measured. The biggest advantage by AFM is that most samples can be investigated in their natural state, including biological samples (even in an aqueous environment), which is otherwise impossible by electron microscope methods. In the case of repulsive force procedure, the tip physically touches the sample (with a sensitivity of nN force) and traces across the sample without damaging the morphology. In AFM, the sample moves under the cantilever, which is the main reason that the resolution by AFM is somewhat less than that by STM. The images of HOPG are almost of the same resolution as by STM. Furthermore, in some studies where STM and AFM have been compared, images of the same quality have been reported. This has been especially true in the case of calibration grids. However, resolution is highly dependent on the characteristics of the sample (i.e., surface texture and even the degree of hydration) as well as on sample preparation. This point is not settled completely, and one should be careful in reaching any conclusions at this stage.[21] In all of these measurements, the sensor (tip or the cantilever) is moving at molecular distances from the substrate. This means that the images will also depend on such conditions as moisture content or any other impurity in media (air or fluid).

Force versus distance curves are described in much detail later, but for now, an introduction is warrented. Most AFM instruments have the capability to move the tip at a fixed point over the substrate while measuring force. This software facility is available in almost all commercially available AFMs. Essentially, one turns off the z-feedback, and the sample is made to move in and out of the force region. The force is then estimated from the spring constant of the cantilever multiplied by the deflection. The force will then provide information as described in all physical chemistry textbooks. Furthermore, because these AFM force curves can also be executed under fluids, an added dimension to surface and colloid chemistry is provided.

2.6.1 Forces in AFM

It is obvious that in such a setup where two bodies are in close proximity, various kinds of forces, such as van der Waals, hydration forces, and electrostatic forces, will be present.[3,26]

In a force versus distance experiment, the tip moves from a large distance from the substrate toward the substrate, until the tip touches the substrate, and thereafter, it retracts to its original position. These forces can be characterized as follows:

Force	Distance between Tip and Substrate
No force	Large distance (μm)
Prior to touching	Attraction forces/tip bends downwards toward substrate
Almost touching	Repulsion forces/tip bends upwards
Touching	Contact/tip bent downwards

The magnitude of force is proportional to the cantilever spring constant multiplied by its deflection. In other words, AFM allows for the measurement of these forces in a simple manner, as compared to other force apparatus. Prior to this AFM feature, forces at such close distances could not be measured under ordinary conditions. However, in the current literature, a need for investigating these areas of application of AFM is found to be needed.

2.6.1.1 Van der Waals Forces

All bodies exert van der Waals forces when in close proximity, i.e., almost of molecular dimension. The magnitude of van der Waals forces is generally proportional to distance, r_d, as $1/r^6$, and is also called dispersion force.[2,3] In this case, the tip, with radius R_{tip}, is at a distance r_d from the substrate, and the F_{vdw} (vdw = van der Waals) force is given as:

$$F_{vdw} = A_{ham} R_{tip} / (6 r_d^2) \tag{2.5}$$

where A_{ham} is the Hamaker constant and is dependent on the polarizability of the materials.[2,3] This relationship is not time retarded, hence, only valid for distances up to 10nm and for tip radius R = 100 nm.

2.6.1.2 Electrostatic Force

An electrostatic force will exist between the tip and the substrate as a function of distance:

$$F_{el} = (q_1 q_2)/(4\pi\varepsilon D_d^2) \tag{2.6}$$

where q_1 and q_2 are the two charges, ε is the dielectric function of the medium, and D_d is the distance between the charges. Assuming a flat substrate, the electrostatic energy, W_{el}, between the two charges and with capacitance, C_c, with a voltage, U_v, can be given as follows:[2,3]

$$W_{el} = -1/2 C_c U_v^2 \tag{2.7}$$

$$F_{el} = -\nabla W_{el} \tag{2.8}$$

In the limiting case, $D_d << R_{tip}$, one gets:

$$F_{el} = -(\pi\varepsilon R_{tip} U_{el}^2)/D_d \tag{2.9}$$

2.6.1.3 Hydrophobic Forces

The special long-range force between hydrophobic macroscopic surfaces in water is called the hydrophobic force or effect.[20,97] This interaction is not completely defined, and there are probably more than one reason why hydrophobic macroscopic surfaces attract each other, as many suggestions exist as regard to its origin. What seems perfectly clear, however, is that it is the water molecule and its unique properties that are the key components in this interaction phenomenon.[98] The hydrophobic attraction has been measured for distances up to 90 nm using a surface force apparatus. This attraction differs from short-range attraction between small hydrophobic molecules due to its anomalous long range of action.[99] The interaction between hydrophobic moieties in water (i.e., hydrophobic interaction) plays an important role in the forces that stabilize the self-assembly of organized structures such as micelles, lipid bilayers, surface films, protein–lipid complexes, and biological membranes.[20,100] The direct force measurements have revealed, somewhat unexpectedly, that the hydrophobic force is long range, for example, attraction beginning from distances greater than 50 nm is commonly measured. Despite a large number of studies reported in current literature, much remains to be investigated. One important aspect that has not been extensively investigated is the manner in which the surfaces are prepared.

2.6.1.4 Double-Layer Force

Double-layer interactions take place between any two bodies with charge surfaces when they approach each other in a liquid.[2,3] This interaction is, in most instances, repulsive, and it is dependent on the ionic strength of the aqueous phase and on the surface potential (surface charge density) of the two bodies. The interaction force is actually the charge reflection toward the solid surface and the aqueous media.

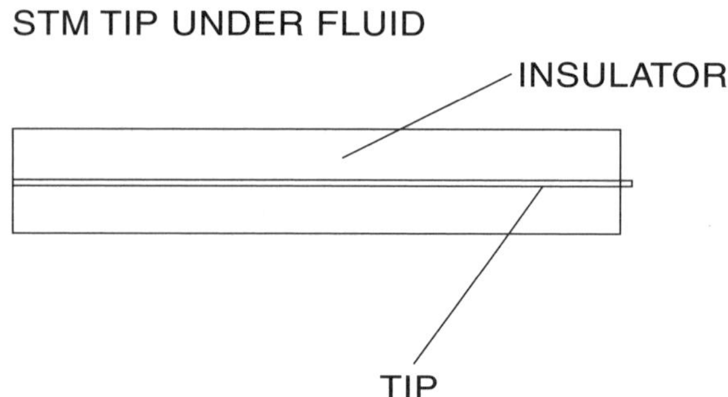

FIGURE 2.8 STM tip modified (partially covered by a nonconducting material, such as wax) for operation under fluids (schematic).

2.7 STM AND AFM STUDIES UNDER FLUIDS

The biggest limitation of electron microscopy has been that in spite of obtaining molecular information, these studies could not be carried out under fluids. For example, even though detailed images of biological molecules have been revealed by x-ray, these were only carried out after drying and other treatments. In some cases, the biological material has been treated with such fluids as glycerol, in order to maintain a certain degree of fluidity under electron microscopy. The later treatment may or may not affect the molecular structure. In STM and AFM, no such extra treatment is necessary.

```
TIP------------------------v
FLUID=====================
SUBSTRATE__________________
```

The STM can be modified so as to be able to function under fluid, as described in the literature.[101,102] The tip is covered by some suitable insulator (wax or Teflon), except at the end of the tip, through which the tunneling current is measured (Figure 2.8). A rather simple method of modifying the commercially available AFM is shown in Figure 2.9. The principle is to glue a glass slide just over the cantilever. A drop of fluid is then added under the glass plate such that the sample is covered (10 μL is often enough volume). The laser beam can pass through the glass slide and the fluid in order for AFM to operate for fluids.

A different method is shown in Figure 2.6.[103] For some of the apparatus, such cells are commercially available. In this report, living renal epithelial cells in an aqueous environment were studied by AFM.

In all kinds of everyday systems, one would like to have the knowledge of molecular information under dynamic conditions and indifferent media (e.g., gas or liquid). The STM and the AFM can be modified with a so-called fluid cell, which allows one to measure under almost any kind of fluid. This invention is the greatest advancement in the modern microscopy field.

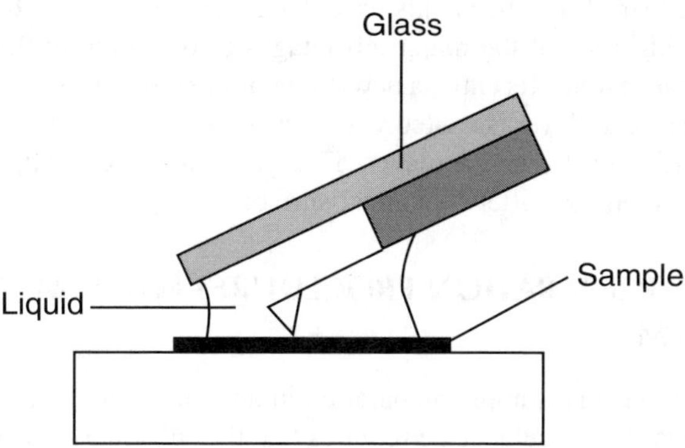

FIGURE 2.9 A simple modification for fluid cells with a glass slide over the cantilever is depicted for AFM. (From Birdi, K.S., *Handbook of Surface and Colloid Chemistry,* CRC Press, Boca Raton, FL, 1997.)

COVERED TIP

================

----------------------------- NAKED TIP

================

A simple method of modifying the commercially available AFM is shown in Figure 2.9. The principle is to glue a glass slide just over the cantilever. A drop of fluid is then added under the glass plate such that the sample is covered (10 μL is often enough volume). The laser beam can pass through the glass slide and the fluid in order for AFM to operate under fluids. It is also easy to suck out the fluid and to replace with another fluid or composition. In the author's laboratory, this cell has been exenisvely used.

Another method is shown[103] in Figure 2.9. In some apparatus, such cells are commercially available. The cantilever is fixed on a glass slide that forms a closed fluid cell in conjunction with another glass slide and a rubber spacer ring. More elaborate fluid cells are now available commercially with the possibility of adding or circulating fluids (Figure 2.10).

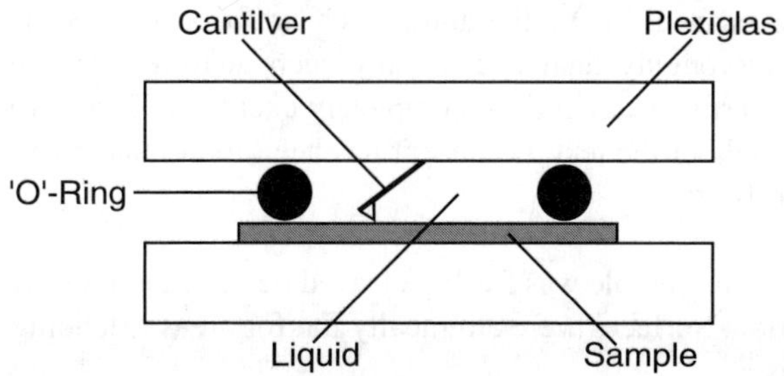

FIGURE 2.10 Schematic of the AFM fluid cell. The O-rings are used to contain the fluid.

A great number of advancements have been made as regards the control of the probe under fluids. One of the major advantages of operation in fluid media is the relative lack of adhesion. Recently, a new mode of operation was reported (magnetic AC mode).[104] The cantilever is coated with a magnetic film, and the solenoid is used to apply force directly to the cantilever. This gives more gentle tip movement and is found to be useful for softer biological surfaces.

2.8 SAMPLE PREPARATION PROCEDURES FOR STM AND AFM

It should be obvious that sample preparation in any microscopy procedure is an art. To obtain images at molecular scale requires that the substrate have almost no effect on the images of the molecules under analyses. Consider that the end image will consist of information that may not be easily discernable to the eye. The image will need further analyses and comparison. The preparation procedure is well developed in the case of electron microscopy (which is many decades old). One of the most stringent requirements is having well-defined sample requirements. However, when one considers molecular microscopy, these demands become even more critical. Because the procedures used are dependent on each sample and its characteristics, only general procedures are given in this section. The procedures needed are relative to the substrate and to the test substance. The selection of the substrate becomes very important in those cases where the test substance may not exhibit adhesion to the substrate. These various parameters are described throughout the text.

As is well known from current electron microscopy literature, sample preparation is the basic determining factor for obtaining high-resolution images. Accordingly, one may use the same well-known procedures for STM and AFM analyses, with appropriate modifications. Some of these procedures will be given in the following. However, in most of these procedures, surface phenomena are mainly used for better sample preparations, as described below.

2.8.1 SUBSTRATES

The range and variety of substrates used in STM and AFM studies is large, and sometimes, extensive preparations are involved. The first criteria of a substrate are that it must be clean and as free as possible of any defects that may turn out as artifacts in the images. In the literature, such artifacts have been mentioned, but these are not thoroughly analyzed. In any such analyses, self-controls must be included so that artifacts are almost completely excluded. These procedures will be mentioned throughout the text, because it has been found that STM and AFM need such self-controls.

- Mica — The sample was freshly cleaved before Langmuir–Blodgett (LB) deposition. Surfaces were atomically flat for areas extending over tens of microns.[105]
- Graphite — It was recognized at an early stage that atomic resolution on HOPG could be achieved. HOPG is commercially available. The surface

is almost molecularly flat (as seen from the *z*-axis scale in Figure 2.2). The distance between carbon atoms is used for calibration for STM and AFM. HOPG has proven to be the most useful for STM and AFM. The surface of freshly cleaved HOPG (by peeling with a tape) provides a highly molecularly smooth and clean surface. The same specimen can be used many times, thus allowing for few apparatus adjustments, as delineated herein. In a recent study, the *ab initio* periodic Hartree–Fock calculations for interpretation of the STM images of graphite were reported.[106]

- Gold surfaces — Highly flat gold surfaces are commercially available. However, pure gold plate (24 carat) when heated just below its melting point has been found to give a highly flat surface.[21] This is a useful substrate, because it is readily available and can be easily cleaned. Gold represents an excellent substrate, because it has no surface oxide and remains clean for long periods of time. High-resolution images have shown steps of 0.25 nm.

2.8.2 Diverse Substrates

Because a large variety of substrates was used in the literature, these substrates will be described in more detail throughout the text. Lead dioxide and galena (PbS) have also been used as substrates.[3]

2.8.3 Langmuir–Blodgett (LB) Films

As described in detail elsewhere,[20,100] monolayers of lipids and biopolymers are spread on liquid interfaces to form self-assembly structures. In order to study the strictures of these structures, monolayers must be transferred to solid substrates. In studies of lipids and biopolymers, the LB layers are prepared according to the standard literature procedures.[100] Transfer of amphiphile molecules [e.g., fatty acids, lipids, or macromolecules (synthetic polymer or biopolymer)] monolayers is generally performed at a low speed of 1 cm/min at a constant surface pressure of the lipid film. The substrate used for the transfer can be a freshly cleaved HOPG or any other suitable material (gold sample, mica, etc.):

Formation of LB Film from Monolayer on Water
LB FILM
MONOLAYER ON WATER
TRANSFER TO SOLID

In the study of lipids and biopolymers, the LB layers are prepared according to standard literature procedures.[20] Transfer of a lipid or biopolymer monolayer is generally performed at a low speed of 1 cm/min at a constant surface pressure of the lipid film. The substrate used for the transfer can be a freshly cleaved HOPG or any other suitable material (gold sample, mica, etc.). Because of their applications in the areas of nonlinear optics, molecular electronics, and biosensors, LB films have a very important role.

2.8.4 Biopolymer Samples

The analyses of biopolymer samples are varied, because these materials cannot be treated by the same procedure. This arises from different reasons. The concentration of biopolymers cannot be easily detected if samples with nanogram/mL concentrations need to be used. Another difficulty is the need to use buffers. These electrolytes that would be present in much higher concentrations than the biological sample may contribute background problems in image analyses. The major efforts used in biological analyses are therefore expended in avoiding these electrolytes in the sample preparation. These different procedures will be described in detail herein.

The analyses of DNA have been made as follows.[108] Mica was soaked in magnesium acetate solution (2 h) and, thereafter, sonicated in water. After drying in air, a drop of DNA solution (50–300 ng/20 μL) was applied. DNA was also investigated after fixing on gold surfaces by 2-dimethylaminoethanthiol. Images of entire plasmid molecules were obtained.[109]

In another procedure, a drop of dilute solution of biological macromolecule (DNA or proteins) was imaged by AFM after water had evaporated overnight.[20]

Other procedures used for sample preparations are as follows:

1. **Solution evaporation or spray:** In most cases, merely a dilute solution (10 μg/mL) was applied as a drop (a few microliters) to the substrate (HOPG or gold or mica), and STM or AFM was carried out after evaporation. The aqueous solutions of proteins or peptides were performed by this method. Because the volume of sample used is so minute, the chance of finding nanometer particles by microscopy may be vanishingly small. A procedure was developed where a drop of protein or virus solution (1–10 μL) was used.[110] The graphite sample was barely allowed to touch the surface of the drop of the solution, thereby, a monolayer of material (biopolymers, virus, cells) was attached to graphite. However, this can only be useful in the case of surface-active substances.

   In order to analyze images, it is important that the sample under study be as well defined as possible, i.e., free of any unwanted impurities. This is achieved in most cases by dilution, because minute impurities are diluted further. The evaporation procedure has been found to be the most useful in all kinds of SPM studies.

   This procedure has been found to be the most useful method. For example, in the case of studies of mixtures, mixed solutions of varying ratios can be studied, and the evaporated mixture when studied provides useful information in regard to packing and other phase equilibria. The molecular interactions can be investigated. Examples of such studies are given below.
2. **Drop-substrate contact (only for surface-active compounds):** Although until a decade ago electron microscopy was useful for biological systems, it had large drawbacks. For example, the volume of a negatively stained sample that could be examined by electron microscope is so small, that the chance of finding particles can be vanishingly small. This is a typical

problem associated with samples sent to diagnostic virology analyses. Hence, in order to obtain dense distribution of virus particles on a grid for electron microscopy, viruses must first be concentrated from dilute suspension, for example, by centrifugation. Virus particles suspended in a drop of water tend to concentrate at or near its surface, with the air.[110] This surface activity arises from the presence of proteins.[20] The concentrated and probably more purified particles may then be collected on a film-coated grid for negative staining and electron microscopy.[20] This is a useful method, more simple than others [e.g., high-speed centrifugation, lyphogel, or precipitation by $(NH_4)_2SO_4$] that are used to process clinical specimens for diagnosis, where virus particles may be too dilute in the original sample. It was shown, by freeze fracturing for electron microscopy, that most of the virus particles accumulate at the surfaces of drops. The freeze-fractured drop can be visualized as follows:

.......VIRUS PARTICLES AT SURFACE...
.......DROP OF VIRUS IN BUFFER................O

Virus particles were found at the edge of the freeze-fractured rim due to their surface-active characteristics.

2.8.5 STM AND AFM ANALYSES OF ELECTRON MICROSCOPE GRIDS

Electron microscope procedures are much more advanced than in the case of scanning force microscopes. In the author's laboratory, the grids as used for the electron microscope have been analyzed by AFM and by STM (if a conducting layer is present). In fact, this procedure allows for comparison of these different techniques. However, STM and AFM provide three-dimensional images, which are more useful.[3]

2.9 CALIBRATION AND IMAGE ANALYSIS OF STM AND AFM

Calibration of any instrument, such as a microscope, is necessary. The calibration of STM and AFM in the *x–y* direction is accomplished by using HOPG or suitable grids available commercially. The calibration in the *z*-direction has been a somewhat more difficult problem. This also arises from the fact that the *z*-direction calibration needs to be performed individually on each apparatus. This calibration control is sometimes more demanding in the case of SPMs than in the case of other microscopes. At this stage, the best procedures are based on the following:

1. Use of gold samples with steps of one layer of gold
2. Use of lipid layers deposited as LB films[3,20,112]
3. Use of macromolecules with known dimensions[3,20,113]
4. Use of HOPG

The observation of surface corrugation of HOPG by STM in air has been reported.[111] These procedures are delineated in the review herein.

Image analysis is varied and covers a wide range of possibilities. Filtering techniques include high- and low-frequency as well as two-dimensional Fourier transform filtering. However, scrupulous preparation and extreme care will yield excellent images.

In the case of STM and AFM, one needs to calibrate in the *X–Y* and the *Z* directions. The calibration of STM and AFM in the *X–Y* direction is accomplished by using HOPG or suitable grids available commercially. In Figure 2.11, the images of a grid are shown.

The images can be viewed in two-dimensional and three-dimensional modes. The versatile treatment of these images is a great advantage over ordinary photography procedures. The printout can be of very high resolution. Because the images are digital, extensive image analyses procedures have been reported in the literature. Generally, the images are in TIF (or converted from IMG or similar form) and, therefore, are compatible for analyses by various image software programs. However, other image form files can also be easily processed (such as BMP, CDR, GIF, JPG, etc.). The reader is referred to consult the software that covers this image analyses subject in the literature. However, most software programs can easily handle these image files.

Self-assembled multilayer thin films consisting of alternating layers of ca. 6 nm Au nanoparticles and dithiols have been prepared on glass substrates.[114] They have been studied by UV/Vis spectroscopy, ellipsometry, STM, and temperature-dependent conductivity measurements. Au sols were prepared in toluene. Substrate glass microscope slides were cleaned in pirana solution (H_2SO_4/H_2O_2). Dithiols used were 1,6-hexanedithiol, 1,9-nonadithiol, and 1,12-dodecanedithiol. STM images showed Au particles in the expected range of 6 to 10 nm.

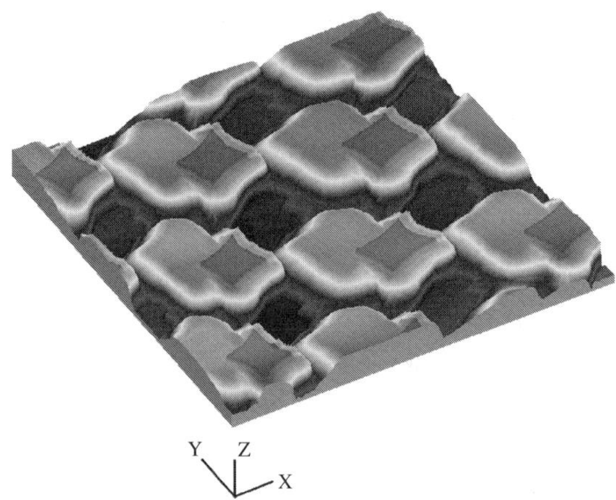

FIGURE 2.11 AFM image of a commercially available grid for calibration of AFM (and STM) (90,000 Å × 90,000 Å — maximum height was 7289 Å).

The z-calibration is also sensitive to the particular system being used and the system under study. Therefore, specific procedures and routines are required in order to be able to determine accuracy. Generally, the images are in TIF form and are, therefore, compatible for analyses by various image software programs. It is the experience in the author's laboratory that calibration of each microscope must be carried out using well-defined substrates. Further, there must be special experimental procedures that will enable these calibrations to be carried out, as this can ensure that there are no artifacts related to mechanics or electronics of the microscope. One can easily compare this to such other instruments as microbalance, viscosimeters, pH meters, or spectrophotometers.

2.10 COMPARATIVE STUDIES OF DIVERSE MOLECULES BY STM AND AFM

It has been argued that AFM or STM should principally provide the same kind of morphological features as reported by electron microscopy.[115]

This is the case when such a surface as graphite (HOPG) is analyzed. The images obtained are the same by both methods. It may depend on the radius (or rather the shape) of the tip and on the geometry and the physicochemical state of the biological material (for example, adsorbed water). The less corrugated the objects are, the less important the actual tip shape. This is found in the case of near-atomic resolution images in the case of LB films. In the case of STM, better resolution can be obtained by averaging the signal-to-noise ratio.

Even when the contact area is comparatively large, the pressure in the probe is still quite high.[116] It has been calculated that a force of less than 10^{-11} N would be required to avoid specimen deformation. Under these conditions, for a fully hydrated biological material, compression might occur under various AFM studies. Furthermore, the degree of deformation of a macromolecule would also be dependent on the substrate.

3 Lipid-Like Molecules on Solids and SAMs

Lipids and lipid-like molecules are known to play an important role in various aspects of everyday life (biological cell membranes, fats and nutrition, soaps and emulsions, lubricants).[20] Physisorbed monolayers of alkylated molecular species have been extensively investigated. Only little is known about molecular structure and dynamics at domain boundaries of two-dimensional molecular crystals. In the literature, many studies have been carried out on long-chain hydrocarbon molecule adsorption on graphite surfaces as a model system for determining the forces responsible for the adsorption processes.[117–120]

Thermodynamic analysis has indicated that the hydrocarbon molecules adsorb with high affinity to graphite surfaces and that the heat of adsorption increases with chain length,[119] which suggests that the alkane molecules are adsorbed with their long axis parallel to the graphite surface, as also confirmed by STM studies.[120–122] During the past decade, increased interest has been given to a specific system, the self-assembly monolayer (SAM) of lipid-like molecules on gold surfaces.[20]

SAMs are found to play an important role in many areas of industrial applications as well as in biological and pharmaceutical systems. This has been noticed from the extensive number of studies reported on SAMs in the past few decades. The SAM studies have been supported by investigations by SPMs that have provided information about structures at molecular scale. This means that in the future, the application of SAMs in diverse areas will increase extensively based on these investigations. The combination of monolayers on water, SAMs on solids, and SPM studies, thus leads to an almost complete molecular picture of such systems.

Two-dimensional molecular patterns were obtained by the adsorption of alkanes, alcohols, fatty acids, and a dialkylbenzene from solutions on graphite and from STM.[123]

For example, SAMs formed on Au surfaces after immersion in solutions of benzeneselenol (BzSe) and diphenyl diselenide have been studied by STM.[124] It is well known that organosulfur substances chemisorb to gold surfaces to form SAMs.[20] SAMs were formed on mica from diethyl ether solutions of alkylselenols. STM images were recorded in constant current mode. SAMs obtained from BzSe were identical to those found by STM analyses, whether exposed to air or oxygen-free atmosphere. It is known that colorless solutions of BzSe rapidly oxidize in air to produce yellow solutions of diphenyl selenide.

In recent years, due to the advent of *nanotechnology*, there has been considerable interest in the production of metallic nanoparticles and nanoparticulate films. These nanoparticles exhibit unique properties as compared to larger particles. The latter characteristic is in the case of charged particles, the ratio charge/particle being much different than in the case of larger (macrosize) particles. In a recent study, the gold nanoparticles were prepared by ultraviolet irradiation of LB films of octadecylamine, hexadecylaniline, and benzyldimethylstearylammonium chloride deposited from aqueous $HAuCl_4$ subphases.[125] AFM data agreed with TEM analyses, as regard to size and shape of gold nanoparticles. The size of these platelike particles was on the range of 20 to 800 nm across.

The SAM structures of dodecanthiol on Au have been investigated by x-ray, contact angle, and STM.[126] The comparative analyses by x-ray and STM were in agreement. Thermal stability was investigated by STM. SAMs were prepared by immersion of Au surfaces into freshly prepared 1 mM solutions of dodecanethiol in ethanol for 2 h at room temperature. The first process that occured upon immersion of the gold surface into alkylthiol solution was the formation of a *domain*-like growth that competes with gold erosion. These domains then grow with different sulfur lattice positions and different chain tilt azimuths. This process leads to three different kinds of domain boundaries on the surface. These domains were found to be dependent on temperature. At temperatures around 50°C, the chain tilt mismatch azimuth disappears and leads to much larger domains. At higher temperatures,100°C, the top layer of the gold becomes mobile and the depressions fuse with terrace edges.

The main recent interest in alkanethiol SAMs arises from the ease of preparation of these highly reproducible and molecularly well-defined SAM structures.[20] In a recent study, the STM was used to obtain information on the self-assembly process and the surface dynamics of SAMs.[127] STM tips were prepared from polycrystalline tungsten wire using a DC etch. A single-crystal Au was prepared by chemical etching and flame annealing method. The clean Au crystal was incubated in 0.001 M butanethiol ethanol solution. Time-dependent sequence images were obtained after 5, 9, 12, 20, 30, 56, 72, 100, and 127 h. All analyses were obtained from the same surface region. These data showed the formation of simultaneous vacancy-island or pits defects in the top layer of Au. In these images, a time-dependent evolution of ordering domains phenomena was observed. The relationship between pit ripening and molecular ordering was analyzed. The plots of number density and the fractional coverage of SAM were used. During the coarsening phase, the data were found to fit a phenomena logical equation (with a power-law time-dependence as expected from a random-walk-mediated process):

$$N(t) = N_o/(1 + B_p t^{0.5}) \tag{3.1}$$

where N_o is the initial pit number density and B_p is an adjustable parameter. The time, t, at which the number density saturates, corresponds with completion ordering.

These data establish a correlation between facile Au migration and the presence of liquid-phase SAM.

Epifluorescence microscopic studies of monolayers containing dioleoyl- and dipalmitoylphosphatidylcholines (DPPC) domains were observed in pure DPPC monolayers at relatively low surface pressures. These domains grew with increasing surface pressure.[128] Only liquid expanded phase repetitive compression and expansion of the monolayers containing DPPC:DOPC:NBD-PC 49:50:1 at an initial rate of 3.2 Å/molecule produced monolayers with visual properties consistent with there being a preferential exclusion of the unsaturated lipid from the monolayer.

3.1 COLLAPSED LIPID MONOLAYERS (SELF-ASSEMBLY)

Self-assembly monolayer (SAM) structures are an important molecular phenomenon which has only recently been extensively studied.[20] The combination of molecular interactions in certain molecules, especially lipid-like, leads to SAM packing. All lipids (under given temperature and pressure) when spread on an aqueous interface form stable monomolecular films.[23] These monolayers have been used to investigate two-dimensional assemblies, to make organized arrays, and to model more complex biological membrane structures.

Monolayers of lipids at the air–water interface provide ideal systems for the study of two-dimensional phase transitions for such assemblies.[20] Phase equilibria in such SAMs are thus of importance for thermodynamic analyses. All lipids, when spread on aqueous interface, form stable monomolecular films. These monolayers have been used to investigate the two-dimensional assemblies, to make organized arrays, and to model more complex biological membrane structures. As described elsewhere, a lipid film on compression exhibits a collapsed state, followed by a change in compressibility (i.e., the slope of the surface pressure versus area isotherms changes) (Figure 3.1).

In most cases, a drop in the surface pressure, π, is observed. This collapse state (Figure 3.2) is described as arising from the movement of the bilayer to slide over the monolayer.[20]

LIPID MONOLAYER.....AT COLLAPSE PRESSURE.......
....TRANSITION TO COLLAPSE STATE

The magnitude of π after the collapse state remains constant until the surface is completely covered by a trilayer (or multilayer). After this state, the π again rises but with a somewhat lower compressibility. It has been observed that above the equilibrium surface pressure, the π versus A isotherms are in a state of supersaturation, at which a monolayer held at constant π or A shows a definite relaxation phenomena.[129]

The collapse state has remained neglected, because there is no direct or indirect information on the exact molecular structures. The equilibrium near the collapse state exists between a monolayer and a multilayer (generally, a trilayer).[20,113]

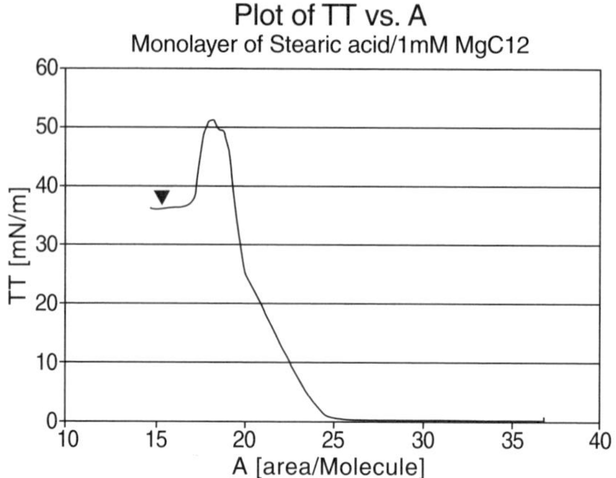

FIGURE 3.1 (a) Surface pressure, Π, vs. area (A) isotherm for Mg-stearate. (From Birdi, K.S. and Vu, D.T., *Langmuir*, 10, 623, 1994. With permission.)

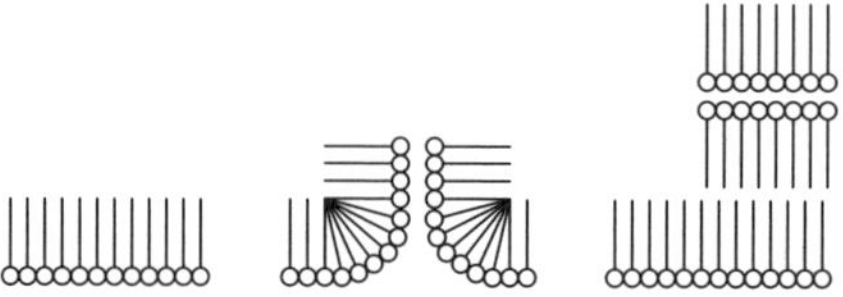

FIGURE 3.2 Schematic collapse state structure (see Figure 3.1). (From Birdi, K.S. and Vu, D.T., *Langmuir*, 10, 623, 1994. With permission.)

3.1.1 Mg-Stearate Films

Collapsed Mg-stearate films were transferred to HOPG as LB films. AFM studies of collapsed monolayers on HOPG (Figure 3.3) clearly showed steps of heights that correspond to the trilayer of Mg-stearate$_2$.[113]

It is of interest to mention here, that clean graphite substrate (HOPG) does not show such step images. These studies were carried out using a commercially available AFM (DME A/S, Denmark).

The AFM analyses in Figure 3.3 show that collapsed film partially has broken up into islands under mechanical handling. However, large parts are perfect in structure. This information is useful when such LB films are to be used in the electronics industry or for biosensors. It is important to mention that no domains were observed in these LB films.

The most advancement with STM and AFM is the possibility of obtaining three-dimensional images, under ambient conditions or other (such as UHV). However, it seems that the literature is not exhaustively complete with this kind of information. One may presume that the research has advanced somewhat faster for the full application of this tool to the data found in literature. These studies showed that a trilayer (step height/length of the molecule = 70 Å/23 Å = three layers) is formed as LB film after the collapse state. This finding agrees with the proposed structure in Figure 3.2. The morphology also indicates that the trilayer is a perfect state. This

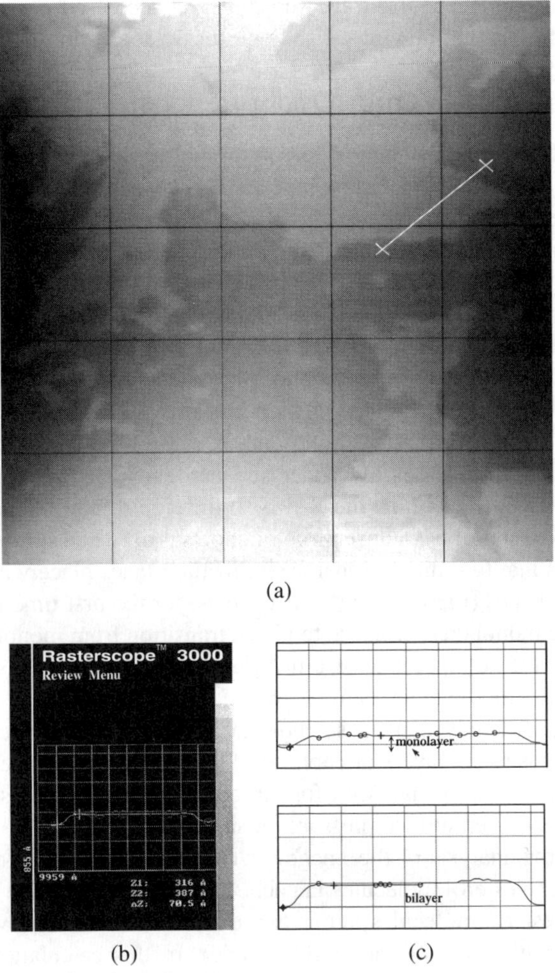

FIGURE 3.3 (a) AFM images of Mg-stearate collapsed films 20 as LB films on HOPG (50,000 × 50,000 Å); (b) step height analysis: Z = 70.5 Å, the light area corresponds to a step height of 70 Å; and (c) molecular model of Mg-stearate (length = 23.4 Å). (From Birdi, K.S. and Vu, D.T., *Langmuir*, 10, 623, 1994. With permission.)

means that equilibrium exists between the monolayer and the collapsed state. It will be shown below that equilibrium is dependent on the lipid molecule. This structure can thus be used as a means of calibration of the AFM apparatus.

AFM studies[20] of collapsed monolayers as LB films on HOPG (Figure 3.2) clearly show steps of heights that correspond to the trilayer of Mg-stearate$_2$.

The height analysis shows that these films consist of multilayers, because the height is consistently 70 Å. This value is much larger than the fully extended chain of Mg-stearate, found from molecular modeling to be 23 Å.

During these measurements, no damage to the LB films by the probe tip was observed. However, changes in AFM pictures are sometimes observed only in those cases where the tip distance from the substrate has decreased, i.e., under increased resolution conditions that may give rise to closer tip separation from the substrate.

3.1.2 Cholesterol and Other Oxidized Cholesterol Films

The cholesterol molecule is the most important biological lipid of interest in the cell membrane structural phenomena.[20] Cholesterol is known to be related to different diseases (cancer).[20]

Bilipid membranes (BLMs) are mainly composed of phospholipids and cholesterol. The main difference between these two groups of lipids is that while the former carry ionic charges, the latter are neutral. In other respects, both kinds of lipids form SAMs on the surface of water. On the other hand, while phospholipids readily form vesicles, the cholesterol molecule cannot and even hinders vesicle formation of phospholipids when mixed (depending on the mixing ratio). However, oxidized cholesterol analogs are even more important, because these molecules are reported to be related to such diseases as cancer and hardening of the capillaries. It was, therefore, of interest to examine the SAMs of different oxidized species of cholesterol. The AFM data of the collapsed film of cholesterol (when spread on the surface of water) shows that two-dimensional crystallization takes place with characteristic *(half) butterfly* shapes (Figure 3.4).[113] This shows, for the first time in the literature, that not all lipid monolayers collapse to give a transition from monolayer to trilayer. This shows that the collapse state is a *two-dimensional crystal* phase, as should be expected from thermodynamic analysis.[20]

The step height analysis of these cholesterol two-dimensional crystals is 90 Å, which corresponds to six layers of cholesterol (90 Å/length of cholesterol molecule = 15 Å). This is different than seen for the collapse films of Mg-stearate, as well known from monolayer surface pressure versus area isotherms.[112,113]

From the area analysis of the image (as obtained from the software of AFM), the number of cholesterol molecules in each "half-butterfly" can be calculated. The number of cholesterol molecules in this two-dimensional six-layer was found to be ca. 60 10^6 molecules (0.0005 ng = 10^{-16} mole). In this calculation, the area per molecule cholesterol was assumed to be 40 $Å^2$ (as found from surface pressure versus area isotherm).[20] This is a new kind of application of AFM, whereby the detection of very small amounts (<nanogram) of materials can be carried out.[113] Further, the morphology can provide information about the composition as well.

Another oxidized homolog of cholesterol, i.e., cholestane, showed collapsed films (Figure 3.5) with two step heights.[113]

This observation shows that a two-dimensional crystal with a narrow size distribution will self-assemble into ordered structures. The explanation is that cholesterol molecules are able to self-assemble under the influence of the molecular forces to form such nanocrystals. In regard to the self-assembly characteristics, there are many molecules with amphiphile characteristics possessing this property.

This indicates that in the collapse state, the trilayer is able to form higher-order two-dimensional crystals, i.e., six layers. These analyses show that in the collapsed monolayers of different lipids, the following phase equilibria (with equilibrium constant, K) exist.[20]

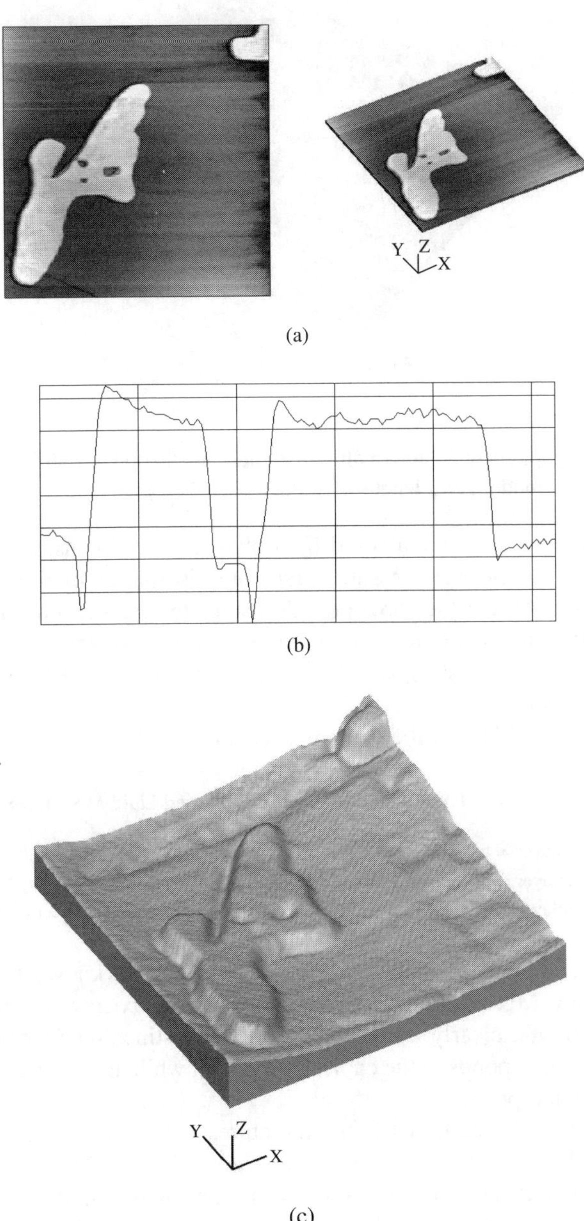

FIGURE 3.4 (a) Morphology of collapsed films of cholesterol (50,000 × 50,000 Å); (b) step height = 103 Å (molecular model gives length = 16 Å); and (c) three-dimensional image.

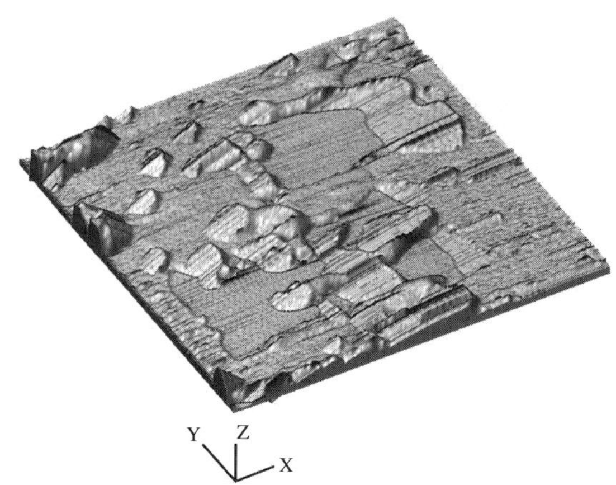

FIGURE 3.5 Morphology of collapsed films of cholestane (50,000 Å × 50,000 Å); step height = 103 Å (molecular model gives length = 16 Å).

The differences between collapsed lipid phases (two-dimensional crystals) are easily seen from the surface pressure versus area isotherms of lipids.[23] It is thus found that such AFM analyses now provide complete three-dimensional structures that can be useful in explaining these differences in the collapsed states. Analyses of the stabilizing forces in such monolayer assemblies as van der Waals forces have been given elsewhere.[20,100]

Lipid-Phase Equilibria in the Collapsed Lipid Monolayer:[20]

Lipid	Phase Equilibria in the Collapsed Lipid Monolayer
Mg-stearate	Monolayer = trilayer, K = (trilayer)/(monolayer)
Cholesterol	Monolayer = hexalayer, K = (hexalayer)/(monolayer)
Cholestane	Monolayer = trilayer, K_1 = (trilayer)/(monolayer) = hexalayer, K_2 = (hexalayer)/(trilayer)

The phase equilibria of oxidized lipids in bilayers are known to be of much interest in biology. Mixed cholesterol plus cholesterol epoxide were investigated by AFM (Figure 3.6). It is clearly seen that there are two distinct, separate phase regions. The flat plateau corresponds to the cholesterol phase, while the light globular shapes indicate the epoxide phase.

It is important to remark that in the literature, monolayer structures have been studied where fluorescent probes have been added. The AFM studies have been carried out with the need of such probes, as the latter might affect the phase equilibrium. Furthermore, the break up of collapsed lipid monolayers is detected by AFM, which will be useful for other more complicated systems. Because the two-dimensional packing of the monolayer on the aqueous subphase can differ from that observed after transferral as a LB film to a solid substrate, the molecular packing in the film in the two systems must be known.

These results show that monolayer structures of cholesterol and oxidized species are complicated assemblies with equilibria in two-dimensional phase. It is for the

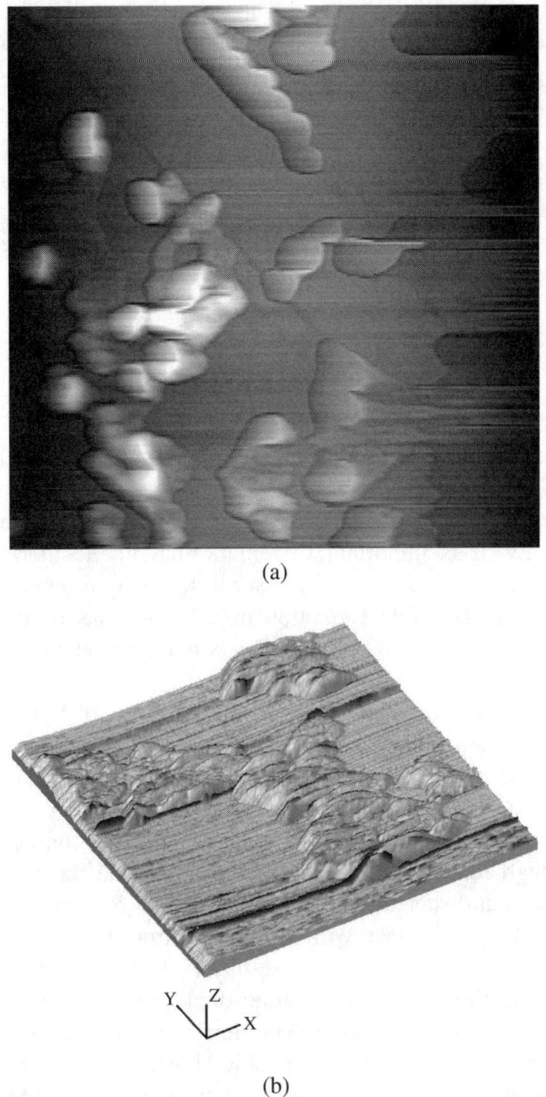

FIGURE 3.6 Mixed collapsed films of cholesterol plus cholesterol epoxide AFM images: (a) two-dimensional image (60,000 Å × 60,000 Å) and (b) three-dimensional image (60,000 Å × 60,000 Å).

first time in the literature that such two-dimensional phase equilibria have been studied. The AFM data of LB films have added much useful information. Especially, these monolayer and multilayer equilibria are of thermodynamic significance.[3] Furthermore, this is the first example of a combination of monolayer–LB film–AFM that provides information of such self-assembly monolayers (SAMs).

In another study, relaxation phenomena were analyzed in detail.[129] The analysis was based on a nucleation-growth theory that describes the two-dimensional monolayer to three-dimensional phase. The growth was related to the interfacial tensions at the center–water–air three-phase contact. The rate of growth was given as:

$$(A_o - A)/(A_\infty - A) = 1 - \exp(-Ct^{3/2}F(k_n t)) \quad (3.2)$$

where A is the total area per molecule, A_o is the initial area per molecule, $A\infty$ is the area per molecule for time $t \to \infty$, k_n is the nucleation rate constant, and k is the growth rate constant. The function $F(k_n t)$ was approximated by series expansion. All the constants were given as:

$$C = 4/3(\Pi(Mm/\rho G_f)^{1/2}\, d_m^{3/2}\, k^{3/2}(N_{max}/n_{cinf}) \quad (3.3)$$

where N_{max} is the total number of nuclei, n_{cinf} is the total number of molecules transferred from monolayer phase to three-dimensional phase, d_m is the molecular diameter, and G_f is a geometry factor.

The phase equilibria of oxidized lipids in bilayers are known to be of much interest in biology. It is important to remark that in the literature, monolayer structures have been studied where fluorescent probes have been added.

AFM studies of crystallite formation in LB films has been reported of Cd-arachidate monolayers.[20,130] This observation is not completely investigated at this stage in the literature.

Many efforts have been made to study the LB films of lipids by STM. It has been reported that a nonconducting material on a conductive surface can be imaged at atomic resolution by STM.[131] The Cd-arachidate LB film is nonconductive with a resistance of 10^9 to 10^{15} Ω cm, which is considered to be large for STM. However, on substrates such as gold, images have been obtained. It can be that the organic SAMs have enough conduction for STM, for example, defects in LB films. It may also be that at certain distances, Cd-arachidate film is 5.6 nm, the current may be sufficient for STM. The distance will be less if current was operative to the COO group. At this stage in the literature, these observations are not completely understood.[131] These investigations showed rather thick organic SAMs (5 nm). In these studies, LB films were placed on HOPG or calcite coated with 100 Å layer of gold. In order to deposit LB films, the hydrophobic HOPG surface was shock oxidized. This was carried out by electrochemical treatment in 0.1 M Na_2SO_4 solution with a current pulse of 2 to 5 sec duration that could be investigated by STM. The magnitude of the bilayers of Cd-arachidate and DMPA were 55 Å and 24 Å, respectively, which corresponds with the lengths of the molecules. HOPG images showed the regular hexagonal lattice with C–C bonds of 1.42 Å and 2.46 Å distance between the nearest spots along one axis.

A diverse range of metal ions has been incorporated into monolayers and LB films.[20] In a recent study, monolayers of arachidic acid on subphases containing PtII and PdII amine complexes were investigated.[132] The metal fatty acid films were deposited on gold-coated glass and mica substrates as LB films. The reactions of these films with H_2S were investigated using UV-visible grazing angle infrared and x-ray spectroscopic and AFM.

As is well known, fatty acids are examples of water-insoluble amphiphilic compounds that self-assemble in aqueous solution and at water surfaces.[20] Traditionally, the information on the organization of the monolayers has been obtained from surface pressure-area isotherms. During the past decade, more sophisticated methods have been developed, such as fluorescence microscopy,[133] Brewster-angle microscopy (BAM),[20,134] and synchrotron x-ray diffraction,[135] which provide a more detailed picture of the molecular organization in the monolayer. However, these methods show a limited lateral resolution of structures existing in the plane of the film. By transfer of the monolayer to a solid substrate, thereby creating a so-called LB film, further techniques become available for the investigation of the monolayer. Transmission electron microscopy (TEM) and Fourier transform infrared spectroscopy (FTIR) have been used to investigate monolayer domain topography, but none of these techniques can provide direct information of heterogeneous domains as well as nanometer-scale structure and surface properties of heterogeneous multicomponent monolayers. AFM is a surface imaging technique with angstrom-scale lateral and normal resolution that operates by measuring the forces acting between a probe and the sample.

In recent literature, one finds many studies on the model membrane structures reported by monolayer and AFM investigations.[136]

Biological membranes have complex organizations that are essential for their functionality. Phospholipids are the main constituents in many biological membranes, and many studies of their monolayers have been reported.[137,138] Particularly studied are the lipid bilayers existing between the corneocytes in the horny layer of the skin.[139] Here, phospholipids are rare, and the main classes of lipids are ceramides, cholesterol, and free fatty acids of different chain lengths and saturation.[140,141] For a complex biological mixture, it is hard to identify individual components from the topographic images. By studying each component separately and then the formation of simplified models, more information of the complex system can be achieved. The lateral phase separation and the formation of domains on the nanometer scale can be determined by studying monolayers of two fatty acids of different chain lengths and their mixtures. Therefore, AFM measurements were carried out on LB films transferred from the air–water interface at controlled surface pressures.

Monolayers of palmitic (C16:0) and lignoceric acid (C24:0) and their equimolar mixture were transferred to a hydrophilic mica substrate, as LB films, at various surface pressures, and were investigated by means of AFM in contact and lateral force modes.[142] The first-order transition of lignoceric acid gives a plateau region, representing a liquid expanded to liquid condensed phase transition in the pressure-area isotherm. Theoretical analyses of these isotherms have been extensively described in the literature.[20] This was seen in AFM images as stripes of a condensed phase within the expanded phase, exhibiting a small height difference and a significant difference in the magnitude of the friction force. The corresponding phase transition of the palmitic acid was continuous, and no changes of the LB films with respect to surface pressure were observed with AFM. Surface pressure versus area isotherms and the direct observations of *domains* of irregular size and shape using the AFM showed that lignoceric and palmitic acid were immiscible. The height difference between the domains was found to be 1.1 nm, corresponding to the difference in hydrocarbon chain length of the two fatty acids. Solutions of single fatty acids and their equimolar mixtures (1:1) were used in this study. LB films were made on sheets of freshly cleaved mica dipped into the subphase (0.1 M acetate buffer, pH 4.0). All lipids were dissolved (1 mg/mL) in chloroform and spread on the subphase. After the solvent had evaporated after 20 min, the monolayers were compressed until the desired surface pressure was achieved. These monolayers showed transfer ratios of values almost unified.

The LB films were investigated under constant force AFM and lateral force AFM (LFM),[142] with a 10 μm × 10 μm × 2.5 μm scan range. Microfabricated square pyramidal-shaped tips of silicon nitride with a bending spring constant of 0.12 N/m were used as received. The scan rate was 2 Hz, and the applied force was of the order of 1 to 10 nN. The imaging artifacts were checked using scan direction variation. The surface pressure–area isotherms of the systems were studied before depositing the monolayers as LB films on the mica support. The latter procedure is important in such studies. The monolayer isotherm provides the information about the lipid structures. Using these isotherms, in combination with AFM studies, one can interpret the images with a much higher security than without. The pressure-induced phase behavior was studied for single fatty acid monolayers and for mixed monolayers [isotherms of lignoceric acid (C24:0), palmitic acid (C16:0), and the equimolar mixture at 19°C]. The isotherm of the lignoceric acid exhibits a transition from a liquid expanded to liquid condensed state at a surface pressure of approximately 8 to 10 mN/m and an area of 25 to 22 A^2/molecule. This transition state is extended and rather flat, and a large decrease in head-group area takes place at equilibrium between the condensed and expanded phases at ideally constant surface pressure. The transition occurs with a small variation in pressure, and this slight deviation from the ideal behavior of a first-order phase transition may be considered an indication that a true macroscopic phase separation is not taking place in the monolayer. Nonequilibrium *nanometer-sized domains* or aggregates on the surface can result in nonhorizontal transitions in the monolayer isotherm.[20,143]

If one eliminates the possibility of any impurity effect, the most likely cause of the nonideal phase transition is the formation of small domains in the monolayer.[20] The corresponding phase transition in the palmitic acid monolayer is continuous, with a small decrease in head-group area, around 22 mN/m and 22 $Å^2$/molecule. These data are in agreement with literature studies.[20,144] The mixed monolayer of lignoceric acid and palmitic acid gives a pressure-area isotherm, where the phase transitions for both components are seen at unchanged surface pressures. This is an indication of immiscibility between the fatty acids, which is in agreement with previous studies of mixed fatty acids with several hydrocarbons difference in chain length.[20]

The nonhorizontal transition of lignoceric acid was found to be more pronounced in the mixed fatty acid isotherm, where the transition takes place under the same surface pressures (8 to 10 mN/m), but the reduction in head-group area is smaller (25 to 23 $Å^2$/molecule). The transition from gaseous to liquid expanded state of the monolayers was clearly observed. LB films of lignoceric acid were studied, and the height and friction AFM images were simultaneously obtained, where the height images originated from normal forces and the friction images from lateral forces. Frictional measurements can give information on heterogeneities in samples that are not caused by height differences and are, therefore, a good complement to the topographic measurements. In regard to the interpretation of these data, absolute values of the friction measurements are not reliable, while relative values of friction within an image are more accurate.[145] Isotherms that exhibited the expanded and the condensed phases were present. Flat areas were observed and some crude areas of corrugated form with stripes were also observed in the transferred monolayer at this surface pressure region. The stripes have various orientations within the sample, and the distance between them is typically 150 nm. The height of the stripes is only about 0.1 to 0.2 nm and would be difficult to detect with methods other than AFM. In the friction image, these stripes are much more pronounced, showing two to three times higher friction relative to the flat areas within the same sample. The orientation and size of these stripes are the same when zooming in and out of an area examined and changing scan direction, strongly indicating that the stripes reflect a property of the system rather than one of the imaging processes. Lignoceric acid monolayers deposited at higher surface pressures, corresponding to a liquid condensed state of the monolayer, were found to be flat with no visible stripes. These films were also found to be robust, because there was almost no sign of film rupture by the tip. For low surface pressures, well below the phase transition pressure, the transferred monolayer shows stripes. However, these stripes differ from those in the previously described sample in shape, and they exhibit much lower relative friction. In the low-pressure sample, one can also see inhomogeneities and a few holes or cracks. The depth of the holes is about 2 nm, and no internal structure could be observed. In the liquid expanded state of lignoceric acid, the head-group area is relatively large, which allows a less ordered hydrocarbon chain organization. This can explain the irregularities observed within the monolayer. The film also seems to be sensitive to the force applied by the tip on the sample. In these studies, the films were deposited on mica at a surface pressure corresponding to a well-defined Π. The difference between the phases was clearly seen in the friction images (5 nm; for the friction images, 0.05 V).

In transferred LB films of palmitic acid, no stripes were observed at any surface pressure. Furthermore, palmitic acid monolayers were found to be not as robust as the lignoceric acid monolayer, and they seemed to be affected by the tip. The condensed palmitic acid monolayer was more resistant to the tip than the expanded monolayer, and no signs of new defects caused by the tip were observed. There was little difference measured between the expanded and condensed palmitic acid films. At the first-order phase transition, liquid expanded and liquid condensed phases coexist in the monolayer. This is demonstrated here as a two-dimensional phase separation in the transferred film, visualized as stripes in the lignoceric acid monolayer. The height images may not be expected to prove that the small topographic irregularities are due to the phase transition. A new procedure was used, where a combination of height and friction images were produced, and a significant difference between the two samples can be shown, where the frictional fluctuations can be related to variations in the crystalline properties within the samples. From this, it was concluded that these features are due to coexistence of liquid expanded and liquid condensed lignoceric acid monolayers and are not the same as the features that lack this significant difference in friction. Phase transition is not observed in the case of palmitic acid monolayer, which is due to the continuous transition without a coexistence region. Inhomogeneities within lignoceric acid monolayers have also been reported but are not assumed to be the same as reported in these studies.[146] The internal inhomogeneous and triangular structures of lignoceric acid visualized by phase contrast microscopy are several magnitudes larger in size than those observed.

3.1.3 Mixed Lipid Monolayers

Mixed lipid monolayers haves been extensively described in the literature in regard to the degree of miscibility of lipid monolayers.[20] In all AFM images of the equimolar mixture of lignoceric and palmitic acid, phase separation was observed with distinct domains of respective fatty acid.[145,146] The relative height difference between the domains was measured at ca. 1.1 nm. For eight carbon atoms in the all trans conformation of a normal alkane, their estimated length was $8 \times 0.127 = 1.02$ nm. A thinner continuous phase of palmitic acid was noticed as large domains of a thicker phase of lignoceric acid. These domains were of irregular shape, with the boundaries consisting of nonuniformly connected straight stretches on a nanometer length scale. When surface pressure increases, the interfaces become even less rounded. The sizes of the domains are about the same at increased surface pressure. Small domains are observed within the lignoceric acid phase. The relative height difference to the lignoceric acid domains is the same for these small domains as that for the continuous palmitic acid phase. From these data, it was concluded that the small domains consist of palmitic acid. These palmitic acid "lakes" within the lignoceric acid domains were found to be almost circular, with a diameter of ca. 200 nm. Area ratios between the two different lipid phases correlated with the composition of the sample. The palmitic acid phase was less sensitive to the tip at increased surface pressure, which is consistent with the results for the pure palmitic acid monolayer. The height differences between lignoceric acid and palmitic acid domains was measured to be 1.1 nm, consistent with the difference of eight methylene units. However, almost twice the chain length difference for the same system measured with AFM was reported.[147] This deviation is claimed to be due to adhesion forces. On the other hand, it has been reported that from AFM study, the height differences in monolayers of mixed chain length fatty acids are in good agreement with the difference in methylene units.[148,149]

The height differences between Mg-stearate and cholesterol collapsed monolayers also showed satisfactory agreement with the expected values of lengths of the molecules.

In the case of the lignoceric acid domains, specific features were observed for samples prepared at pressures corresponding to the phase transition of lignoceric acid. These features were not easily detected in height images, while they were easily detected in friction images. These data did not show any regular size or shape, although they often resembled stripes. At pressures below the phase transition of lignoceric acid, no signs of these features were observed. The interpretation given is that the small features in the lignoceric acid domains correspond to the stripes in the pure lignoceric acid monolayer. As a result of the large height difference between the two fatty acids, compared to the internal height difference in the lignoceric acid, the features in the lignoceric acid domains are less pronounced in the height image of the mixed samples compared to the single lignoceric acid samples. Hence, the relative frictional differences within the lignoceric acid phase are of the same order of magnitude as the relative frictional differences between the two fatty acid phases. At surface pressures below and above the phase transition of lignoceric acid, no frictional differences within the lignoceric acid phase are detectable. This was observed for the lignoceric acid samples and the mixed samples, with the exception of the weak stripes. One may conclude that the frictional measurements in these samples exhibit the coexistence of liquid-condensed and liquid-expanded phases in the lignoceric acid monolayer. However, more studies are needed at this stage, before making any general conclusions.

3.2 DOMAIN PATTERNS IN MONOMOLECULAR FILM ASSEMBLIES

As shown in Figures 3.1 through 3.3, the monolayer film on compression will have to undergo some kind of two-dimensional rearrangement after reaching the collapse point (= collapse surface pressure).[20] The self-assembly characteristic will, however, give rise to three-dimensional structures. If one compares the monolayer film with the soap bubble, one can expect that the collapse state will be comparative to the break up of the soap bubble. The monolayer film could undergo an abrupt transition (less than a millisecond). These monolayer films on liquid interfaces are the most important assemblies for industry and biology. The rearrangement that must take place at the collapse region is dependent on different forces that stabilize these structures. In recent literature, two major different procedures used to investigate these phenomena were reported. One has been based upon the use of an ordinary microscope with fluorescence probe. This allows changes in fluorescence of the labelled lipid during such collapse state to be observed. The sizes of domains are of the order of micrometer. The second procedure has been to study the LB film using STM or AFM methods. The size of domains in the latter can be of the order of nm. Because these sizes are different, we will designate these as *macrodomains* and *molecular-domains*, respectively. In current literature, these two types of domains, unfortunately, are generally not recognized. By both procedures, it was found that *domain* structures were observed only for some lipid self-assemblies under definite experimental conditions. For example, stearic acid monolayers show no domains, while cholesterol and oxidized cholesterol exhibit domains (Figure 3.7).[20,150]

In macrodomain studies, concentrations of dye used are as low as 0.2 mol%. However, this gives rise to some drawbacks, such as the fact that there are only a small number of molecules fluorescing in the visible. Additionally, even such small amounts of dye molecules can affect the data due to nonideal film mixed behavior.[20] The spontaneous formation of domain assemblies in monomolecular films of amphiphiles at air–water (oil–water interface needs to be investigated) interface has evoked great interest. Considerable evidence seems to suggest that these two-dimensional assemblies arise in response to competing interactions.

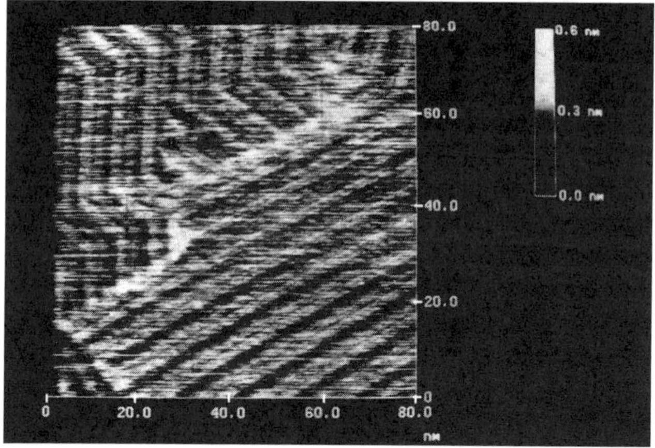

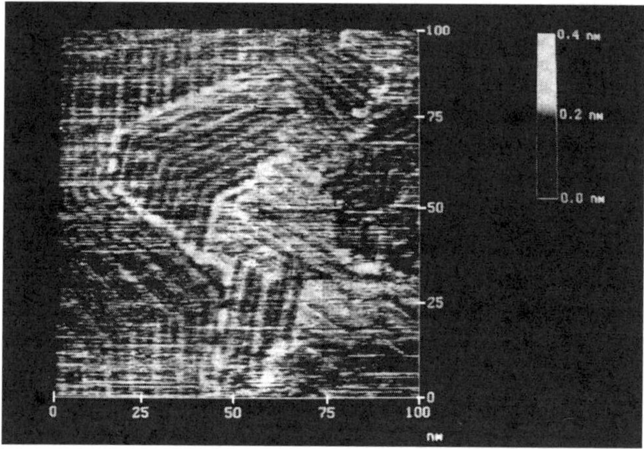

FIGURE 3.7 STM images of 1-docosanethiol in phenyloctane adsorbed on graphite. The bright spots (dispersed) in the image are attributed to the SH functional groups. Tunneling conditions: 1550 mV/150 pA. (From Venkataraman, B. et al., *J. Phys. Chem.*, 99, 8684, 1995. With permission.)

3.2.1 Macrodomains

Typical procedure is to use 2% of a fluorescent lipid analog that renders visible a domain (micrometer range) pattern. This assumes that the fluorescence moiety has no effect on the assembly structure. This aspect has not been extensively investigated. DPPE macrodomains as formed by these mixed lipids were studied by ellipsometry method.[20,151] It must be stressed that these domains are valid for mixed lipid film assemblies and may be different in the case of pure lipid films (i.e., without the fluorescence moiety).

In order to obtain more information on the surface monolayers, a new method based on fluorescence was developed. It consisted of placing the monolayer trough on the stage of an epifluorescence microscope, with doped low concentration of fluorescent lipid probe.[152] Later, ordered solid–liquid coexistence at the water–air interface and on solid substrates was reported.[152] The effect of cholesterol on domain shape has been also reported.[152,153] The theory of domain shapes has also been extensively described by this method.[154]

Molecular domain structures of phospholipid monolayer LB films have been investigated by AFM.[150] The domain structures of phospholipid [di-palmitoyl-phosphatidylcholine (DPPC)] were studied as LB films. Star-shaped domains of thiolipid monolayers transferred by the LB technique onto gold surface were studied by AFM.[153] The dimension of the stars was 10 to 15 μm. At this stage, in the literature, there is no theoretical description of these domain shapes. One may imagine that this could be a two-dimensional crystal formation, analogous to the three-dimensional phenomena.

The domain microcharacteristics of stearic acid, cholesterol, and oxidized cholesterols have been investigated by LB film analyses using AFM[3] (Figure 3.6).

In the case of *macrodomains*, a theory was presented that explained the thermodynamically controlled strip-like shapes of two-dimensional solid crystal domains as observed in phospholipid monolayers in the presence of trace amounts of cholesterol.[154]

It must be stressed here that these *macrodomains* are images with no information about thickness. In this theory, dipole–dipole repulsions between lipid molecules were assumed to favor the elongation of the domains into long strips, but this elongation was supposed to be opposed by increasing interfacial free energy (line tension) associated with the parameter of the domains. An expression was derived for the dependence of line tension on the concentration of cholesterol in the monolayer, and this relationship was used to compare the predictions of the theory with experimental measurements of the width of domains as a function of monolayer compression. The theory was found to agree well at high compression, but there were found to be deviations from experimental data at low compression, where the domains were found to be short and end effects neglected in the theory became apparent.

As mentioned earlier, SAMs of alkanethiols on gold surfaces have been found to have a variety of potential applications, such as sensors, corrosion inhibitors, wetting control, and biological devices.[155] The formation of SAM of decanethiol (DETH) on Au was investigated in dilute solution *in situ* STM in real time. Although solvent ethanol was used in many cases, heptane was preferred due to the absence of leak current. All STM measurements were made with a bias of 700 mV and a tunneling current of 30 pA in constant current mode. Although it is generally believed that the SAMs of alkanethiols are formed as a result of the chemical bond formation between substrate atoms and sulfur atoms of thiols and the hydrophobic interaction between alkyl chains, their growth process is still not absolutely clear. The alkyl–alkyl chain attraction may be much more important in such SAM once sulfur–gold bond is formed. Data of STM images were obtained after 3, 7, 12, and 56 min. These are the only images reported *in situ*. It was found that gold surface is completely covered after 56 min.

3.2.2 Theoretical Analysis of Domains (Macrodomains)

The lipid-phase transition at the air–water interface was investigated using fluorescence microscope. At the transition between the fluid (low density) and solid (higher density) phase, two-dimensional *domains* of solid lipid (phospholipids) coexist with regions of two-dimensional fluid.[152,156] The coexisting phases form a variety of domain shapes. For obvious reasons, these data cannot give information about the thickness of the domain structures. These domains were reported in most cases to be of circular shape. This was in accordance with the fact that a circle minimizes the solid–liquid interfacial energy. Later, it was found that the addition of cholesterol to DPPC films gave rise to a thin stripe of uniform width.[152] Furthermore, it was found that chiral DPPC gave different shapes than did the racemic form. The mixed films of DML plus cholesterol were analyzed to determine the dynamics of domain shape.[153] These analyses were carried out by using epifluorescence microscopy of monolayers of lipids containing 2 mol% of fluorescent lipid. The size of domains was in the range of 10 to 30 μm. These macrodomains were analyzed based upon a competing interactions model.[157]

The charge effect was observed because DMPA domains were different than those of DPPC.[158]

The solid strip of width, w_w, is surrounded by fluid phase. The dipole density of solid and liquid is ρ_s and ρ_l, respectively. The magnitude of w_w in these studies varied from 1 to 3 μm. The free energy that determines the size of the strip is of two parts (interfacial forces and electrostatic repulsion forces):

$$F = 2\lambda l + F_{el} \tag{3.4}$$

where λ is the line tension for the solid–liquid interface, *l*, is the length of the lipid strip, and F_{el} is electrostatic free energy. If the lines are long, and neglecting end effects, one can write for the electrostatic interaction energy, U_{ll}, between two lines:

$$U_{ll} = U_l \, l/d_s \tag{3.5}$$

where U_l is for each strip, and d_s is the square for each dipole.

In order to estimate w_w, one needs to minimize F with respect to w, by assuming that the total area of the monolayer, $A = N/L$, is fixed; and the ratio of the solid area to total area, $\Phi = A/A_o = w_w/l$, is fixed. This minimization procedure gives:[156]

$$w_w = d_s \, e^C e^{\varepsilon\lambda/\Delta u\verb|^|2} \tag{3.6}$$

where $C = 0.5772$ (Euler's constant), and ε is the dielectric constant of the medium.

Similar considerations for the array of alternating solid and fluid strips give w_w:

$$w_w = d_s e^C e^{\varepsilon\lambda/\Delta u\verb|^|2} \, (\Pi\varphi/\sin(\Pi\varphi) \tag{3.7}$$

the last term determines the interactions between the strips and is close to unity for $\varphi < 0.3$.

These theoretical considerations were based on the following:

1. Dipole–dipole repulsions within solid domains that favor elongation
2. Dipole–dipole interactions between regions of solid separated by fluid regions that oppose elongation
3. Line tension at the solid–liquid interface that opposes elongation
4. Tilt anisotropy in the solid (which indicates that the magnitude of width, w_w, is determined by electrostatic and interfacial force balance)

The addition of cholesterol was found to arise from among other effects due to lowering of the line tension. This means that an increase in cholesterol content caused narrowing of the domains.

The shape stability of the macrodomain of a rectangular shape was analyzed.[154] The straight line of the domain was stabilized by the line tension, λ, and destabilization was related to the harmonic shape distortions by long-range dipole forces.

These considerations are, however, devoid of information about the thickness of the domains. As is well known from physics, most materials contract in all dimensions when pressure is applied, i.e., volume compressibility ($-dV/V\ dP$), area compressibility ($-dA/A\ dP$), and linear compressibility ($-dL/L\ dP$) are positive. Self-assembly monolayers, when undergoing higher surface pressures before the collapse state, would undergo these compressibility phases. The π versus A isotherms indicate a buckling state, i.e., where the isotherms show spikes before entering the collapse state.[20] The buckling state can lead to the detachment of the lipid assembly with a kind of droplet formation in two dimensions.

The softening of lipid bilayer membranes undergoing a gel to fluid phase transition with temperature was studied by computer simulation.[159] The presence of nonhorizontal isotherms of lipid monolayers was suggested. As described earlier, the effect of temperature is high on these films. The computer simulation has thus not been successful in analyses. The main criticism that has been made is the omission of water (i.e., polymorphism) in such simulations.

3.2.2.1 Domains (Macro- and Nano-Size) Shape

Equilibrium of the phase-separated domains would be expected to be of circular or at least regular shape.[160] The shapes of the domains in the mixed fatty acid monolayers exhibit a nonequilibrium state and incomplete phase separation. In these experiments, the monolayer was left for 20 min at a determined pressure before depositing. To investigate the equilibration for the domain formation more thoroughly, the mixed monolayer was left for 12 h at a constant pressure of 22 mN/m before deposition. AFM images of this sample showed domains of comparable size and shape, as for the earlier experiments. Finally, a spread monolayer was left at zero surface pressure for 12 h before compression to 22 mN/m and deposition onto mica. This monolayer showed that domains of lignoceric acid are squared and have smooth borders to the palmitic acid and few "lakes" of palmitic acid within the lignoceric acid. It was concluded that phase separation in two dimensions is slow. Perfect phase-separated structures were not obtained after 12 h at zero surface pressure. The slow phase separation allows for the study of monolayers transferred at nonzero surface pressures with reproducible results.

Investigations of transferred LB films generally aim to increase the understanding of the monolayer structure at the air–water interface. Even if there is a general belief that the structure of a monolayer at the air–water interface resembles the structure of the monolayer transferred to a solid substrate, one has to be aware of factors that may influence the molecular arrangement during and after the deposition.[149] Samples were, therefore, prepared with varying deposition dipping speeds in the range of 2 to 10 mm/min, showing no visible differences. Some of the transferred samples were reexamined after 1 and 2 days, and no significant changes could be noticed. To ensure that the routine for the monolayer preparation gives representative and reproducible results, the monolayer at the air–water interface was compressed and decompressed in isocycles six times before relaxing at constant pressure and then depositing onto mica. The compression speed was varied in the range of 10 to 200 cm^2/min. No detectable differences were observed in these samples compared to films made under normal conditions. The effect of varying the pH of the subphase gave rise to better film quality for monolayers prepared on a buffer of pH 4 than on one of pH 7. Because the fatty acids in the monolayer effectively titrate at a higher pH than in bulk, it is assumed that all lipids should be in an undissociated state at pH 4, resulting in a more homogeneous monolayer when no lipids self-assemble to form aqueous phases.

From these studies, it was concluded that domain formation in single and mixed LB films of free fatty acids could be studied by AFM.[160] The method clearly revealed domains in the lignoceric acid monolayer at the liquid-expanded phase to liquid-condensed phase coexistence region. These are particularly apparent using the friction mode of the AFM device. Similarly for the mixed palmitic lignoceric acid system, separate domains of the two components are easily seen, and they show a difference in thickness, reflecting the difference in chain length of the two acids. The domains appear over a substantial variation in the procedure for preparing the monomolecular film, and one could conclude that the domains are also present in the parent film at the air–water interface. The domains are most likely not equilibrium structures, but they form generically, and their presence can explain the pressure variation in the coexistence region for liquid-expanded and liquid-condensed phases. The irregular shape of the domains in the mixed lignoceric acid–palmitic acid system shows that not only is the equilibration slow with respect to the formation of large, macroscopic domains, but also, the relaxation of the shape appears to be slow under experimental conditions. These data need to be compared with other *domains* observed for such lipids as cholesterol (consisting of 60 million molecules).[3,20]

The nanometer-scale aspects of molecular ordering in nanocrystalline domains at a solid interface were investigated. The role of liquid crystal–surface interactions was studied by STM.[161]

SAM of thiols stimulated a series of computer simulation studies, based on models with varying degrees of sophistication.[162] In the molecular dynamics model, the system consisted of 90 alkanethiol molecules in a rectangular cell. However, these model studies are premature, and more studies are needed before any useful information is obtained.

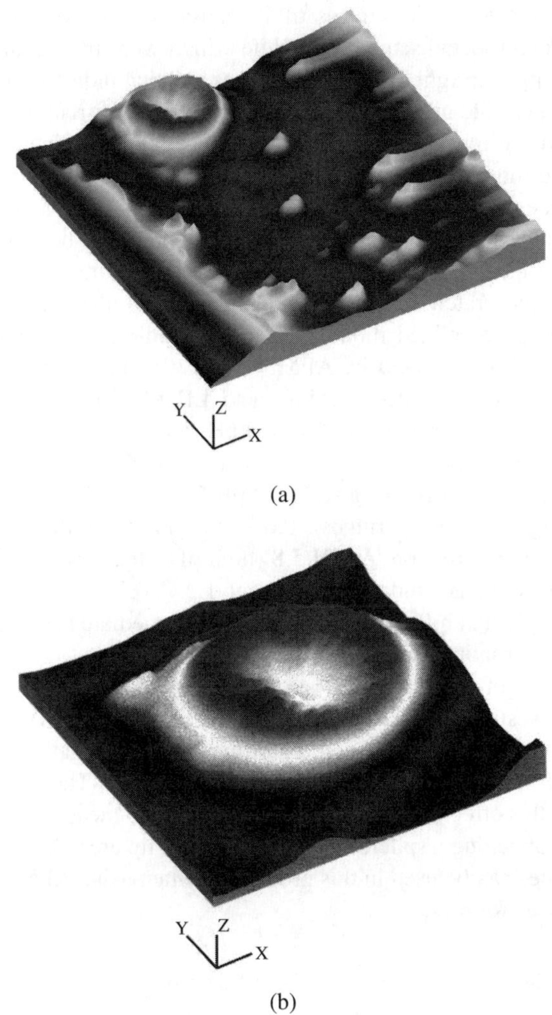

FIGURE 3.8 AFM image of vesicle of DPPC on HOPG. (a) Image is depicted in embossed state for clarity (90,000 Å × 90,000 Å), and (b) an enlarged view of a single vesicle (45,000 Å × 45,000 Å).

3.3 MIXED LIPID MOLECULE ASSEMBLIES

The adsorption of mixed 1:1 triacontane/tricontanol mixtures in solutions were investigated by STM (Figure 3.8).[163] STM images of 1-docosanol, 1-docosanethiol, didocosyl disulfide, and 1-chlorooctadecane were compared on HOPG substrate.[164] The images showed bright spots arising from sulfur, S, atoms. This suggests that S atoms in STM images can be used as a chromophore. The alcohol region was observed as "zigzags" in the top left corner, and the alkane region as "straight" rows in the lower right. Tunneling conditions were 1200 mV and 60 Pa. The data for a mixture of 1:10 showed STM images of 1-docosanol, 1-docosanethiol, didocosyl disulfide, and 1-chlorooctadecane on graphite which were investigated.[165] The S–H and S–S groups appear bright in the STM images and that indicates that the presence of an S atom on graphite gives rise to a higher tunneling current when the tip scans as compared to the current over a carbon, C, oxygen, O, or chlorine, Cl, atom (Figure 3.7). This finding might be useful in using S atoms as chromophores.

The resonance and environmental fluctuation effects in STM currents through adsorbed molecules have been analyzed.[166] AFM has also been used to modify structurally lipid bilayers in a controlled procedure.[167] The images showed that after the lipid tubule was scratched, the molecules rearranged after time (24 h).

The morphology of mixed monolayers of arachidic acid–cadmium arachidate LB multilayers was investigated by AFM.[168] It is well known that the fraction of cadmium salt incorporated into arachidic acid LB films deposited from a dilute $CdCl_2$ subphase increases from 0 to 1 over the pH range of about 4.8 to 6.2.[20] In this study, a systematic change was reported in the surface morphology of such LB multilayers over this pH range using AFM. At pH 5.0 (low pH), the surface displayed increasing coverage of stripes (ridges), 0.6 ~ 0.2 nm above the surrounding area, aligned in the dipping direction. At pH 5.8 (high pH), the surface was pockmarked with irregular but compact indentations about 1.2 ~ 0.3 nm deep (in addition to numerous monolayer and bilayer deep holes). At intermediate pH values, the surface was covered by alternating stripes and deep holes.

The adsorption of SDS on gold surfaces covered with SAMs of hexadecyl mercaptan was investigated by probing the surface charge.[169,170] This surface charge was measured from the force between a modified (with a negatively charged silica sphere) tip of AFM and the surface as SDS was adsorbed. The negative charge of SDS gave rise to this effect. The quantitative procedure to measure forces from AFM was achieved by attaching a sphere to provide a larger tip area. A virtually unlimited choice of substrates can be used in this procedure, when using AFM. The setup used can be described as follows:

(CANTILEVER)---GLUE---Silica sphere (negative charge).....----------------

....

....

....Aqueous SDS phase...

....

.....SSSSSSS.........SDS adsorbed on substrate with SAM ..SSSSSS..........

AFM images gave silica surfaces with a mean roughness of 1.2 nm/μm^2. Gold was vacuum deposited on freshly cleaved mica surfaces.

All natural biological cell membranes are composed of mixed lipids besides other molecules (proteins, etc.).[20] The monolayer cell membrane model system has therefore provided much information. However, until recently, there were no procedures available that provided image data of mixed lipid films. As described in detail elsewhere,[20] two- or three-component lipid films are complex structures. The mixed films of lipids at air–water interface can be further investigated using the LB film method. The π versus A isotherms of distearoylphosphatidylethanolamine (DSPE) and dioleoylphosphatidylethanolamine (DOPE) and mixed (1:1) films were investigated.[171]

These isotherms showed that at 25°C, DSPE is solid-like while DOPE is liquid-like, as one should expect as seen from their molecular structure. At $\pi = 25$ mN/m, the compressional modules are 183 and 79 mN/m for DSPE and DOPE, respectively. The LB films on mica of 1:1 mixed monolayers showed well-defined elliptical domains ca. 5 to 10 μm size, embedded in a continuous matrix. The higher-level domains were assigned to DSPE. The space-filling model gave lengths for DSPE and DOPE as 3.3 and 2.9 nm, respectively. The step height measured between the two phases was 1.3 nm, which is 0.9 nm larger than the height difference expected from the space-filling models. The monolayer thickness of DSPE can be estimated by dividing the volume of the molecule by the area (41 $Å^2$ /molecule) occupied at the air–water interface. Assuming a volume for a saturated chain in the solid state of 27.4 + 26.9 nC $Å^3$ per n_C-carbon chain, and a head group volume of 243 $Å^3$, gives a monolayer thickness of 3 nm. These differences were related to the force used in the AFM analyses. It could also be that the tilt angle is different than expected from simple geometrical packing assemblies. As mentioned earlier, AFM is a powerful technique, but the interpretation of the images (as is also the case in electron microscopy) at a nanometer scale is sometimes complicated. The force versus distance curves of DOPE and DSPE were found to be different. The adhesion pull of forces over DOPE (10.5 nN) was greater than that over DSPE (6 nN).

As mentioned elsewhere herein, the height analyses of collapsed lipid films, however, was in accord with space-filling models (in the case of fatty acids and cholesterols) (Figures 3.2 and 3.4). It seems that more studies are needed in this area.

The effect that a single fluorine substitution in a hydrocarbon chain of stearic acid has on the molecular order and the effect that the fluorides have on the STM contrast of an alkyl chain physisorbed on graphite were reported.[172]

Epifluorescence microscopic studies of monolayers containing dioleoyl- and dipalmitoylphosphatidylcholines domains were observed in pure DPPC monolayers at relatively low surface pressures. These domains grew with increasing surface pressure.[128] Only liquid-expanded phase repetitive compression and expansion of the monolayers containing DPPC:DOPC:NBD-PC 49:50:1, at an initial rate of 3.2 Å/molecule, produced monolayers with visual properties consistent with there being a preferential exclusion of the unsaturated lipid from the monolayer.

Stearic acid thin films prepared by the hot-wall technique were observed, and barrier heights were measured using STM, which is operated in air.[163] Three kinds of substrates were used to prepare stearic acid films: highly oriented pyrolytic graphite (HOPG), gold thin film, and indium-tin-oxide.

3.4 HOLES IN LB FILMS OF SELF-ASSEMBLY MONOLAYERS

In some of the AFM images of various lipid LB films, in the current literature, "*holes*" were reported, as shown in Figure 3.4. However, no holes were observed in some lipid films, such as Mg-stearate. Therefore, no clear relationship exists between hole formation and lipid monolayer assemblies. At this stage in the literature, there is no plausible description as to the physical reasons for formation of such nanostructures in SAMs. The shapes and sizes of the holes have been found to be of high regularity.

The nature of the *holes* in self-assembled thiol monolayers as found in STM studies[173] is elucidated with unprecedented high tunneling resistance, 1 T Ω. The molecules were found to order in $\sqrt{3} \times \sqrt{3}$ domains separated by different types of missing row structures. These regions are neither openings (pinholes) nor regions of disorder in the monolayer. The spontaneous organization of molecular quasicrystalline structures, termed SAMs, is responsible for fundamental processes, such as cell membrane formation. The holes were clearly visible as empty regions. These images can be described as follows:

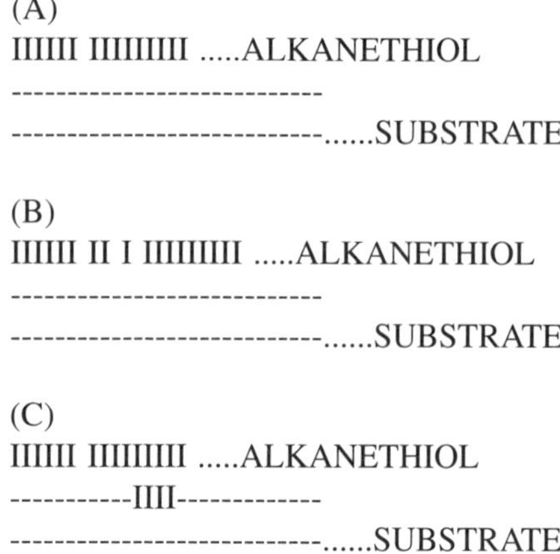

The schematic drawings in (A) and (B) were excluded from direct visualizations. It was concluded that the apparent holes in the STM images were depressions in the substrate (Au) surface layer of one unit cell depth originating from an etching process. However, it seems that these conclusions are not extensively studied. In the author's laboratory, holes were observed in the case of some lipid LB films (cholesterol), but no holes were observed in the case of other lipid LB films (Mg-stearate).

3.5 VISUALIZATION OF VESICLES BY AFM

Many lipids exhibit different kinds of characteristics found to be of much importance. The self-assembly characteristic is one of the most important phenomenon that leads to vesicle or liposome formation.[20] The study of liposome structure and properties was extensively reported in the current literature. The behavior of liposomes (*in vivo*) in clinical applications (phospholipid-based vesicles) has been investigated since 1970 as a possible system for the delivery or targeting of drugs to specific sites in the body.[174] While in monolayers on water surfaces lipids self-assemble in two-dimensional phases; in vesicles, the lipid molecules form spherical multilayer assemblies. The structures of vesicles are related to lipid structure and to experimental conditions.[20]

When a lipid molecule is dispersed in water, it may form a vescicle if the experimental conditions (such as temperature) are correct:

LIPID MOLECULE + WATER
VESICLE ASSEMBLY IN WATER PHASE

Interactions between phospholipid vesicles and bilayers play a central role in cell physiology, enabling secretion, signaling, and intracellular transport. In many instances, these processes require fusion between membranes. Considerable attention has been paid to acquiring the structural details of proteins that mediate fusion,[175] however, less is known about the structure and organization of lipids during this process. In addition, the interaction between lipids and proteins is known to be important for fusion. Evidence for this comes from the fact that when fusogenic proteins are reconstituted into lipid bilayers, their activities are often a sensitive function of lipid type.[176] Substantial evidence also implicates membrane cholesterol content in the ability of some enveloped viruses to fuse with and enter liposomes.[177–179] Because of this, the physical mechanisms of membrane fusion in reconstituted lipid systems have received intense study with a variety of physical methodologies that includes fluorescence spectroscopy and microscopy,[180] light scattering,[181] and surface forces apparatus.[182] These studies revealed that membrane fusion propensity depends on several parameters, including fluidity, lysis tension, and bending modulus, and they produced a wealth of quantitative data on the mechanical properties of bilayers formed from different lipid compositions.

Vesicles formed by surfactants containing amino acidic groups are of interest in considering membrane interactions with proteins.[183]

AFM is particularly well-suited to the characterization of reconstituted complexes of biological macromolecules, because of its ability to operate under buffer, in real time, and at nanometer length scale. Thus, much effort has been made to use AFM to study lipid vesicles, planar bilayers, and other self-assembled lipid microstructures. Similarly, AFM would be an attractive tool with which to study structural intermediates in membrane fusion. Several investigators presented AFM images of purified synaptic vesicles and reconstituted phospholipid vesicles on solid substrates. The former studies provided useful information about the mechanical properties of synaptic vesicles as well as how those properties changed in the presence of biochemical effectors.[184–186] The latter studies used receptor–ligand interactions to tether the vesicles to substrates such as gold and mica[187] or focused on the kinetics and mechanisms of formation of supported bilayers from phospholipid vesicles.[188,189] These studies helped elucidate the various stages of membrane assembly. Nonetheless, definitive visualization of individual vesicles has been limited. Presumably, this is due in part to vesicle deformation during scanning and the presence of artifactual contributions from the AFM tip. Even less is known from AFM about the association of vesicles with a planar bilayer, including structural details of the processes of adsorption, wetting, and fusion. The same is true of vesicle–vesicle interactions; while aggregates of small unilamellar vesicles (SUVs) are well within the limits of AFM's spatial resolution, freeze-fracture electron microscopy remains the most widely used method for directly visualizing interactions between SUVs. Obtaining high-resolution images of vesicles in contact with each other and with planar membranes would serve as an important first step toward introducing AFM to the study of membrane fusion intermediates. Recent light microscopic studies with giant vesicles suggest that membrane fusion *in vitro* can proceed through an orderly sequence of adsorption, adhesion, wetting, and merger, any or all of which might be directly captured by the AFM.

AFM images of vesicle–bilayer complexes formed by adsorbing SUVs onto mica to form a continuous membrane and allowing excess vesicles to settle onto the membrane were recently reported.[190] Depth measurements showed that one can successfully form a continuous supported membrane, and mechanical and structural evidence was found that the adsorbed structures were individual lipid vesicles. The morphology of the adsorbed structures was found to be a function of lipid composition.

Vesicles were prepared from mixtures of DPPC and cholesterol dissolved in chloroform, 1.0 mg of total lipid. The solution was dried from stock solution for 1 to 2 h under a gentle stream of nitrogen followed by treatment with vacuum over desiccant for 2 to 6 h. In some experiments, the lipid was weighed directly as dry powder, with no noticeable differences in results. The dry lipid was then resuspended in 1 mL of 20 mM NaCl buffer and sonicated under nitrogen until clear (usually 30 to 60 min). This suspension was centrifuged at 16,000 g for 30 min to remove any large aggregates or contaminants.

Vesicle–bilayer complexes were prepared using a method based on the vesicle adsorption technique.[191–193] After centrifugation, 50 μL of supernatant was pipetted onto a freshly cleaved mica substrate mounted onto a magnetic metal disk. The drop was confined by a silicone O-ring affixed by a minimal amount of vacuum grease. After allowing the vesicles to adsorb at room temperature for 30 min, 100 μL of excess buffer was added. AFM images and force measurements were obtained with a multimode microscope using a glass fluid cell. All measurements were made with silicon nitride cantilevers with lengths of 200 μm and nominal spring constants of 0.02 N/m. In these studies, AFM was operated in contact mode. To produce elasticity maps, force volumes were obtained using suitable software, and individual force curves were subsequently converted to plots of force versus distance to hard contact.[193] The curves were analyzed with appropriate software to obtain relative elasticity maps using the force integration to equal limits (FIEL) method.[194] Force volumes were collected in relative trigger mode, and surfaces were constructed by mapping the piezo position at each point of relative trigger on the surface.

DPPC vesicles were also studied by AFM in a procedure where a vesicle suspension was placed (10 μL) on HOPG. After water evaporation, the vesicle of DPPC could be clearly seen (Figure 3.8). For comparison, the vesicles of dioleyl lecithin were investigated by AFM (Figure 3.9). These vesicles are regular in size (ca. 4500 Å diameter).

3.5.1 DPPC-Cholesterol (1:1 Molar) SUVs

Small unilamellar vesicles (SUVs) were adsorbed onto mica to form a supported bilayer and allow excess vesicles to settle to the membrane to form vesicle–bilayer

FIGURE 3.9 AFM images of vesicles of dioleyl lecithin (on HOPG): (a) two-dimensional and (b) three-dimensional (45,000 Å × 45,000 Å).

complexes. This gave rise to a surface covered by a high density of dome-like protrusions that emerge 10 to 50 nm from the supported bilayer. These dome-like structures coexist with saucer-like structures, so called, because they resemble saucers resting face down on a flat surface, in which the central portion of the structure appears raised and rounded relative to its periphery. A topographic profile along such a structure demonstrates that the height difference between periphery and center is approximately 10 to 20 nm. The characteristic feature of a saucer is a central protrusion of greater height than the edges. Underneath these vesicular structures, is a flat, defect-free surface. A simple interpretation of these images is that the flat surface is a continuous lipid bilayer, and the dome- and saucer-shaped structures are vesicles that have stably adsorbed to and wet the membrane. Some of the structures seen here are smaller in height than 20 nm, which is below the typical size range of sonicated vesicles. The reduction in apparent height may be explained by vesicle deformation due to wetting of the bilayer and compressive forces from the AFM tip.[187] Another possibility is that the smallest structures in the image are nonvesicular structures, such as cholesterol micelles. This may account for some of the smallest structures and these adsorbates to be present at significant numbers, even at cholesterol mole fractions below 0.1. As found from mixed monolayer studies of cholesterol,[20] these findings are acceptable. Height images and topographic profiles of vesicle–bilayer complexes made from 1:1 molar mixtures of DPPC and cholesterol were studied. The dark background corresponded to the bilayer, and the lighter areas were material adsorbed onto the bilayer. Topographic profiles across several dome-like structures and saucer-shaped structures were found beneath the images. The top portions of some vesicles were found to be labile to AFM imaging, which was indicated by streaks that emerged along the scanning direction from one side of the vesicle. The surface underlying the adsorbed vesicles was related to a lipid bilayer, as was found from the image that was scanned over a 1 μm × 1 μm region of a sample with a high force (normal to the surface) to scrape through to the mica substrate and measure the bilayer thickness. The height difference between the lipid surface and the underlying mica substrate was approximately 4.8 nm, around the expected thickness of a phospholipid membrane. This thickness was also consistent with values measured in solution with cholesterol-DPPC lamellae by x-ray diffraction.[195]

In these studies, images were obtained at minimal scanning forces, (>50 nN) and scan rates (>20 Hz) for 10 min, thereby scraping through to the underlying substrate. The topographic profile along the line drawn in the image is shown below. The height difference between the bilayer surface and the mica surface was approximately 4.8 nm, consistent with the expected thickness of a phospholipid bilayer.

The adsorbed particles were found to be of the expected size and shape of individual lipid vesicles complexed with the bilayer. However, they were also of the same approximate dimensions and shape as the AFM tip. This could suggest that the vesicular structures may not be vesicles but rather convolutions of the AFM tip scanning objects with smaller radii of curvature (e.g., contaminants, surface heterogeneities).[196] These tests, which may provide information beyond simple recognition of size and shape, are needed to demonstrate that these are actually vesicles. One such test exploits the ability of the AFM to make indentation measurements. Lipid vesicles are expected to be far more soft and compliant than the underlying membrane, which is mechanically tightly coupled to the rigid mica substrate. Therefore, experiments were performed of micromechanical mapping on the surface by force volume imaging.[197] However, as indicated elsewhere, these results are not conclusive at this stage.

An isoforce image was constructed from a force volume of a 1:1 DPPC/cholesterol surface, and an elasticity map was produced using the FIEL method. The areas of low elasticity corresponded in general to the positions of the adsorbates. This difference in elasticity was more directly seen in the individual force curves collected on the flat part of the membrane and on the adsorbates. There was essentially no deformation over the bilayer until 1 to 2 nm before hard contact. This small region of deformation may be due to the intrinsic elasticity of the bilayer. In contrast, there was a prolonged deformation region of at least 15 nm over a typical vesicle. Therefore, the adsorbed structures are far more compliant than the supported bilayer. The fact that the z-distance over which the vesicles are elastically deformed is around the same size as the vesicles suggests that the vesicles are soft over their entire height. In addition to excluding a tip-shaped artifact, this softness also makes it unlikely that the adsorbates are particulate contaminants in the buffer. Finally, this deformability is consistent with elasticity maps of cholinergic synaptic vesicles and indentation measurements on biotinylated phospholipid vesicles bound to avidin-coated surfaces.[187]

The elasticity measurements of adsorbed vesicles were estimated from force volumes. It was found that a positive cantilever deflection prior to zero distance was indicative of sample deformation.

A second test of whether these structures are truly vesicles was concluded from the images of adjacent vesicles. In all cases, the vesicular structures were found to be flat, mutually deforming interfacial contact. This type of interface is characteristic of two adherent vesicles and was observed by freeze-fracture electron microscopy of egg-phoshatidylcholine SUVs[198] and light microscopy of giant unilamellar vesicles.[180] An additional feature of these adsorbates is that the highest points are soft and somewhat labile to AFM imaging, resulting in streaks. These streaks suggested that the adsorbates were far softer or more mobile than the underlying bilayer, a finding consistent with the interpretation that they are bound vesicles. It was found that even at modestly high scanning forces, the adsorbates were readily dislodged from the surface. This is in contrast to the underlying bilayer, which at high cholesterol content proved quite resistant to tip-induced abrasion from the mica surface.

The AFM images of vesicular structures were obtained in adhesive contact mode. In some images, it was found that the AFM tip had smeared away a piece of a vesicle or dragged it along the surface. By using the same method, vesicle–bilayer complexes with SUVs made of pure DPPC were prepared. These analyses showed that the surface was no longer covered by bound vesicles; instead, larger aggregates were found atop the first bilayer. These larger aggregates may be partial formation of a second bilayer or may result from multilamellar structures settling on the bilayer. The absence of individual vesicles for pure DPPC was consistent with a previous AFM report of multilayer formation by DPPC on mica.[199]

Why cholesterol appears to stabilize the bound vesicles and prevent multilayering is likely a result of the changes cholesterol is known to confer on phospholipid bilayers. These changes are well documented from micropipet aspiration studies and include increases in toughness, fluidity, bending and area expansion moduli, and lysis tension.[200] Interestingly, fluorescence spectroscopic studies of vesicle–vesicle fusion *in vitro* indicate that cholesterol has a biphasic effect on the ability of distearoyl-phosphatidyl-choline vesicles to fuse with mixed phospholipid vesicles: total fusion increases at mole fractions less than 0.1 and then falls markedly to 0.45.[201] Thus, one explanation for the unusual stability of bilayer-associated vesicles seen with these systems may be the reduced fusion propensity of cholesterol-rich vesicles coupled with their increased mechanical rigidity. However, more studies are needed before the effect of cholesterol on the characteristics of the vesicle–bilayer complexes can be explained. Furthermore, when DPPC-cholesterol SUVs were used to form the complexes, one repeatedly observed two distinct vesicular structures (dome shaped and saucer shaped). This may be interpreted as that the two structures are simply different types of stable intermediates that form when a vesicle wets or partially fuses with a bilayer. Whether this is a consequence of heterogeneities in vesicle size, lipid composition, or some other parameter is unclear. A second interpretation of the saucer-like structures relates to lamellarity. The saucer-like morphology may result from the adsorption and spreading of a multilamellar vesicle onto the surface; as the inner shells of the multilamellar structure wet the surface during adsorption, some combination of surface tension and steric confinement by the outer shells arrests spreading. In this model, the edges of the saucer come from the outermost shells wetting the bilayer. In the literature, it was reported that sonication and centrifugation may yield predominantly unilamellar structures. One possibility is that the few multilamellar structures that remain adsorb more stable to the membrane surface than unilamellar structures. Because AFM allows the simultaneous acquisition of structural and mechanical data under aqueous conditions, these data showed that one may be able to probe these structures at very high resolutions.

In a recent study, the interaction of DHP (dihexyadecyl phosphate) vesicles with increasing quantities of guanidinium-type counter-ions was investigated.[202] This has relevance to the use of many natural antiviral compounds, such as guinidine. Vesicles were prepared by adding 2 mL solution of DHP (0.003 M) to 8 mL of water (70°C) under constant stirring. Vesicles were hereafter forced through a 3 μm filter. AFM images were taken by placing a drop of vesicle solution, after solvent was evaporated on mica. Vesicles thus formed were clearly visible in the AFM images. Spontaneous vesicles are reported to form in aqueous solutions of mixtures of cationic and anionic surfactants.[202]

The vesicles formed in mixed solutions of cetyltrimethylammonium toluenesulfonate (CTAT) of odecyltrimethyl ammonium bromide (DTAB) with the branched chain anionic surfactant sodium dodecylbenzylsulfaonate (SDBS) were investigated by AFM and other methods. The vesicles of these systems are shown in Figure 3.10.

3.6 LB FILMS OF LIQUID CRYSTALS

Liquid crystals are of importance in different industrial applications (flat-panel displays, temperature devices, etc.) and in biological systems (lipids). As in practice, mixed liquid crystals have been widely used but empirically used in flat-panel displays without precise information on the substrate alignement. It was therefore realized that the anchoring phases of mixed systems needed to be investigated.[203] Mixed liquid crystals of 4-*n*-octyl-4-cyanobiphenyl (8CB) and 4-*n*-dodecyl-4-cyanobiphenyl (12CB) were studied on molybdenum disulfide (MoS_2) by STM. A drop of solution with mixture was placed on a freshly cleaved MoS_2 after heating to 100°C. The STM images in Figure 3.11 show the patterns as a function of mixing ratio of 8CB:12CB. The images show homogenous single-row structures only for

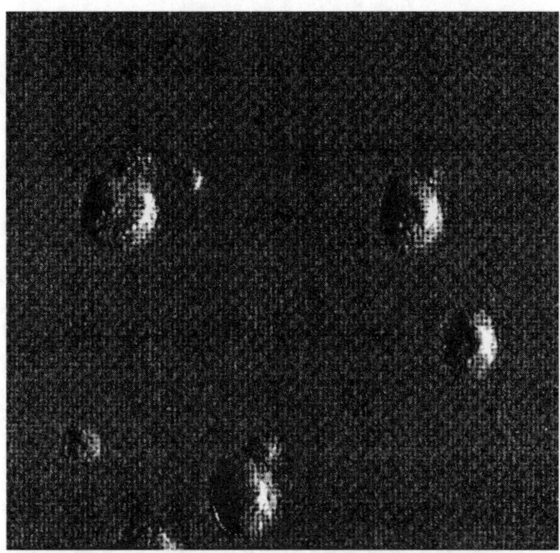

FIGURE 3.10 AFM image of vesicles deposited on mica (1 μm × 1 μm). (From Morgan et al., *Langmuir*, 13, 6447, 1997. With permission.)

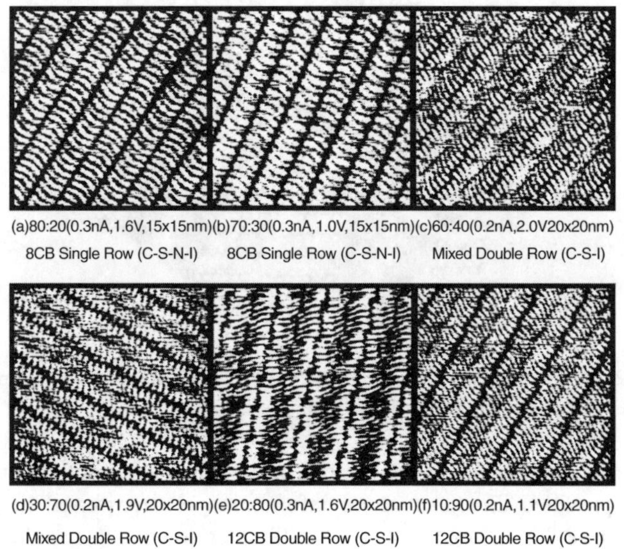

FIGURE 3.11 STM images of the mixtures of 8CB and 12CM. The mixing ratio is indicated in the text. (From Iwakabe et al., *Langmuir*, 10, 9, 1994. With permission.)

mixtures with ratios 80:20 and 70:30. On the other hand, individual patterns are found for other mixing ratios, as depicted in Figure 3.12 as models. These structures show the anchoring mechanism in liquid crystal mixtures.

Several studies demonstrated that AFM can provide an important new view of ultrathin, well-ordered organic multilayer films.[204] With very small loadings, on the

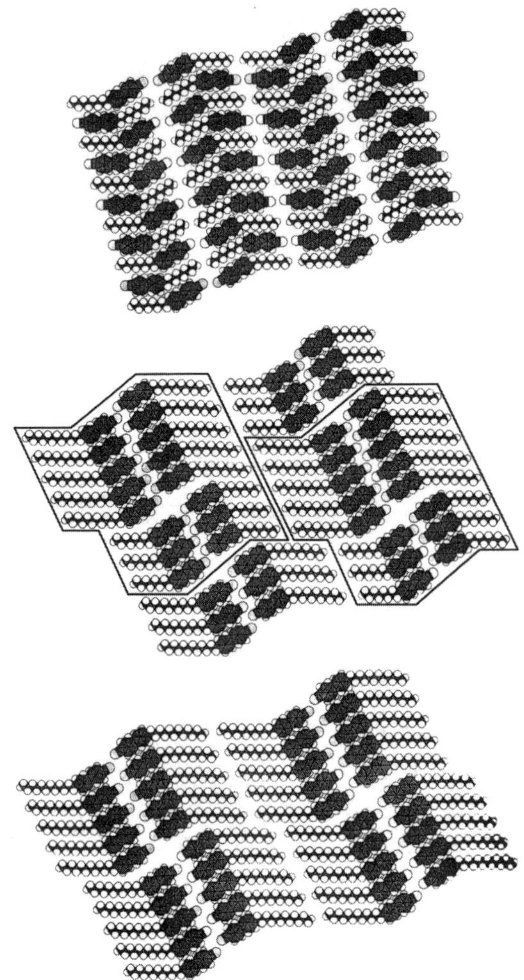

FIGURE 3.12 Models showing anchoring structures from STM images in Figure 3.11. (From Iwakabe et al., *Langmuir*, 10, 9, 1994. With permission.)

order of 10^{-9} to 10^{-8} N (1 to 10 nN), it has been shown to be possible to image LB films. AFM provided direct structural information on transferred liquid crystals, ranging from morphology on the scale of millimeter down to intermolecular spacing on the angstrom scale. Step heights as observed by AFM were shown to be in agreement with x-ray data. It must be mentioned that while AFM operates under ambient conditions, x-ray always operates under high vacuum.

Alkyltrichlorosilanes and 2(perfluorooctyl)-ethyltrichlorosilane (FOETS) mixed monolayers were prepared on the water subphase and were covalently immobilized onto the silicon wafer surface.[205] AFM observation of the mixed monolayers of octadecyltrichlorosilane (OTS)/FOETS revealed that the crystalline OTS formed circular (macro) domains of ca. 1 to 2 μm diameter. These domains were surrounded by a sea-like amorphous FOETS at OTS molar fractions even above 75%. The domains were higher by 1.2 nm than the flat surrounding matrix.

The controlled preparation of micro- and nanometer-size features in materials, by physical or chemical methods, has been recognized as an important technical invention. Monolayer isotherms of the polymerized surfactant 2-pentadecylaniline (2PDA) as well as detailed study of morphology were reported.[206] Through thermodynamic and AFM studies, it was shown that this system exists as a phase-separated monolayer before the polymerization is initiated. Microdomains with diameters ranging from 2 to 5 μm were observed.

3.7 STM AND AFM STUDIES OF DIVERSE MOLECULES ON SOLIDS

The STM has been found to provide information on atomic scale about the motion and interactions of adsorbates on solid surfaces. The ability to examine the local molecular chemistry at specific surface sites is one of the important areas of applications of using STM for dynamic interactions. The lateral motion of molecules in surfaces is important in all kinds of surface phenomena.

3.7.1 STUDIES OF DIVERSE SMALL MOLECULES

Diverse small molecules can be transferred to an air–water interface. They can be transferred to LB films of stable quasicrystalline nanomono- or multilayers. Applications of the LB films are expected in different areas, such as: nonlinear optico- (second- and third-harmonic generations), bimora, lubrication of magnetic tape.[207] The results obtained with a new stand-alone AFM integrated with standard optical fluorescence microscope are discussed. Imaging a molecule in real space is an elegant way to obtain information of its exact structure, i.e., its size, shape, mode of function, and the nature of its bonds with surrounding molecules.[208] This was investigated by using the constant-current STM image of a Cu-phthalocyanine molecule adsorbed on a Cu surface.

The adsorption of different organic molecules on electrode surfaces in aqueous solutions has been a well-known area of research in electrochemistry. The STM image of iodine-modified Pt in 0.1 $MHClO_4$ is given in Figure 3.13.[208] The image clearly shows the areas with defects.

3.7.2 SURFACTANT MOLECULES STUDIES BY AFM

AFM analyses of surfactant molecules on solid surfaces are useful in providing information on the adsorption process of these amphiphiles. The orientation of adsorbed molecules is the determining factor of the properties of the system. For example, a surfactant molecule adsorbs with its polar end toward a glass (or similar polar solid) surface. On the other hand, on a nonpolar solid surface (such as graphite), it adsorbs with its alkyl chain. In a recent study, the adsorption of hexadecyl-trimethylammonium bromide (CTAB) was studied by AFM.[209] The mica substrate was allowed to reach equilibrium in the solution of interest. Imaging was performed at a repulsive force with the tip separated from mica by about 3 to 4 nm. Under these conditions, the force

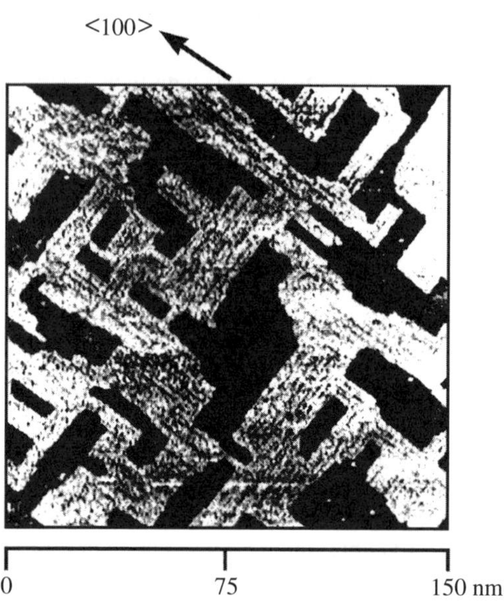

FIGURE 3.13 STM image at iodine-modified Pt surface in 0.1 M $HClO_4$. (Potential = 0.6 V, tunneling current = 1.0 nA). (From Sashakat et al., *Langmuir*, 14, 2896, 1998. With permission.)

on the tip is dominated by the surfactant film, and therefore, the image provides information about the surfactant film. The concentration of CTAB used was twice the *cmc* (critical micelle concentration) at 25°C. In pure CTAB solutions, the mica surface was almost flat, as seen from AFM images. On the addition of HBr or KBr, cylindrical features were seen, which indicated large micellar structures. The height analyses showed features that were ca. 0.2 nm, however, the aggregates should be 4 nm in height. This discrepancy was explained by the possibility that the tip was not able to penetrate the cylinders due to tip curvature. Similar images were obtained in the case of anionic detergent, sodium dodecyl sulfate (SDS) (Birdi, to be published).

In obtaining AFM images of SDS, a drop (10 μL) of SDS solution (1 mg/L) was allowed to evaporate on HOPG. AFM analyses was carried out after 24 h (Figure 3.14). SDS images showed that almost perfect flat layers of SDS molecules form the structures, with step height almost equal to the fully extended SDS molecule, i.e., ca. 15 Å. These results agreed with the findings that such amphiphiles form self-assembly structures (micelles and bilayers).[3]

LB films of surface-active azo dye, *p*-tert-octylphnol yellow amine poly(ethylene oxide) were measured by AFM.[210] Two-dimensional phase separation was observed in these LB films. The orientation of dye molecules was related to the surface pressure of the monolayer of films.

3.7.3 C60 Monolayers

Fullerene as well as fullerne derivatives monolayers were prepared and characterized using a variety of procedures.[211]

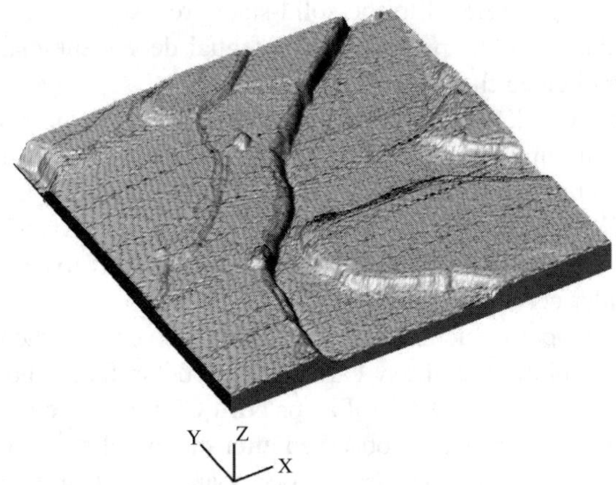

FIGURE 3.14 AFM images of SDS on HOPG. A drop (10 μL) of SDS solution (1 mg/L) after evaporation of water (90,000 Å × 90,000 Å × 1572 Å).

In a recent study, the monolayers of bipyridine of C60 were investigated STM.[212] C60 molecules were investigated by STM.[37] The diameter of C60 was found to be 1.1 nm.

There has been considerable interest in electrochemical processes that allow the direct conversion of one form of a solid material into another reduced or oxidized form, especially for the case of Buckminster-fullerene, C60.[213] The driving force for these experimental studies has arisen mainly from the need for new technology for specialized energy storage devices, interest in new superconducting materials, and the fundamental study of the energetics and kinetics of processes at phase boundaries in contact with an electrode.[214,215] As reported in the literature, the C60 appears randomly distributed over silicon substrate, although at least five typical adsorption sites have been defined.[216]

The reduction of C60 occurs in solution as well as in the solid state[8–15] via several consecutive one-electron steps.[212–216] In the latter case, the charge neutrality for the solid-state process has been shown to be maintained by the reversible insertion of countercations such as NBu_4+, Li+, Cs+, etc., from the contacting solution phase.[217] A recent review covered many of the advances and problems reported in this field.[218]

The electrochemical redox-cycling of solid C60 was reported under a wide range of experimental conditions, including various deposition procedures such as solvent casting,[219] gas phase,[220] and electrochemical deposition, and a range of contacting solvent systems and electrolyte salts.[221] The interaction of the solid with neutral solution species and also catalytic processes have been reported.[222]

The most extensive work has been published on the redox-cycling of solid C60 deposited on an electrode and immersed in an acetonitrile solution containing NBu4+ cations. For this system, the starting material, C60, as well as the products from the first two reduction steps, were sufficiently insoluble for the process to be dominated by a solid–solid pathway, because the loss of material from the electrode into the solution phase was small. In the complex reaction scheme, the dissolution of C60 and the solid-state conversion were proposed to occur as parallel processes.[218] However, mechanistic details related to the solid-state process are not well understood, and phenomena due to "film" porosity and potential-dependent changes of conductivity further complicated the system.[219]

The reduction of C60 deposits has been shown to involve factors such as counter-ion (cation) uptake and transport into the deposit, significant structural rearrangements on reduction and reoxidation, resistivity changes, and dissolution on the basis of EQCM and elegant SECM studies.[218–221] On the basis of mass spectrometry studies, it was suggested that the NBu_4+ cations are likely to be partially retained in the deposit under certain conditions.[223]

The electrochemical reduction of solid C60 abrasively attached in the form of microcrystals to graphite and glassy carbon electrode surfaces and then immersed in acetonitrile containing 0.1 *M* NBu_4PF_6 or NBu_4ClO_4 as the electrolyte has been studied.[224] Voltammetric responses observed after the initial stages of redox cycling experiments, when redistribution processes occur, are consistent with those observed previously for so-called "film" deposits. The characteristic "steady-state" shape of the solid-state voltammetric response with a large potential gap between reduction and reoxidation responses is shown by chronoamperometric experiments to be associated with the presence of a nucleation and growth-type mechanism. The initial three reduction processes of C60 attached to the electrode surface, which lead to the chemically reversible formation of only slightly soluble C60- and C602- and finally to the loss of the soluble C603- have been followed by *ex situ* SEM and *in situ* AFM experiments. The extent of the electrochemical conversion is shown to depend strongly on the crystal size, with larger crystals being affected only at the solid–liquid interfacial region. Evidence for stochastic processes further supports the proposed nucleation and growth-type mechanism. Crystal redistribution processes also are identified by AFM measurements.

In contrast to previous studies in which thin "films" of material of differing density have been applied to electrode surfaces, in this study, the direct use of microcrystalline C60 was used. For films of C60, cracks and imperfections may be the key factors in allowing electrochemical processes to occur. Experimental evidence for high porosity of C60 deposits formed by solvent casting comes most evidently from SECM and hydrodynamic voltammetry.[217] Direct use of a microcrystalline solid is therefore a suitable approach for mechanistic characterization of the process, because it eliminates the possibility of complications due to the presence of entrapped solvent molecules. The direct use of solid C60 mixed into a polyethylene oxide matrix was reported.[225] The reduction of microcrystalline C60 abrasively attached to suitable electrode surfaces was found, from electrochemical experiments and other studies, to proceed through a nucleation and growth-type mechanism. The electrochemical experiments were carried out using a computer-controlled electroanalytical system or a bipotentiostat controlled by an AFM system. The electrode used was a 4.9 mm diameter basal plane pyrolytic graphite disk mounted in Teflon, a 3 mm diameter glassy carbon disk, or a 3 mm diameter platinum disk. A platinum wire served as the counter electrode, and the reference electrode was a saturated calomel electrode (SCE) (with E1/2,ferrocene = 0.38 V versus SCE) in conventional voltammetry and an Ag/AgCl pseudoreference (with E1/2,ferrocene = 0.38 V versus Ag/AgCl) in *in situ* AFM studies. Samples were gold coated in a sputter coating unit prior to microscopy.

For *in situ* electrochemical AFM, an AFM, operating in the contact mode, was employed. Electrochemical experiments with C60 abrasively attached to a glassy carbon electrode were performed in a special cell. The experimental details for the electrochemical AFM cell were described elsewhere.[218,226] Pyramidal silicon nitride tips were used in these studies. The images were obtained with a resolution of 300 × 300 data points per image, and the scanning rate was about 5 to 10 Hz. All AFM and voltammetric measurements were made under ambient temperature conditions of 20°C. To keep the solution phase inside the electrochemical AFM cell free of oxygen and to minimize the contamination with oxygen diffusing through the latex membrane, freshly degassed solution was flowed through the cell before and after each potential cycle experiment.

Procedures as described in the literature were employed to characterize the morphological differences between C60 films prepared by solvent casting and by abrasive attachment of microcrystalline C60.[219] For the solvent evaporation deposition of C60 on glassy carbon and platinum, the electrode surface was polished and then coated with 5 to 10 μL of a solution of 0.01 M C60 in toluene or a saturated solution (0.36 mM) of C60 in dichloromethane. The procedure for abrasive attachment of C60 consisted of spreading 2 to 5 mg of the solid onto a filter paper and gently rubbing the electrode surface onto it. Glassy carbon electrodes were chosen for the *in situ* AFM experiments because of better adhesion of the C60 microcrystals compared to platinum or gold electrodes. Before use, the glassy carbon electrode was polished with a succession of diamond lapping compounds from 1 μm to 0.1 μm particle size and then sonicated for 20 min.

During a typical experiment, an electrode area of between 1 μm × 1 μm and 50 μm × 50 μm was initially imaged at open circuit potential. Subsequently, a potential of 0.0 V versus Ag/AgCl was applied to the C60 modified glassy carbon electrode, and then the topographic changes occurring during about 300 s were recorded. Finally, topographic images at different locations on the electrode surface were obtained after cycling the potential over the range between 0.0 V and –1.0 V, –1.2 V, or –1.6 V versus Ag/AgCl.

3.7.4 Preparation of C60 Deposits for Electrochemical Studies

Many applications and properties of thin films or deposits in electrochemistry depend strongly on their morphological (grain shape, contact between grains and the electrode) and structural (defects) properties. Surprisingly, relatively little is known about the morphology and the structure of the C60 deposits employed in earlier voltammetric studies, despite the fact that a range of techniques such as SEM, mass spectrometry, and quartz crystal microbalance have been applied in conjunction with the electrochemical measurements. The deposition of C60 at the electrode surface by solvent evaporation has been by far the most widely used technique in electrochemical studies.[219] The most commonly used solvents for this purpose have been benzene, dichloromethane, and toluene.[218] Differences in the morphology and porosity of the deposits may be expected due to the rate of nucleation and crystallization, which are related to the rate of solvent evaporation, the solution concentration, and the roughness of the electrode surface. These factors, together with the deposition procedure, are believed to influence the morphology, topography, and the mechanical stability of the resulting deposit.

In the following studies, AFM images of C60 deposits as prepared by using different procedures were investigated. The images showed the typical morphology of the C60 microcrystals formed on a glassy carbon electrode by evaporation of 5 μL of 0.01 M C60 in toluene solution. The images showed that the deposit did not completely cover the substrate surface. It also indicated that it is composed of various differently shaped single crystals and star-like agglomerates. These data were in accordance to that reported for a nonuniform C60 deposit grown by a sublimation procedure.[227,228]

AFM images are obtained in air of microcrystals of C60 formed on glassy carbon by evaporation of a 5 mL solution of 0.01 M C60 in toluene and by evaporation of a 10 mL saturated solution of C60 in dichloromethane. An AFM image of a glassy carbon electrode surface modified by mechanically adhered C60 was shown after immersion in acetonitrile/0.1 M NBu_4ClO_4, and the effect of cycling the electrode potential over 10 cycles (scan rate 0.1 V s^{-1}) between 0 and –1.0 V versus Ag/AgCl is shown.

Nanometer scale microcrystals and a much more uniform thin layer of C60 were formed on glassy carbon surfaces by evaporation of 10 μL of a saturated solution of C60 in dichloromethane solution. The fact that the deposits formed from an organic solvent of relatively low boiling point such as dichloromethane are more homogeneous may account for the better reproducibility of the electrode.

3.7.5 Voltammetry of Solid C60 Mechanically Attached to a Graphite Electrode

A marked difference is usually observed between the first and subsequent scans during potential cycling voltammetric experiments with drop-coated films and with vapor-deposited C60 film, irrespective of whether only the first or the first two reduction processes were studied.[217,229] The first two reductions were proposed to overlap during the first voltammetric scan recorded for a relatively thick C60 deposit but became resolved on subsequent scans. Similar behavior was observed for compact vapor-deposited films and for electrochemically grown films.[229,230] These reports suggested that the preparation method and the initial film structure affect the initial electrochemical behavior of the film. It was suggested that the initial reduction needs an overpotential for the incorporation of the cation into the C60 structure, and this initial activation process may facilitate the redox reaction in subsequent potential sweeps.

In these studies, cyclic voltammograms obtained for solid C60 mechanically attached to a basal plane pyrolytic graphite electrode were obtained. It was found that in the initial 10 potential cycles between 0.0 and –1.0 V versus SCE, a somewhat broad reduction signal developed into the relatively stable "steady-state" voltammetric response characteristic of the C60/NBu_4C60 system.[230]

The initial potential cycles involve more complex processes affecting the first (A_{red}) and especially the second (B_{red}) reduction signals. The exact nature of these processes, probably related to "breaking-in" or "redistribution" is not well known.

The data of cyclic voltammograms were obtained for the reduction and reoxidation of solid C60 mechanically attached to a basal plane pyrolytic graphite electrode (in acetonitrile/0.1 M NBu_4PF6) (scan rate 0.02 V s^{-1}, $T = 20°C$).

The characteristic features and the complexity of the observed voltammetric responses obtained for the reduction of microcrystals of solid C60 attached to basal plane pyrolytic graphite electrodes after repetitive cycling of the potential, appear to be similar to those described previously. The data suggested that processes attributed in previous studies to "films" are probably more accurately described as processes associated with reactions of arrays of solid particles or crystalline deposits. The voltammetric data observed for the reduction of solid C60 abrasively attached to basal plane pyrolytic graphite or glassy carbon electrode were compared to data obtained for the reduction of C60 deposited by dichloromethane evaporation onto a platinum electrode. The peak width at half-height of the voltammetric responses is, in most cases, small compared to the peak width of 91 mV expected theoretically for a "thin film" of noninteracting redox centers.[231] The peak potentials, especially for the first chemically reversible reduction redox couple, $A_{red} - A_{ox}$, can be seen to exhibit a wide separation or "inert zone" of $E_{pox\text{-}red} = 0.52$ V, which is characteristic for a wide range of solid-state electrochemical processes. This behavior has been attributed to structural changes associated with the process of cation insertion and expulsion into or from the solid or other mechanisms. Recently, systems with large peak separations have been interpreted in terms of a nucleation and growth mechanism.[232] Compared to the similarly related cases of reduction of solid 7,7,8,8-tetracyanoquinodimethane (TCNQ) and the oxidation of tetra-thia-fulvalene, this mechanism seems preferable to other schemes that were reported.[232]

In the case of C60, strong evidence for the presence of nucleation-growth-type kinetics in the interconversion of C60 microcrystals is provided by chronoamperometric (double potential step) experiments. The data of two successive potential step experiments for the reduction of C60 mechanically attached to a basal plane pyrolytic graphite electrode immersed in acetonitrile/0.1 M NBu_4PF6 after 10 redox cycling experiments were analyzed. It was seen that the reduction at $E = -0.850$ V versus SCE and the reoxidation current-time profile at $E = -0.205$ V versus SCE exhibit "rising" current transients rather than the monotonic decay of the current usually observed with diffusion or electron transfer controlled systems. This means that the current-time transients include well-defined peaks. The existence of peaks in response to potential steps is highly characteristic for solid-state processes associated with nucleation-growth kinetics. According to this mechanism, the slow formation of nuclei is followed by almost "explosive" growth at large overpotentials, which explain the large separation of peak potentials and narrow widths at half peak height. The calculations for this process showed that from an energetic point of view, the most likely location is at the triple interface between solid, electrode, and solution phases.[232]

A chronoamperogram was obtained for the reduction and reoxidation of solid C60 mechanically attached to a basal plane pyrolytic graphite electrode and immersed in acetonitrile/0.1 M NBu_4PF6. Potential steps are from 0 to –0.85 V for reduction and –0.85 to –0.205 V versus SCE for reoxidation after 10 initial potential cycles between 0 and –1.0 V versus SCE.

A study reporting the morphological changes that occur upon reduction of a film of C60 vapor deposited onto a platinum electrode and immersed in 0.1 M NBu_4PF6 in acetonitrile led to the conclusion that crystalline products may be obtained upon reduction to a NBu_4C60 film. In agreement with mass spectroscopy and EQCM evidence, results implied that these crystalline-reduced products can be formed irreversibly. It has also been reported that the extent of formation of the microcrystals depends on the film thickness, with the thicker films resulting in a higher coverage of crystals.[220]

SEM images were obtained for C60 deposited by CH_2Cl_2 evaporation onto a basal plane of pyrolytic graphite electrodes before and after electrochemical reduction at $E = -1.0$ V versus SCE. In agreement with the AFM images, the dichloromethane evaporation method leaves behind an array of solid, almost shapeless, particles deposited on the basal plane pyrolytic graphite rather than a film. After electrochemical reduction of the deposit at $E = -1.0$ V versus SCE for 60 s in acetonitrile/0.1 M NBu_4PF6, removal from the solution, air drying, and sputter coating, the morphology of the deposit clearly changed, and a crystalline product identified as NBu_4C60 was formed. Some of the initial C60 material was not converted to the well–defined microcrystalline form. However, the crystalline product observed after reduction resembled that observed by data obtained by *ex situ* experiments in the absence of the solvent (electrolyte) phase and were treated with care.[220]

The studies of redox cycling experiments of solid C60 by *in situ* AFM were carried out. The formation or dissolution of solid materials at electrode and solution interfaces by *in situ* AFM techniques was developed and applied successfully in recent years.[233] Further, it was shown that the electrochemical solid–solid conversion of mechanically attached solids, e.g., TCNQ, can be monitored directly by *in situ* AFM.[234]

The original C60 material adhered by abrasive attachment was found to be composed of small particles between 200 and 300 nm in size. These, combined with larger-sized single particles and micron-scale conglomerates of C60 particles, were suggested to give rise to areas of low and high coverage. Hence, monitoring the topography changes that occurred as the potential was scanned proved to be difficult because of the time it takes to record a good quality image over a wide area of surface and probably also because of formation of mechanically sensitive reduction products. Therefore, the monitoring of morphology changes was undertaken after redox cycling experiments for a specified number of cycles. Images showed (at different levels of magnitude of the same electrode surface at 0.0 V after 10 redox cycles between 0.0 and –1.0 V) that the surface of the larger conglomerates became smoother, and the smaller particles spread and lost their sharp boundaries. The extent of conversion of C60 to NBu_4C60 solids and vice versa during the course of each redox cycle is not known in detail, but it is likely that especially for the larger conglomerates, only partial conversion is achieved. However, even for these larger conglomerates, it is interesting to note that the surface appears to be affected by the change in topography, which indicates that probably a surface-based process occurs, leaving a considerable amount of original C60 material "buried" under the "active" layer.

In these studies, *in situ* AFM images obtained at 0 V in acetonitrile/0.1 M NBu_4ClO_4 showing the changes in the morphology that occured when C60 mechanically adhered as a crystalline solid to a glassy carbon electrode was subjected to redox cycling experiments (scan rate 0.1 V s^{-1}): C60 microcrystals after 10 potential cycles between 0 and –1.0 V, after two further cycles between 0 and –1.2 V, after eight further cycles between 0 and –1.2 V, and after three additional cycles between 0 and –1.6 V versus Ag/AgCl. Under the redox cycling experiments (0 and –1.2 V), which were measured with a switching potential corresponding to the case of the first two reduction steps, further substantial changes in topography were observed. These data showed adherence to C60 after 10 cycles between 0 and –1 V, and two cycles between 0 and –1.2 V, and 10 cycles between 0 and –1.2 V versus Ag/AgCl. The degree of roughness was found to increase with the number of redox cycles. In areas of low coverage, no significant change in topography was observed. This suggested that the redistribution of material occurs to achieve a lower energy configuration after adding two electrons. The redox cycling over a potential range that encompassed the third reduction-oxidation step (between 0 to –1.6 V versus Ag/AgCl) gave rise to material loss from the surface of the electrode. Thus, it could be concluded that the height of material observed was only a fraction of 1 to 3 μm. A thin film of material remained on the electrode surface even after the voltammetric activity was lost. This may not be due to the dissolution of C60, as it has been suggested, but may be due to the formation of a film of nonelectrochemically active material.[235] It has also been reported that the presence of a resistive film on the electrode surface after C60 lost its electrochemical activity.[219]

To compare the above results with the morphology changes of the C60 deposits formed by the solvent casting process, the topography changes of C60 microcrystals formed by toluene solution evaporation on a glassy carbon electrode surface were studied. It was found that the microcrystals become smaller with each redox cycling, and the roughness of the surface simultaneously increases, suggesting the occurrence of a surface reaction. Areas of low and high reactivity can be identified, and the presence of stochastic nucleation-growth-type kinetics may be responsible for this pattern.

The morphology changes in C60 microcrystals deposited on a glassy carbon electrode were studied, after evaporation of a 5 μL solution of 0.01 M C60 in toluene during the course of redox cycling experiments between 0 and –1.0 V versus Ag/AgCl. The images were taken after different cycles: (a) fresh deposit, (b) after one potential cycle, and (c) after four potential cycles in acetonitrile/0.1 M NBu_4ClO_4 (scan rate 0.1 V s^{-1}).

These studies showed that the longer time scale processes involved in the electrochemical reduction and reoxidation of solid C60 mechanically attached to electrode surfaces closely resembled those observed for deposits formed by solvent evaporation or gas-phase deposition. In other words, after a few initial cycles of the potential, the same voltammetric features were observed, irrespective of the method of the material deposition.

3.7.6 Benzene and Phenyl Radicals on Metal Surfaces

In order to understand the adsorption phenomena on solids, the imaging of benzene molecules and phenyl radicals on Cu (III) was investigated by STM.[236] This system was chosen, because Cu is highly suitable as a substrate for STM. Benzene molecule was selected due to its special characteristics. Imaging with the STM is relatively slow. All experiments were performed in an UHV at 77/K. The Cu (III) was prepared by electrochemical polishing, followed by sputtering and annealing. The resultant Cu surface showed clean flat crystal surface with ca. 100 nm terraces. In benzene atmosphere, low-temperature STM images at 4 and 77/K showed that step edges on the Cu surface were populated with benzene molecules. In images of areas 37 Å × 37 Å, one could clearly see rows of benzene molecules at the edges of terraces. Phenyl radicals were found to have adsorbed on the Cu terraces and at the step edges. Because benzene is a pollutant under given circumstances, this technique can be adapted to the nanoanalytical technique.

3.7.7 Other Diverse Systems

The co-adsorption of sulfur and carbon monoxide (CO) on platinum single-crystal surfaces was studied by STM.[237] In catalysts, submonolayer amounts of sulfur are frequently added to surfaces in order to modify these surfaces. The sulfur molecule is also known to hinder the adsorption of CO. These studies are of relevance to pollution control and prevention.

3.8 STM Studies on the Effect of Functional Group

In STM, there is still discussion about the mechanism by which nonconducting material can be investigated. In a recent study, the tunneling mechanisms of STM were carried out by varying the functional group of the molecules adsorbed on solid surfaces.[238] Experiments were carried out to study the spatial overlap between the STM probe and the functional group and electronic structure, the coupling between the energy levels of the adsorbate and the surface fermi level. STM studies were carried out at the liquid–solid interface. HOPG was used as the substrate. A few microdroplets of the solution containing the derivatized hydrocarbon of interest were pipetted onto HOPG, and the samples were mounted on the STM. The variation of the tunneling effect on the changes of constant-current topographs as a function of bias voltage for three functionalized hydrocarbons was studied. The materials studied were octadecanamide (ODA), 1-bromodocosabe (1BDO), and octadecyl sulfide (ODS). Solutions of these substances (1 mg) were added to 1 mL of phenyloctane. Height versus voltage measurements were obtained by sequentially varying the bias voltage, in 0.1 V increments, applied to the sample in the range from –1.6 to –0.6 and +1.6 to +0.8 V, while recording STM topographs as a function of tip height. In these studies, the brighter areas correspond to topographically higher regions. ODA was found to form a two-dimensional thin film comprised of molecules arranged in an all-trans configuration. From these images, it could be concluded that the molecules orient in a head-to-head configuration where the functional group of one molecule lies adjacent to the functional group of the next molecule. The angle between the molecular axis and the trough was found to be 68°, which was in accordance with a preferential packing arrangement of hydrogen bonds between the adjacent amide groups. Similar image analyses of 1BDO showed that the bromide end group was clearly distinguishable from the hydrocarbon backbone. The two-dimensional thin film of ODS revealed that hydrocarbon arms extended linearly from the sulfur atom positioned at the center of the molecule. The sulfur molecules were found to align with their molecular axis oriented at approximately 90° with respect to the troughs. As already mentioned, the tunneling mechanism for insulating molecules adsorbed on conducting surfaces is still the subject of much interest. In the literature, various mechanisms have been put forward to describe the STMs' ability not only to distinguish between the insulating adsorbates and the atoms of the surface but also to differentiate between atoms within the adsorbate. From these analyses, it was found to be useful to describe these data in two different categories, including weak-resonant tunneling or strong-resonant tunneling coupling on the basis of the degree of electronic coupling within the tip–adsorbate–substrate system. These studies were based on the fact that not only is STM able to distinguish between different molecules at surfaces but also to differentiate between atoms within the adsorbate. For example, it was found that Xe molecules adsorbed on Ni surfaces appear in the STM images as 1.5 Å protrusions.[239]

The STM tip is known to exert some force (nN) on the adsorbate, which requires the fixing of adsorbed molecules on the substrate. By covalently attaching porphyrin to a gold substrate, STM images were obtained.[41]

In some experiments, it is observed that the tip might give rise to moving or rearrangement of the substrate molecules. Under controlled conditions, this gave rise to the possibility of writing on substrates at atomic scale (nanolithography), as described later. In the case of other molecules, one even observes assemblies.

The interaction of such small molecules as H_2 and O_2 with Cu and Ni surfaces has also been investigated using STM.[240]

4 Biopolymers and Synthetic Polymers Structures by STM and AFM

Even though SPMs (mainly STM and AFM) have existed for only a few decades, by any standard, they are still in the early-level experimental phase. Prior to SPMs, all molecular level images of biological systems were known from microscopic conditions much different than their ambient environments. For example, all x-ray microscopy is carried out under vacuum. All biological processes take place in aqueous environments, and therefore, STM and AFM have been found to be useful for these systems. For biological applications, the major aspects have developed extensively, including instrumental development, such as different operation modes, and the control of imaging environments.[116]

During the past few decades, x-ray diffraction has contributed much useful information in the understanding of the structure of fibrous and globular proteins. Structures such as α-helix and β-structures are some of the most important analyses reported.[241]

The literature contains an extensive number of studies in which biopolymers were investigated with the help of SPMs. These reports will be described in detail here.

The synthetic polymers (such as polystyrene, polyethylene, polymethyl acrylate, etc.) are known to play an important role in everyday life. The molecular structural studies of the macromolecules have been extensively investigated by SPMs. Detailed molecular information was obtained from these SPM studies as described in the following.

The most recent investigations involving a single polymer chain, are the most important innovations for SPM applications.

4.1 DNA STRUCTURES BY STM AND AFM

DNA is one of the most important molecules for biological (genetical) evolution and mankind. DNA has attracted a great deal of interest from the SPM scientists since the invention of STM. The sequence of DNA is the most basic information stored in this molecule. The current methods of sequence analysis are time consuming and not always perfect. Furthermore, the amounts needed are larger than those sometimes available. It was hoped to accurately image the individual bases by STM or AFM, so that the sequence of a long nucleic acid polymer could be directly read. However, at present, it has become evident that this high-resolution demand cannot be realized, but various studies reported some success.

In DNA, the helical structure and the kinks are thought to play an important role in DNA–protein interactions, but this has not been directly observed until recent data was obtained by using AFM. By using AFM, the 168 base-pair axially stained circles were observed.[242] DNA images had an average half-height of 3.5 nm, which indicates an instrumental broadening of 1 nm. An AFM image of DNA on HOPG is given in Figure 4.1. In spite of low resolution, the helical structure of the DNA molecule in these clusters can be seen. This might be due to the fact that DNA was dissolved in water. As shown elsewhere, the AFM images of other polymers, such as polyglycine, more clearly show helical structures.

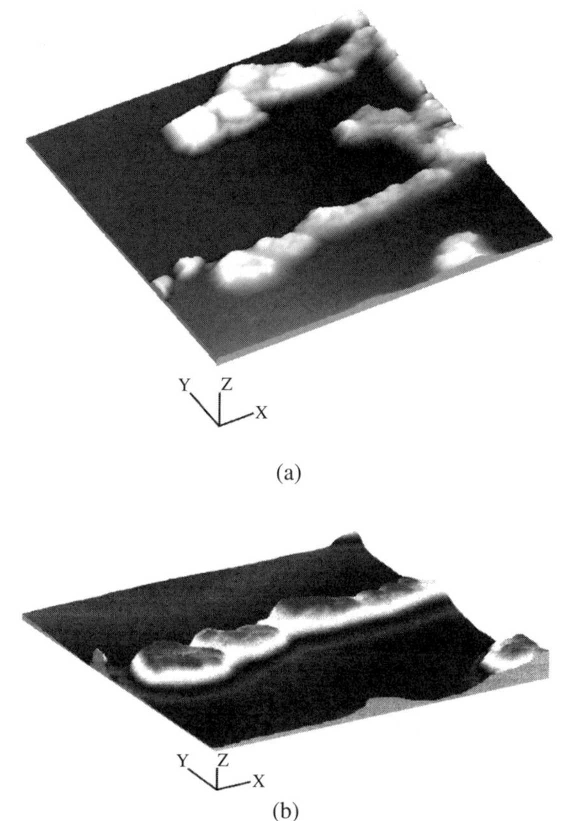

FIGURE 4.1 AFM image of DNA on HOPG (dissolved in water): (a) 50,000 Å × 50,000 Å; (b) three-dimensional 30,000 Å × 30,000 Å.

Adenine and thymine molecules when placed on graphite were investigated by STM, and one could easily distinguish the two molecules (thymine contains a single hexagonal ring and the adenine molecule has a double-ring structure).[243] These results lead to the possibility of using STM and AFM to obtain the sequence of DNA.

In current literature, the visualization of circular DNA molecules by AFM to generate atomic resolution images of biological DNA molecules was reported. AFM images of DNA, using AFM imaging of DNA under propanol, have been shown to give better resolution and narrower apparent widths than imaging in air.[244]

The AFM to DNA has the ability to obtain high-resolution images routinely in ambient conditions. The AFM was used to image double-stranded DNA and showed an open circular shape without drastic contortions and contour lengths that measured within 1% of the calculated length.[245] Images of single-stranded DNA showed two distinct forms. Double-stranded DNA adsorbed on Ba (11) treated mica. The sample prepared by adsorbing DNA (single and double stranded) on treated mica showed isolated strands. AFM was used with a force constant of mN 0.06 S/m.

Three different advances involving DNA in AFM were reported.[246] First, an HEPES-Mg buffer was used that improved the spreading of DNA and provided good DNA coverage at 5 μm with as little as 200 to 500 picograms per sample. Second, the new "tapping" mode was used to improve the ease and resolution of AFM imaging of DNA in air. Finally, AFM images were presented of single-stranded $X-174 virion DNA with the gene 32 t single-stranded binding protein. A summary of the current state of the field and of the methods for preparing and imaging DNA in the AFM was described.[247]

The need of a universal buffer is a necessity in all kinds of STM and AFM analyses. One-Phor-AII buffer plus (OPA+) was reported to be excellent for preparing DNA samples for AFM.[246] DNA molecules bind to the mica substrate strongly in OPA+ buffer, even though OPA+ contains 50 mM K+, which has been reported to inhibit DNA binding to mica. The availability of a universal buffer is useful for researchers who use a variety of restriction and modifying enzymes. For AFM of DNA, however, it was found that it is a universal buffer.

The application of STM to image the internal structure of biological molecules was pursued.[248] Samples were studied under UHV for STM of biological molecules. Various molecules of biological importance that were studied included hydroquinone; oligomers including pentaphenyl ether, enkephalin, and SDS; large homopolymers of lysine and glycine; calmodulin, tropomyosin, and IGG; and double- and single-stranded plasmid DNA. During the evaporation process of charged molecules from a drop of solution placed on a solid, molecules will concentrate at the solid–liquid interface. Charged groups are often essential for the aqueous solvation of large molecules. These charged molecules in solution can be neutralized by reattachment of counterions during the final stages of evaporation, or they may be neutralized, along with a remote counterion, when both attach to the Pt substrate. Upon evaporation, it is highly probable that naturalized, surface-bound counterions will accumulate near molecules as a result of the partially ionic character of the nominally neutralized macromolecule. STM images of 550-mers of polylysine (0.05 μg/mL) appeared as compact clusters, which were suggested as single molecules. The DNA images showed heights of 5 nm, which agreed with models.

Single-stranded DNA were found to be circular in shape. It was concluded that STM can be useful in providing high-resolution images. In the case of macromolecules, some degree of blurring in images may be observed. This may be due to overlapping or three-dimensional structures. It was concluded, as mentioned elsewhere, that such STM studies may lead to the possibility of DNA sequencing.

The AFM method was used to study various biopolymers, such as DNA on substrate mica, etc.[249–255] However, the resolution of AFM is known to be limited by the sharpness of the tip, and the interpretation of the image has been related to the geometry of the tips.[256] For example, due to the finite radius of the tip, AFM images of DNA are, on average, seven times broader than the known 2 nm (20 Å) width of DNA. This result requires description of the resolution of AFM images, especially when comparing images of AFM with electron microscopy. Some investigations on this subject have been reported.[20]

AFM images of some substrate molecules might be subject to artifacts, such as the broadening of structures and ghost images of tips, due to finite size and shape of the contacting probe. This means that the shape or radius of the tip used in AFM measurements should be known. However, these observations have not been systematically or exhaustively investigated. Therefore, at this stage, no clear description can be given. Recent analysis exhaustively described these observations for a variety of polymers under varying experimental conditions.[3]

4.2 SPM STUDIES OF THREE-DIMENSIONAL PROTEIN STRUCTURES

Biological activity of proteins and enzymes is strongly dependent on molecular configuration. Such macromolecular structures as α-helical, β-sheet, and random-coil are well-defined, three-dimensional structures.[3,100,257]

Although electron microscopy has been very useful in determining such structures, the advent of STM and AFM has added much information in recent years.[3,258] In fact, the first image of DNA (double-helical image) by STM was a highly acclaimed result.[259]

In general, in all cases, the need for three-dimensional protein structures becomes urgent, as with the bacterial proteins (for example, cholera toxin, which is known to produce the disease cholera). In the latter case, these investigations have been going on since 1977 using protein crystals and x-ray diffraction methods.[260] This knowledge is essential in order to produce new kinds of vaccines.

The AFM has become a well-established and valuable tool in the three-dimensional topographical analysis of biopolymers.[261–267] The three-dimensional structure of most of the biopolymers is within a range of 4 to 10 nm (40 to 100 Å) in lateral dimension by 1 to 5 nm (10 to 50 Å) in height. Unfortunately, in this size range, the finite size of the probe tip leads to relatively large systematic enlargements. However, this observation needs extended investigation.

In the structural analysis fields, electron microscopy has long been successful as a well-established imaging application, and it is still improving in terms of resolution quality for imaging the three-dimensional structures of biological molecules.[249] However, observation with an electron microscope must be carried out in a vacuum, resulting in difficulty when imaging biological specimens under physiological conditions. Three-dimensional structures are observed as two-dimensional projections, often distorted by the drying processes in the electron microscope. In addition, to obtain sufficient contrast in electron microscope analysis, the specimen usually must be treated by heavy-metal stain, which easily induces denaturation of the biological molecules.[268] The two-dimensional crystal of the membrane protein bacteriorhodopsin (purple membrane) was studied, as deposited on mica, silanized glass, or lipid bilayers, and the two-dimensional Fourier transform of AFM images was reported, which was in agreement with results of electron diffraction measurements. In the case of STM images, high-quality images of bacterial HPI (hexagonally packed intermediate) layer have been reported.[269] Some difficulties were encountered when trying to obtain unambiguous AFM images of biological macromolecules. These macromolecules must not be packed into dense aggregates but should be distributed in a two-dimensional plane. To demonstrate the first trial observations of "artificially formed" two-dimensional arrays of water-soluble protein molecules, AFM studies of monolayers of ferritin molecules electrostatically bound to a charged polypeptide layer of poly-l-benzyl-L-histidine (PBLH) on a silicon wafer were carried out.[270] Individual water-soluble molecules of the protein ferritin have been imaged an a silicon surface in pure water at room temperature with the AFM. The ferritin molecules formed an ordered monolayer by binding to a charged polypeptide monolayer of PBLH spread at the air–water interface. The film, fully wetted with water, was horse spleen ferritin, and was dissolved in pure water and fractionated by ultracentrifugation several times at 200.000 g for 40 min. The heavier fractions were diluted with a solution of sodium chloride (<10 mM) to a final concentration of 100 mg/mL. The ferritin solution was diluted with phosphate buffer (5 mM, pH 5.3) and prepared to a concentration of 30 μg/mL. Horse spleen ferritin has an isoelectric point of 4.5 and is negatively charged in a solution at pH 5 or above. AFM images showed the existence of channels, the size of which were in agreement with x-ray crystallography data. The STM images of PBLH revealed details of the molecular structures with the α-helical conformation, as expected. Images obtained by STM were compared with the macromodel, and these agreed with the images of STM.

4.2.1 Catalase

Globular features were reported from studies of catalase by using STM[271] studies. Individual catalase molecules could be observed from STM images. The repeat distance of 1.1 nm was obtained. STM images of catalase showed clearly individual molecules with diameter of 10 nm and height of 6 nm. This corresponds to the x-ray crystallographic data (10.5 nm × 10 nm × 6 nm). The effects of hydration and dehydration were also measured.

4.2.2 Other Protein Molecules

Gelatin was studied by AFM after placing it on HOPG [a 10 μL drop of solution (1 g/L) was placed on HOPG]. After water evaporation, the AFM image was obtained. A three-dimensional AFM image of gelatin is given in Figure 4.2. The

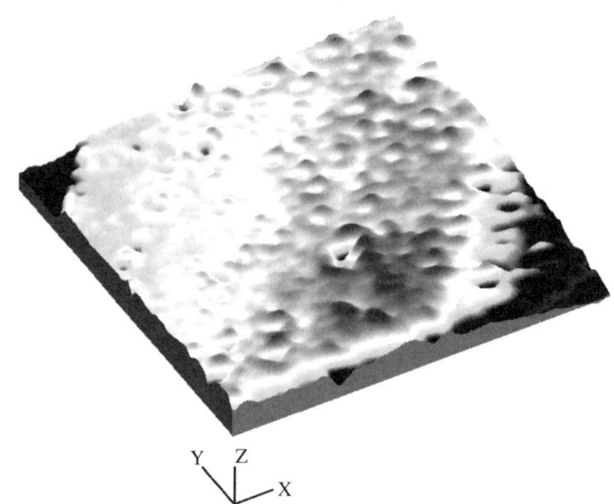

FIGURE 4.2 Three-dimensional AFM image of gelatin on HOPG (30,000 Å × 30,000 Å).

AFM images were further analyzed by software that allows for manipulation of the image data.

The molecular resolution images of β-helices were obtained by AFM studies of pentadecavaline films.[272]

The AFM images[273] of the von Willebrand factor, which is a plasma glycoprotein, in hemostasis and thrombosis of blood were reported. The size of the glycoprotein as measured by AFM was some two to three times larger than that reported by electron microscopy (EM). Other authors have reported similar divergences for other systems, which may be ascribed to such factors as ambient humidity.

4.2.3 Pectin AFM Analyses

AFM images can be useful in analytical applications for determination of purity of substances, such as pectin. The image of pectin (Figure 4.3a) showed that there might be two or more different kinds of aggregates. Some data obtained for other pectins showed clearly that the composition was different. In Figure 4.3b, an AFM image of a mixed sample of pectin plus lysozyme (1:1 g:g) is given for comparison. It is shown that lysozyme appears as a smaller molecule (mol. Wt. 14,000) in the image, as expected.

4.3 PROTEIN ADSORPTION STUDIES BY AFM

The phenomenon of protein adsorption on surfaces is an important process encountered in everyday life. The adsorption of proteins on solids by using AFM under water obviously has much importance. This method allows the adsorption mechanism to be studied *in situ*. The adsorption of fibrinogen has been reported.[26] The adsorption of fibrin was reported using the same procedure.[26] Fibrinogen molecules were adsorbed on silicon wafers from 1 μg/mL solutions for 5 min. The silicon

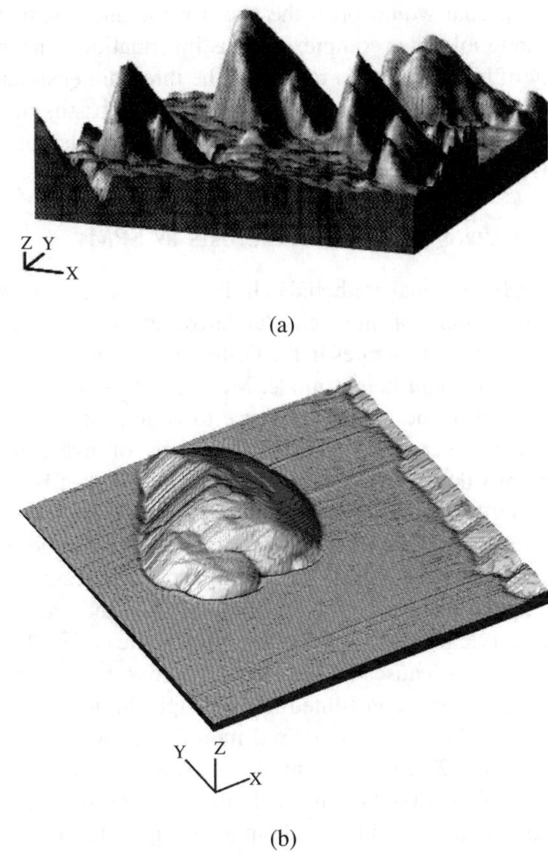

FIGURE 4.3 AFM image of pectin (a) on HOPG surface applied from a solution (1 mg/mL) in water (7500 Å × 7500 Å); (b) pectin + lysozyme.

surface was rinsed and dried before AFM. AFM images showed two forms of topography. One form was trinodular with a length of 60 nm:

TRINODULAR...O-O-O...
LENGTH...60 nm / WIDTH 40 nm...

This value differs from other literature data for trinodular, stated as 45 to 50 nm. At this stage, it is not easy to explain these differences, although, this may be related to tip effects. One must also consider the fact that AFM is operated under ambient conditions.

4.4 BIOLOGICAL MACROMOLECULAR STRUCTURES

It is the goal of high-resolution biological microscopy to obtain images of individual protein and other molecules clearly showing the internal atomic structure.[249] The realization of such a goal would open the way for the analysis of the structure of many molecules and molecular complexes. This information is needed, because all biological activity of biopolymers is related to the three-dimensional structure. It is shown that the contrast in high-resolution electron micrographs of biological macromolecules, illustrated by a study of TMV in ice, falls considerably below the level that should theoretically be attained.

4.4.1 Studies of Virus and Cell Structures by SPMs

In renal physiology, living renal epithelial cells have been imaged by AFM is scaniled the apical plasma membrane of linie cells "under water," i.e., after constantly superfused the cells at high perfusion rates in the buffered solution at 37°C.[274] The images were obtained in the constant height mode. Measurements were made with special care in order to minimize the amount of force to which the cell was subjected. A V-shaped 200 m cantilever was used. AFM images of live renal tubules were obtained. It is obvious that AFM is going to enter the field of biological research, for example, possibly for observing intracellular dynamic processes.

It is well known that communication of a stimulus from one cell to its neighboring cell is mediated by the release of an agonist. In the case of nerve–nerve or nerve–muscle synapses, the agonist (acetylcholine) from the presynaptic nerve cell binds to nicotinic acetylcholine receptors (nAChR) on the cell.[275] The binding of the agonist to the receptor site causes an ion channel across the cell membrane. In this study, the LB properties of reconstituted nAChR-lipid films were investigated. LB films of protein (nAChR) and protein–lipid monolayers were prepared on silicon slides for AFM studies. These AFM studies showed circular features that were attributed to protein molecules floating in lipid film. The protein molecules were 8.5 nm × 11 nm in dimension, which is about 30% larger than expected. It was also observed that in AFM images, protein molecules in the lipid surrounding tend to cluster in a protein–lipid mixture of membrane fragments and lipid domains. The tapping mode of imaging has been used for soft materials, where the probe lightly taps the sample rather than scraping the surface.[276]

4.5 SYNTHETIC POLYMERS STUDIES BY SPMS

The role of synthetic polymers in everyday life is highly important (for example, in housing, cars, food and packaging, heart valves and other transplants, etc.). Because all the properties of these polymers are related to molecular weight, conformation, and other characteristics, the molecular information from microscopy is essential. STM and AFM have been extensively used to study synthetic polymers. These studies provided much useful information that could be related to the structure–property relationship of polymers.

The molecular configuration of isotactic PMMA as LB films was investigated by STM.[277,278] The analysis showed that the polymer was present as a closely packed well-ordered arrangement with a helical and linear chain mixed configuration.

The polymerization of vinyl stearate was followed by the change in the π versus A isotherms.[279] The conversion of monomer to polymer was increased by increasing the π maintained on the film and by annealing the film prior to irradiation. The polymerization in the monolayer was 75%.

The surface properties of polyacryloylacetone (PAcA) were investigated by surface viscosity and AFM studies.[280] The orientation of polymers at interfaces is not very well known in many systems. AFM studies of PAcA were carried out as LB films on freshly cleaved mica plate. The substrate mica was first immersed into an aqueous subphase, followed by monolayer transfer as LB. AFM images were made in the noncontact mode. No microdomains were found at monolayer surface pressure below 6 mN/m. Three-dimensional structures were observed after the collapse pressure at 22 mN/m. This indicated the expulsion of polymer chains from the water phase.

It is known that hydrogen bonding governs the elastic properties of poly(vinyl alcohol) (PVA) in water. The single-molecule force spectroscopic studies of PVA by AFM have been investigated.[281] AFM studies of tethered polystyrene-B-poly(methyl methacrylate) and polystyrene-β-poly(methyl acrylate) brushes on flat silicate substrates have been reported.[282]

Synthetic polymer latexes prepared by emulsion polymerization are important industrial products widely used in the paper, paints, and coatings and adhesives industries. The essential feature of such application is the formation of a continuous latex film. This film consists of polymer particles dispersed in an aqueous medium. It is now well recognized that film formation from latex involves three stages: (a) evaporation of water, resulting in packing of latex particles; (b) followed by their coalescence and deformation; and (c) interdiffusion of polymer chains between adjacent particles. The final stage of interdiffusion of polymer chains across the particle–particle interface eventually leads to the formation of a coherent film. There have been many different methods used to study these films, such as elipsometry and neutron scattering. AFM has been used which has provided much useful information, because it can provide three-dimensional images.[283] In these studies, the AFM was operated over a surface area of 10 μm × 10 μm.The spring constant was 50 Nm^{-1}, in the tapping mode. The AFM of alkali soluble polymer SAA (polystyrene) on the film formation of PBMA (poly-*n*-butyl methacrylate) were reported. The surface morphology of the latex film of PBMA in the presence of post-added SAA was found to be strongly dependent on the amount of SAA.

The morphologies of individual latex particles with AFM were studied after the particles were adsorbed onto mica and then dried.[284] The aim was to verify if the morphology of composite materials correlates with the topography and the shape of the single latex particle on flat substrates, And from the shape of individual particles on mica, information about the behavior of the latex film was obtained. The AFM images showed that when a latex film dries, the volume is reduced, and the spherical particles deform into dodecahedrons of diameter 84 nm.

INDIVIDUAL
LATEX
PARTICLE.............SHAPE OF DODECAHEDRON....84 nm

In the analyses of AFM images, the influence of the tip was investigated. The tip can be modeled as a cone (opening angle 2 α), with a spherical end, radius of curvature, r_{tip}. The model considered was where the tip radius is smaller than that of the latex sphere:

AFM tip.....r_{tip}
Latex sphere.....d_{latex}

Under these conditions, the measured diameter at half maximal height, d_{afm}, is related to the true diameter, d_p, by:

$$d_p = d_{afm} \cos(\alpha) - 2r_{tip}(1 - \sin(\alpha)) \tag{4.1}$$

The radius of curvature of tip, r_{tip}, is 10 nm, as reported by the supplier or analyzed by a microscope. A value of α ca. 20° was estimated from indirect measurements.

Because the initial report that polyacetylene becomes conducting upon exposure to halogen vapors,[285] there has been great interest in the synthesis and characterization of polyacetylene as well as other conducting polymeric materials. Of these new materials, the polyaniline family of compounds is the most important system. The techniques of scanning tunneling microcopy (STM) and atomic force microscopy (AFM) have been used to characterize the initial nucleation and growth as well as the final surface structure of the conducting organic polymer films polyhydroxyaniline and aniline-(3-aminophenyl) boronic acid.[285] STM images were obtained by UHV instruments. With STM scans, individual polymer coils were identified. STM and AFM scans indicated that the surfaces of the domains were smooth, with size around 10 Å.

4.5.1 Dextran Molecule

AFM was used to study dextran under water with high-resolution sensitivity.[286] Dextran is a hydrophilic polysaccharide macromolecule, and it has been used in many applications, such as for a protein-resistant surface coating, as a hydrogel matrix, and in other chromatographic applications. In this study, the hydration of dextran molecules was investigated by AFM of a thiolated dextran monolayer under water and propanol *in situ*. Dextran solutions were allowed to react with gold overnight before making the measurement. AFM images of dextran on gold substrate were taken in air. Later, this sample was incubated in water for varying lengths of time. After 5 min, the images begin to show spherical features. The images were distorted in the direction of the scan after 8 min. This indicated that the tip can deform the dextran molecule after it is hydrated. The force–distance curves of the probe against cantilever deflection (force) are given for different times of exposure in Figure 4.4.

These data showed that a gradual increase in the sample's elastic modulus took place as the degree of hydration increased with time. This led to a decrease in the linearity in the contact region, which indicated deformation of the soft surface by the tip–dextran contact. Similar data was also found for gelatin samples. Exposure to proponal led to dehydration and the reversal of the plots shown in Figure 4.4.

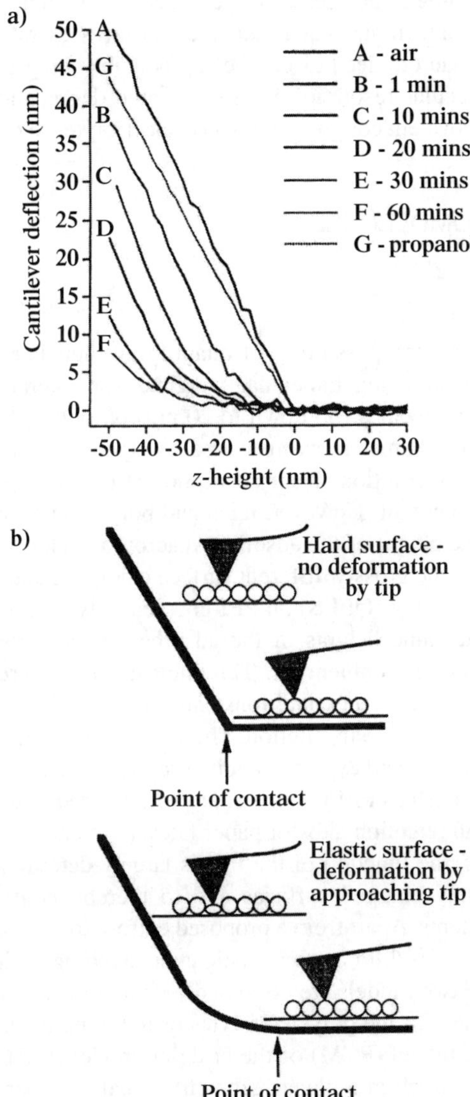

FIGURE 4.4 (a) Force–distance curves as a function of time of exposure to liquid of dextran. (b) A schematic model of the sample elasticity and hydration. (From Frazier et al., *Langmuir*, 13, 4795, 1997. With permission.)

4.5.2 Single Macromolecule Adsorption Studies

It is the biggest aim of researchers to be able to study the characteristics of a single molecule. In solutions, this cannot be achieved for obvious reasons. In a recent study, detection of a *single molecule* adsorption structure of poly(ethylenimine) (PEI) macromolecules by AFM was described.[287] In this study, tapping mode SFM was applied to characterize the distribution and adsorption structure of PEI macromolecules adsorbed onto negatively charged polystyrene latexes as colloidal model systems and onto molecularly flat mica surfaces as reference systems. On both surfaces, PEI macromolecules can be reproducibly mapped by tapping mode AFM, yielding molecular resolution without sample degradation despite the only relatively weak noncovalent coupling of the polyelectrolytes to the substrate surface.

TIP.........v
SINGLE MACROMOLECULE
....
SUBSTRATE

These data allowed the possibility of quantifying their lateral dimensions as a function of the corresponding molecular weights. The lateral dimensions of the adsorbed PEI macromolecules (60 down to 20 nm) on both types of substrates are in agreement with the diameters as measured by dynamic light scattering for the respective molecules in solution. Their adsorption structure is patch-like flat in the dried state under ambient air. However, mica and polystyrene surfaces resulted in a large difference in the height of the adsorbed macromolecules, which may be interpreted as being due to the grossly different surface charge densities of the substrates. Quasielastic light scattering (QELS) on PEI-covered polystyrene latexes in solution yields essentially the same heights of the adsorbed macromolecules as found by AFM in the dried state in ambient air. This indicates that there is no appreciable collapse upon drying at ambient conditions and further backs the notion of a dense patch-like adsorption structure in solution. These findings are discussed with respect to implications for the flocculation mechanism relevant for PEI.

Polyelectrolytes, such as PEI polymers, are widely used in the paper industry as effective drainage and retention aids for paper fines, pigments, fillers, and dyes.[288,289] In these applications, the function of the PEI is largely determined by its ability to adsorb onto negatively charged surfaces and to thereby control the aggregation behavior in these systems. Apart from a proposed entropy-driven sharing of polymers between solid surfaces,[290,291] for a mechanistic understanding of flocculation, besides heteroflocculation,[292] two models are discussed[293,294] that mainly differ in the adsorption geometry assumed for the polyelectrolytes or in the particular case for the PEI, i.e., the patch charge model (PCM) or the bridging model (BM).

Polyelectrolytes adsorb in a planar geometry (small number of loops and tails, many trains) and create, together with uncovered areas of the particle surface, a mosaic of positively and negatively charged patches. In this model (PCM), flocculation is induced by the attractive electrostatic interaction of positive (covered) and negative (uncovered) patches of adjacent particles. There is strong evidence that this model especially holds for highly charged polyelectrolytes and for PEI, in particular.[295] In the BM, the polyelectrolytes adsorb with many loops that, during flocculation, adsorb to adjacent particles, forming bridges. These findings suggest the relevance of this model, especially for polyelectrolytes of high molecular mass and low charge density.[293]

In earlier reports,[296] single molecules of PEI on polystyrene latexes were investigated by chemical force microscopy (CFM). In CFM,[296–300] a specific, desired chemical

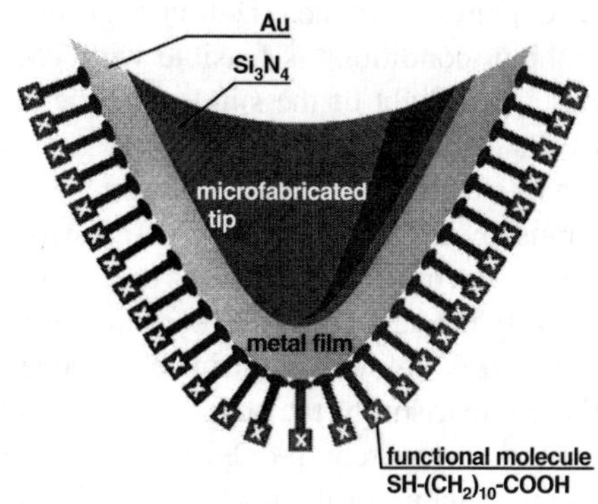

FIGURE 4.5 The modified tip is depicted (schematic). The gold-coated silicon-nitride tip is covered by self-assembling a carboxyl-terminated thiol on the tip. (From Akari et al., *Langmuir*, 12, 857, 1996. With permission.)

sensitivity is conveyed to the tip, and thereby, a chemical- or material-specific contrast is obtained that is superimposed onto the "true" sample topography (Figure 4.5). In CFM,[296] the sample was studied where it was imaged in the attractive regime of the contact mode, i.e., the tip was kept in contact by attractive capillary forces and actually hovered close to "snap out" on the sample. Here, the long-range electrostatic or acid–base interaction of the modified tip and the sample surface was mapped.

The adsorption structure of PEI molecules was characterized by their topography as seen in tapping mode AFM (TM AFM), by mainly probing the repulsive Pauli forces.[143] However, as in the mapping, a repulsive tip–sample contact is involved, and soft polyelectrolytes with a high affinity to surfaces that are not chemically grafted onto the substrate have not been successfully mapped with molecular resolution in this force regime. This is in contrast to reports, e.g., on end-grafted polymer chains,[301] biological macromolecules such as ferritin or catalase,[302] block copolymers of which clusters or micelles of comparable dimensions have been resolved,[303–306] or recent data on stiff and compact polyelectrolyte brushes.[307] Besides mapping in order to get a three-dimensional image of the adsorbed molecules, AFM can yield, similar to the surface force apparatus (SFA), the heights (steric barrier) of the adsorbed structures in solution via the measurement of the interaction forces between a polymer-covered surface and a tip, e.g., modified with a colloidal probe. This was performed on poly(acrylic acid).[308] There were also attempts[309] to image poly(acrylic acid) adsorbed onto mica in water using contact mode AFM, which is difficult, due to expected swelling under these conditions. Consequently, only relatively low resolution could be obtained, the measurement strongly affected the sample, and the results were mainly discussed in terms of layers of adsorbed molecules.

In contrast to earlier studies, TM AFM was used on samples in the dried state. TM AFM is known to minimize the lateral forces the tip exerts on the sample surface but still yield high resolution, which can be considerably higher than in the contact mode[310] on soft materials, as mapping artifacts are avoided. It was concluded that these systems should be studied in the dried state in ambient air, thus minimizing the swelling state.

It was shown that a geometrical characterization of adsorbed polyelectrolytes in the dried state under ambient conditions is feasible with TM AFM and yields the shape, lateral dimensions, and height of the single adsorbed macromolecules. One can compare branched PEIs of different mean molar masses (Mw = 1000, 150, and 37 kD) adsorbed on charge-stabilized polystyrene latexes, i.e., colloidal model systems, and on planar muscovite mica surfaces. These surfaces also carry negative surface charges in aqueous environments, mainly due to the loss of surface K+ ions.[311] These are molecularly flat and are, therefore, suited for the detection of small macromolecules that have sizes below the intrinsic roughness of the polystyrene latexes. Furthermore, the comparison of the adsorption structure on the latexes and on the mica shows that the surface charge density of the substrate has a strong influence on the adsorption structure of the macromolecules, and that this influence persists in the dried state in ambient air. The combination of AFM and light-scattering data gives evidence that the degree of compression of the adsorbed macromolecules is so high, even in aqueous environment, that there is no strong collapse upon drying, which in turn, justifies our approach of working with samples in the dried state in ambient conditions. These findings are discussed within the framework of the flocculation models cited above.

PEI is a weak polybase with a strongly pH-dependent degree of protonation.[312,313] These studies were carried out at high charge density at pH 4 (buffered) and low salt concentrations of 1 mM KCl, as under these conditions, the molecules have maximum diameters, are fairly stiff, and have a high coulombic interaction with charged substrates. Under these conditions, highly branched PEI with molar masses of 150 and 1000 kD, as determined by analytical ultracentrifuge (sedimentation equilibrium run), were adsorbed to mica surfaces under diffusion control from aqueous solutions in the ppm range (14 or 1.4 ppm).[314] Drops of the polyelectrolyte solution were put onto freshly cleaved mica surfaces for well-defined incubation times, and to control the incubation, the substrates (muscovite mica) were rinsed with water and dried in air.

For the AFM experiments, adsorption of PEI molecules with molecuar weight of 1000 kD onto oppositely charged polystyrene latex spheres was carried out on the surface of mica. The polystyrene latexes with a diameter of 1.9 μm functionalized with COOH groups were used. The charge density was estimated from conductometric titration (of the dialyzed latex) as 12 C/cm^2 (0.7 e$^-$/nm^2). The amount of adsorbed PEI was adjusted by the mass ratio of latex and PEI in the adsorption step. The degree of coverage was estimated from electrophoretic mobility measurements. The latex spheres were separated from the solution and precipitated by centrifugation onto a mica surface fixed at the bottom of the tube, before the AFM measurements. Quasielastic light scattering (QELS) was performed on polystyrene spheres upon adsorption of PEI (pH 4 and 0.1 mM KCl) to full coverage, as checked by electrophoretic mobility measurements. Molecular diameters of PEI in solution were determined at pH 7 in 100 mM KCl. The AFM measurements were performed in air using Si cantilevers (35 N/m, 300 kHz, tip radius 10 nm, tapping mode). PEI adsorbed on colloidal systems (styrene latexes) was studied. Branched PEI (Mw = 1000 kD) was adsorbed to polystyrene latexes with a coverage of 0.8 mg/g.

Adsorption was carried out overnight, thus, it may be assumed that adsorption is complete and that the initial stoichiometry of latex to PEI corresponds to the final coverage. Coverage was checked by electrophoretic mobility measurements. This coverage was adjusted to be well below the point of zero mobility, i.e., in the regime of low surface coverage. The point of zero mobility was determined to be at a coverage of 1.5 mg/g. At 6 mg/g, the sample was in the saturation regime that corresponded to a continuous layer of adsorbed macromolecules. The increase in diameter of the latexes due to PEI adsorption was quantified by QELS. For Mw = 1000 kD, the adsorption layer thickness was about 10 nm. The particle analyses was carried out from the combination of data of static and dynamic light scattering, which yield R_g, the radius of gyration, and R_h, the hydrodynamic radius. The ratio R_g/R_h is indicative for the polymer structure and had the value of 1.2 for the poly(ethylenimine), which lies in the range of Gaussian chains, i.e., spherical structures $R_g/R_h = 1.5 - 2$). On the other hand, in the case of ellipsoidal structures, the magnitude of ratio R_g/R_h would yield values >2.[315]

TM AFM images of identically PEI-loaded latex surfaces were investigated. Images were obtained after the surface was covered by aggregates with heights of approximately 10 nm and with lateral diameters ranging up to 60 nm, which corresponds well to the diameters known for PEI macromolecules of the same molar mass in solution. Both of these features and smaller aggregates around 20 nm were observed in the images. PEI was adsorbed on planar surfaces (mica). From these data of height images, it was found that three PEIs with mean molar masses of 1000, 150, and 37 kD adsorbed onto mica. The concentrations used in these studies varied from 2 to 0.3 ppm.

The mean areas and diameters from these images were estimated by using the grain analysis software of the microscope. From these data, it was clearly seen that the mean size of the structures decreases according to the decrease in mean molar mass of the PEI (e.g., mean Mw), heights (h_{AFM}), and diameters (d_{AFM}) of typical aggregates. These data were analyzed using a simple geometrical model, which assumed the AFM tip to be spherical with a radius of curvature of 10 nm and the macromolecules to be the section of a sphere for which broadening of the features due to tip artifacts is not significant for the aspect ratio of the features observed. The volume for the polyelectrolytes calculated from h_{AFM} and d_{AFM} was also based on the assumption that the shape of the adsorbed polyelectrolyte was a section of a sphere. The mean molar mass of the first type of PEI, the density of the adsorbed molecules, was also calculated. This density was used to determine the theoretical volumes (V_{calc}) for the other PEIs. It was found that the calculated volumes (V_{calc}) were in agreement with the measured values. From these data, the cross section of an adsorbed PEI on a latex sphere together with the model were used to calculate the volume of the adsorbed macromolecule. It was approximated by a section of a sphere of height h_{AFM} and width d_{AFM}.

Mean patch diameters (MPD) from the AFM data using the standard software of the microscope were carried out in order to obtain quantitative analyses of these polymers. The adjustment of the height threshold is known to be critical and basically also limits this procedure to the semiquantitative level. The mean patch sizes together with the mean diameters (d_{patch}) were calculated from these data under the assumption that they were spherical. Obviously, the patch size distribution was broad. The d_{patch} values were found to be smaller than the dAFM values of the marked features. In the case of the 150 and 37 kD PEIs, they are significantly bigger than the d_{LS} values from QELS, which were determined at pH 7 in 100 mM KCl, and hence, gave a lower limit for the diameters of the macromolecules that were adsorbed to the mica at pH 4 in 0.1 mM KCl. This is due to the fact that the molecular diameter of PEI slightly decreases with decreasing charge density on the molecule, i.e., with increasing pH, and with decreasing Debye length, i.e., with increasing ionic strength.[316] Altogether, the lateral dimensions of the patches fairly agree with the sizes expected for the macromolecules in solution, and the features observed show a marked dependence on the molecular weight of the adsorbed PEIs. To further strengthen the point of the detection of single isolated macromolecules, phase contrast images were obtained. These data of height image showed a distinct phase contrast as compared to the background. It was concluded (tapping with high amplitude and moderate damping conditions) that the particles are softer than the surface between the structures.[318] This observation allows one to conclude that the areas between the particles are uncovered. And that because the particles are considerably softer than the bare substrate's surface, these cannot be due to inorganic aggregates.[319] Hence, the presence of inorganic aggregates was excluded in these structures.

The data of TM height (400 nm × 400 nm) and phase contrast pictures were measured at 150 kD PEI adsorbed onto mica. The z range in the height image was 2 nm. These data were found to be in accord with literature data.[317] From these analyses, it was safe to conclude that the observed structures were valid for single macromolecules. A correlation of the adsorption structure and charge density of the substrate was estimated as follows:

1. From light scattering of PEI-covered latexes in solution, typical monolayer coverages of 10 nm for 1000 kD PEI were found.[296] By TM AFM, molecular heights in the dried state were found to be on the same order (10 nm).
2. The heights of the same types of molecules (molecular weight = 1000 kD) adsorbed on mica were, by a factor of approximately 10, lower than for the molecules adsorbed to the latex surfaces, although there was no marked difference in the lateral dimensions.

It was concluded that the PEI is adsorbed in an essentially flat geometry in aqueous environment, as there is no appreciable decrease in height or collapse upon drying. Furthermore, the degree of "compression" into a flat adsorption geometry is strongly influenced by the interaction strength with the substrate, which is dominated in this case by the coulombic interaction, i.e., by charge density. The differing adsorption heights could also be explained by the ellipsoidal shapes of the macromolecules. However, light scattering showed that the macromolecules were spherical in solution.

In these reports, a simple procedure for the adsorption of PEI on an oppositely charged surface was carried out. As the macromolecule approached the surface, it started to interact with the surface charges. This interaction may be considered a localized interaction between discrete ionic centers or a delocalized interaction between two multivalent structures.[320] In most cases, it will be a combination, according to the particular local ionic strengths and the distance between the charge centers on the substrate and the polyelectrolyte. After contact with the surface is established, the macromolecule rearranges and, for linear PELs on oppositely charged surfaces, is believed to proceed via unfolding from a structure with many loops and tails to an equilibrium conformation with virtually no loops and tails, where all segments are adsorbed (as depicted below).[315]

(I) ADSORPTION OF PEI.............................
SURFACE CHARGE INTERACTIONS......
(II) PEI UNFOLDS TO ALMOST NO LOOPS

However, as PEI is highly branched, one may presume that it is largely locked in its lateral dimension and does not unfold, which is in agreement with the literature,[315] that the diameters of the molecules in solution and the diameters of the adsorbed molecules agree. Based on these data, a model for the adsorption and reconformation of PEI was presented. The heights in the adsorbed state or segment density distribution are strongly influenced by the relation of the charge densities of the surface and the adsorbed polyelectrolytes. From these data of PEI, charge densities between 11.3 and 5 e^-/nm^2 were estimated.

The charge density of 11.3 e^-/nm^2 is calculated by taking the total charge of one PEI molecule of 1000 kD and dividing it by its area of projection, i.e., R_g^2, according to the adsorption model.

The magnitude of $R_g = d_{LS}/2$ was taken to be 26 nm in these analyses. The total charge per molecule was calculated from the molar mass, the molar mass (42 g/mol) per mole of monomer (24,000 monomers per molecule), and the monomer charge of one. This yields an upper limit and holds for full protonation. Measured degrees of protonation of the free molecules in solution at pH 4 lie between 0.7 and 0.85. In the adsorbed state, due to interaction with the substrate, which decreases repulsion between the segments, full protonation is a good assumption. In many cases, the adsorption may essentially be viewed as an ion-exchange process.[320]

If an "extended" configuration is presumed with close contact to the surface, each ethylenimine unit would cover an area of approximately 0.2 nm^2, which yields, with one charge per unit, a surface charge density of 5 e^-/nm^2, and would yield an increase of the covered area by a factor of two as compared to the first approach. Depending on the respective model, in the case of mica, on the order of 0.2 to 0.4 of the PEI charges nominally compensate the substrate's surface charges. Based on a crude model, the value found was 1.1 to 0.5 mm as an upper limit for the patch thickness, with 0.4 nm being approximately the N–N distance in PEI. In this approach, bond angles were completely free, steric hindrance or the branched structure was neglected, and according to the charge density arguments, every fifth (20%)

or second (50%) amine group was supposed to be localized at the surface, whereas the rest form loops. The same reasoning can be applied to the case of the latexes.

The model yields good agreement with the AFM measurements for the mica surfaces; however, on the styrene latexes, the measured heights are appreciably higher than the calculated ones. The latter observation indicated that the compression of the molecule due to the interaction with the substrate is not as high as one would expect on the basis of the coulombic interactions or that evolution into a more compressed adsorption structure is extremely slow, and monitor equilibrium structures cannot be monitored. Nevertheless, the tendencies predicted by this simple model, i.e., high surface charge means flat configuration and vice versa, are reproduced by the measurements. Of course, a certain degree of swelling in the "dried" state in ambient air, i.e., under measurement conditions, is presumed that keeps the molecule from collapsing completely. This is likely, as under these conditions, it is known that surfaces are covered with a water layer,[317] and this swelling was measured macroscopically by ellipsometry on layers of end-grafted polyelectrolytes.[321]

The heights of the adsorption structures on the latex particles are appreciably smaller than the apparent height of 30 nm determined[296] by CFM, which, however, convolutes the topography and chemical interaction. It was indirectly deduced from electrooptical measurements of hydrodynamic thicknesses of 3 nm for Mw = 20 to 25 kD PEI adsorbed on hematite particles at pH 7.2 to 8.5, which was also in the range of our results.[322] It was shown that the unfolding of PEI molecules occured during adsorption for low coverages. This is to a certain extent equivalent to the extended configuration (charge density of 5 e^-/nm^2) and might also be reflected in the somewhat higher dAFM values for adsorbed molecules as compared to the dLS values determined in solution. The higher difference for the smaller molecules could hint at differences in the molecular architecture, i.e., lower degree of branching. Obviously, these observations, even in the dried state under ambient air, are a marked influence of the surface charge of the substrate on the adsorption structure of the polyelectrolyte. It was further found that PEI forms patches, i.e., molecules obviously do not unfold on the surfaces. This is due to their molecular structure, as they are branched, or due to the establishment of a strong binding to the substrate, which reduces lateral mobility. Furthermore, evidence was given that in solution, the same flat configuration as in the dried state under ambient conditions is obtained. From these data, it was presumed that under the experimental conditions, PEI was monitored in an equilibrium state, i.e., it has sufficient time to reconform from its initial, more extended configuration to a compressed state.[320] With the charge density arguments as given in the last section, a charge reversal on the patches can be expected. The data showed strong influence of charge density at the adsorbing surface on the patch height, as shown by the comparison of mica and polystyrene substrates, which demonstrates that the relative charge density of PEI and substrate is an important factor not only for the formation of electrostatic patches[323] but also for the conformation within the patches at given pHs and ionic strengths. These results should be of general relevance for polyelectrolytes. These data were in accord with literature reports,[323] the point that in its equilibrium configuration, at the investigated pH and ionic strength, PEI adopts a flat adsorption structure, which supports the concept that PEI acts as a flocculent according to the patch charge model. These images showed that it is possible to detect single adsorbed PEI macromolecules adsorbed to polystyrene latexes and to flat mica surfaces. The high molecular mass PEI macromolecules appear on dried surfaces in ambient conditions as disk-like structures or patches. Their lateral dimensions in the adsorbed state lie well in the range predicted by light scattering for macromolecules in solution, and a systematic variation in size as a function of molar mass of the macromolecules with typical lateral dimensions ranging from 60 to 20 nm was observed. The height analyses of the adsorbed macromolecules showed a strong dependence on the type of substrate used, which was interpreted as being due to charge density differences between the substrates. The magnitudes of typical heights were in the range of 1 nm on charged surfaces. On the other hand, on polystyrene latexes, with lower charge densities, heights ranging up to 10 nm were observed, as expected from light-scattering data on the PEI layer. These data suggested that in the dried state under ambient conditions and in solution, essentially, the same flat patch-like adsorption structure was present. It was concluded that PEI acts as a flocculant via the patch charge mechanism, although more studies are needed at this stage.

4.5.3 Latex Particle Analyses by AFM

Core shell latexes are composite particles composed by a core of one polymer covered in a shell of another polymer. These latexes are used in a variety of applications (adhesive, coating, painting, paper, etc.). Surface morphologies of poly(butyl acrylate)-poly(methyl methacrylate) core shell latex were investigated using AFM.[323] The investigation of the detailed morphology of core shell latexes is essential for a deeper understanding of their properties and a better control of their synthesis and design of their structures. AFM in the tapping mode was used to scan these soft polymers. The dimensions measured were as follows:

Method	Tapping	Contact Mode
x-	305	400 nm
y-	310	350 nm
z-	80	50 nm

The lateral dimensions were enlarged due to flattening by the applied forces. The *x* direction was enlarged much more, because this is the scanning direction. AFM images were obtained for individual particles with raspberry morphology. These data showed the great advantages of AFM with which one can get information on surface mechanical properties and stiffness of latexes. The morphology of core shell latex particles was found to depend on the interfacial energy between the polymer phases during emulsion polymerization.

4.5.4 Other Diverse Polymers

High-resolution mapping of functional group distributions at surface-treated polymers by AFM using modified tips was reported.[324]

For example, the single-molecule force spectroscopic studies of PVA by AFM were studied in order to understand hydrogen bonding in the elastic properties of PVA in water.[325]

STM studies of nonpolar surface elements of different materials were carried out in air and when covered by water in order to determine effects of water on the topography.[102] The tunneling barrier depends on the specimen and the tip materials and on the medium in the tunneling gap. The nonpolar surfaces were found to be smoother underwater than in air. This was explained to be due to the larger distance between the tip and the specimen in water than in air.

Polystyrene (PS) and poly-*p*-bromo-styrene (PBrxS) phase separation is well known.[326] This phenomenon occurs when some polymers do not mix, because two phase systems are formed. The spreading condition for a polymer blend is much more complex than that of a simple homopolymer film. Although the two components separately wet the surface, their interfacial tension can destabilize the film. Further, in polymer mixtures, the strong entropic forces are present that determine the mixing equilibria. In a recent study, scanning transmission x-ray microscopy (STXM), AFM, and dynamic secondary ion mass spectrometry (DSIMS) were used to obtain information on the three-dimensional profiles of PS and PBrxS mixtures.[327] Polymer solutions were made up in chlorobenzene. The morphology of mixtures was analyzed in terms of surface tensions of polymers. The segregation of PS was expected, because the surface tension of PS (γ_{PS}) is less than that of PBrS (γ_{PBrS}), where $\gamma_{PS} = 30$ dyn/cm (mN/m), and $\gamma_{PBrS} = 33.7$ dyn/cm. The spreading coefficients, S_{sc}, were given as:[3]

$$S_{sc} = (\gamma_{PBrS} - \gamma_{PS}) - \gamma_{PS/PBrS} \tag{4.2}$$

where $\gamma_{PS/PBrS}$ is the interfacial tension and is approximately 1 dyn/cm. As described in detail in the literature, the interfacial forces in the equation reach equilibrium after an appropriate time.[3] In this system, $S_{sc} > 0$ (i.e., $S = 2$ dyn/cm), which means that PS spreads on PBrS as one would expect, because PS is more hydrophobic than PBrS. Furthermore, on silica surfaces, the following was considered:

$$\gamma_{PS/Si} + \gamma_{PBrS} < \gamma_{\ PBrS/Si} + \gamma_{PS} \tag{4.3}$$

where $\gamma_{PS/Si}$ and $\gamma_{PBrS/Si}$ are the interfacial tensions between PS/Si and PBrS/Si, respectively. During the hydrogen passivation, the magnitude of the dispersive surface tension, γ_{SO}, is slightly hydrophobic, because $\gamma_{SO} = \gamma_{SOd} = 44.7$ dyn/cm. The magnitudes of dispersive surface tensions can be estimated as follows:

$$\gamma_{SLD} = ((\gamma_{SOD})^{1/2} - (\gamma)^{1/2})^2 \tag{4.4}$$

In the present case, assuming PS surface tension arises mainly from nonpolar forces, although PS has a weak polar part, $\gamma_{PS} \simeq \gamma_{PSD}$, one gets $\gamma_{PS/SiD} = 1.35$ dyn/cm. From these assumptions, one gets $\gamma_{PBrs/Si} > 4.6$ dyn/cm, $\gamma_{PBrS/SiD} < 20.6$ dyn/cm, which

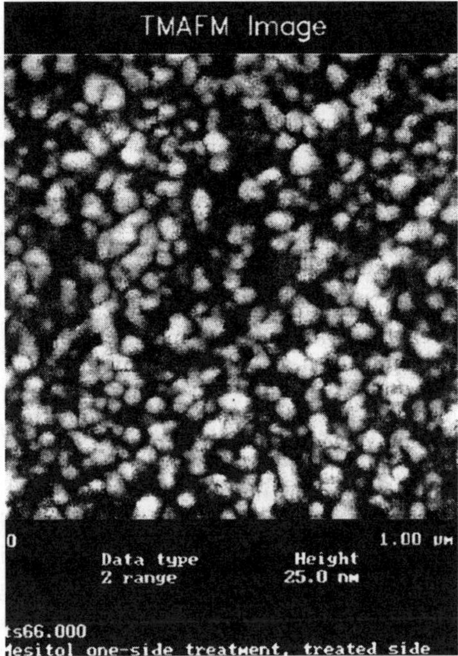

FIGURE 4.6 AFM image of sulfonated aromatic treated nylon-66 film (1 μm × 1 μm). (From Zhang et al., *Langmuir*, 11, 8, 1995. With permission.)

shows that PBrS is approximately 39% polar.[3] These differences in surface tension can explain the polymer mixture's phase behavior as found from AFM images.

Sulfonated aromatic compounds are generally used in the textile and carpet industries to surface treat nylon fibers to produce stain-resistant carpets.[328] The anion forms an anionic bond with the free amine end groups of polyamides, such as nylon. The adsorption of sulfonated aromatics changed the nylon AFM images such that the surface became much more uniform [as found from fast Fourier transform (FFT) analyses (Figure 4.6)]. The size range of the particles was found to be 20 to 62 nm.

Synthetic membranes as used for filtration and microfiltration were analyzed by STM, where 80 nm pores were observed.[329]

4.5.5 Diverse Properties of Synthetic Polymers

The molecular configuration of isotactic PMMA as LB films was investigated[330] by STM . The analysis showed that the polymer was present as a closely packed well-ordered arrangement with helical and linear chain mixed configuration.

4.5.5.1 Determination of Thickness of Spin-Cast Polymer Thin Films

In all polymer systems, the accurate determination of surface film or overlayer thickness is an important aspect in many areas of surface and materials science.

Several authors have proposed methods for measurements of thickness of thin films using ellipsometry or reflectometry.[331–334]

These optical techniques are often used for the film thicknesses in the range from tens to hundreds of nanometers. In contrast, angle-resolved XPS is widely used to study thin layers, i.e., a few nanometers thick. For a thin overlayer on a substrate which both contain a common element, XPS-based algorithms exist that relate the ratios of chemically shifted peaks to thickness, e.g., SiO_2 on Si or Al_2O_3 on Al.[335]

Recently, there were studies conducted on the physicochemical properties of polymer blends and, in one aspect of this work, attempts have been made to relate spin-cast film thickness and property with casting parameters. These investigations were part of the work to develop an algorithm that allows thickness calculations for any uniform film of thickness ($t < 3$) on a substrate using the relative peak intensities of two peaks (one of which is characteristic of the substrate and the other is characteristic of the film) measured at different photoelectron take-off angles. This is the attenuation length of emitted photoelectrons. This method is useful for PS and PMMA films cast from chloroform solution using the C 1s peak of the film and the Si 2s peak of the mica. Atomically flat mica surfaces were used as substrates in order to eliminate the effect(s) of substrate roughness on thin films. In the case of thin films that are relatively soft compared to substrates, e.g., polymer films on mica or silicon, thickness can also be determined by scratching the films with AFM tips. By controlling the loading force of the AFM tip on film surfaces, it is known to be possible to scratch through the film without damaging the substrates. Film thickness is then simply measured by passive AFM imaging of the scratch. Under controlled procedures, these methods have allowed for the creation of nanolithographic features. The atomic-scale wear properties of mica were previously studied by AFM, and it was found that mica wear depended on loading force, number of scans, and tip shape.[336,337]

PS and PMMA thin films (<100 nm thickness) have been spin-cast from chloroform solution onto cleaved mica surfaces (roughness within 0.2 nm).[338] An algorithm for calculating the film thicknesses based on the relative intensities of the C 1s peak of the films and the Si 2s peak of the mica from angle-resolved x-ray photoelectron spectroscopy (XPS) is presented. The film thickness changes as a function of casting conditions. Data from this approach are comparable with thicknesses measured by an AFM tip-scratch method in the range of 1.5 to 5.5 nm. Thicknesses of the films are shown to increase linearly with concentrations of cast solutions. In these studies, *scratch* experiments were carried out on polymer films with the loading force below the mica damage threshold. Thicknesses measured from the angle-resolved XPS and AFM scratch methods are compared for various polymer films cast from different PS or PMMA concentrations in chloroform. Similar scratch phenomena on other AFM samples have been reported. In these investigations, PS and PMMA consisted of weight-average molecular weights of 280 and 350 K, respectively. Films were spin-cast at 293 K from solutions prepared by dissolving a known weight of each polymer in a fixed volume of chloroform (g/mL % concentration). Films were cast from 60 μL aliquots of the polymer solutions onto freshly cleaved mica substrates of about 1 cm × 1 cm area, which were rotated at ~3000 rpm for 2 min. The cleaved mica surface, with roughness that was measured within 0.2 nm, was atomically flat. Film thickness was varied by controlling the concentrations of polymer in solution. The chloroform solvent was completely evaporated from the films, as no chlorine signal was detected by XPS. The x-ray gun was operated at 150 W. An electron flood gun was used to offset charge accumulation on the samples. The XPS survey spectra were taken at five take-off angles (relative to the surface normal) of 0, 30, 50, 60, and 70. The relative photoelectron intensities of C 1s and Si 2s peaks were evaluated from peak areas after subtraction of a linear background. In these studies, AFM was performed with a multimode SPM in contact mode using square pyramid Si_3N_4 probes. Short (nominal length 100 m, nominal tip radius 20 to 60 nm) Si_3N_4 triangular cantilevers with wide legs were used for scratching the films, while longer cantilevers (less stiff) were used for imaging the resulting scratches. Scratches were formed on the polymer films by scanning the tip repeatedly back and forth over the same line with high loading forces. Several different scratches were measured for each film to allow statistical analysis of data. The normal spring constants of the cantilevers were calculated using their measured unloaded resonant frequencies.[339,340]

Individually measured spring constants of each cantilever were used in the estimation of loading forces and the calibration of the instrument. The calibrated values for the short cantilevers were in the range of 0.46 to 0.76 N/m. A loading force of approximately 260 nN (inclusive of the capillary force due to adsorbed water layer), which was evaluated from the calibrated values of spring constants, was used for scratching of all the films. The same force was used to scratch the PS and PMMA films, because the two polymers have similar physical properties and surface stiffnesses. Scratch tests on bare mica substrates using the same cantilevers and force have shown no damage on mica surfaces. It was also found by subsequent imaging in tapping mode, which has been demonstrated to be less intrusive, that contact mode imaging in the low loading force regime (<10 nN) does not significantly damage the polymer films and scratches.[341] The average depth of the scratch below the mean surface plane, corresponding to the film thickness, was measured using the cross-section analysis software. All images of scratched films were acquired in air and were stable and reproducible. The angle-resolved XPS measurements were carried out. The purpose of the following calculation is to determine the thickness of any uniform thin film on a substrate. The thickness measurements are based on the relative peak intensities of the film and substrate at different take-off angles. By assuming that the x-ray intensity does not diminish within the XPS sampling depth, the relation between intensity, Ii, and concentration, Xi, of an element *i* was described as a function of the electron escape depth *i*.[342] The effective escape depth *i* was simply (cos θ), where θ is the take-off angle from the surface normal.

The sampled vertical depth, where 95% of the XPS signal intensity originates from, is 3(cos θ). Taking the ratio of the peak intensities of the film and substrate from these relations, the normalization factor can be calculated from the atomic sensitivity factor (ASF) of the film and substrate elements or be experimentally determined from the bulk standards. The film thickness can be evaluated using the ratio of intensities, if the attenuation lengths of the film and substrate photoelectrons are known. In order to make this analysis, one photoelectron peak from the film and one from the substrate are chosen so that there is no contribution of substrate elements to the film peak intensity and vice versa. For a thin PMMA film ($t < 3$ μm) on a mica substrate, the C 1s and Si 2s peaks are used to calculate the ratio of the photoelectron intensities from the film and substrate, respectively. The Si 2s peak was used instead of the main peak Si 2p, because the Si 2s photoelectrons have their binding energy closer to that of C 1s photoelectrons. The attenuation length of C 1s and Si 2s photoelectrons is effectively the same because of their similar kinetic energies. From these data, magnitude of the film thickness of the PMMA film was calculated as $d = 1.41$ nm. The linearity of the graphs was an indication of uniformity in the film thickness. Imaging with AFM of the spun-cast PMMA and PS films also revealed flat and uniform surfaces with RMS roughness less than 0.3 nm. By applying the same analyses, the thin PS films on mica substrate (cast from 0.02% w/v), the thickness of the PS film was found as $d = 1.33$ nm.

4.5.5.2 AFM Tip-Scratch Method

This method is known to be dependent on different parameters in order to provide an accurate measure of film thickness. Some of these are as follows:

1. The film may be scratched by the AFM tip, with no damage caused to the substrate.
2. The force of 260 nN was found to scratch the polymer film but not to damage the mica surface [which could be detected by AFM in tapping mode or in contact mode with a low loading force (<10 nN)].
3. To check that the AFM tip reached the mica substrate, the films were scratched for various time durations, and the resulting scratch depths were measured by passive AFM imaging (tapping or low force). The scratch depths were found to increase with scratching time up to a limiting value which was reached after a few hundred seconds. Depth values were found to increase with the concentration of the polymer in the casting solutions, as expected.
4. Back-filling of scratches due to film relaxation was examined by imaging the scratches at 90 s intervals after scratching was finished. This resulted in an exponential change in depth with time, allowing an overall relaxation constant for the system to be calculated. By using this method, it was found that the relaxation time (approximately 30 min) is much larger than the acquisition time for one AFM image (45 s). It was assumed that it scratched to the substrate but no further, and that no depth decrease occured due to film relaxation before secondary AFM imaging. Consequently, use of the limiting depth so measured, corrected using the mean surface plane, to avoid inclusion of scratch-edge debris, leads to an accurate measure of film thickness.

The images of the scratch on the PS film, as cast from 0.06% w/v solution were obtained. The ridges formed along the lips of the scratch, which were seen of the sample surface, are due to PS, which is displaced during scratching. This is a general phenomenon. Local depth in one scratch relative to the mean surface plane varies within 8%. In order to improve the accuracy in thickness estimation, the magnitude of depth of the scratch was measured. In these studies, the scratch was formed by an AFM tip at the force of ~260 nN. The average depth of the scratch relative to the mean surface plane was 30.9 nm.

The data of the thicknesses evaluated by XPS algorithm and AFM scratch method were compared. It was found that the XPS method can only be used for very thin films ($d < 6$ nm), which give measurable photoelectron intensities from the substrate. The magnitude of the error of the intensity ratio was found to become larger because of the weak substrate signals from thicker films. Consequently, for measurement of the thickness of the films cast from solutions of between 0.06 and 0.1% w/v, where the mica signals are weak at $q = 0$ and completely disappear at take-off angles greater than 45, the range of angle-resolved measurable signals was reduced. For the films cast from solutions with concentrations of 0.2% w/v (thickness 15 nm) or greater, the mica signals were absent in the normal. Correlation to within 15% between the XPS algorithm and AFM scratch method was observed for thicknesses in the range of 1.4 to 5.5 nm. The differences in the film thicknesses measured by the two techniques can partially be attributed to thickness variations at different locations on the cast films. AFM measurements show a slight decrease (within 10%) in film thickness moving away from the center of the spun samples. (This is probably due to the centrifugal effect of the spinning.) A difference in sampling areas of the two techniques (submicron scale for AFM and tens of microns for XPS) can also result in discrepancies if the film thickness is inhomogeneous. The variation of thickness of PS films with concentration of cast solutions was found to increase almost linearly with the concentration in the range 1.5 to 65 nm. The analyses of the data give a relation d [nm] = 63.5 C, where C is the concentration (% w/v) of polymer solutions. The proportionality factor in the thickness–concentration relation was expected to vary when changing the spin rate of film casting. The film thickness increases almost linearly with concentration, which gave film thickness values for all polymer samples at approximately 6 nm. Data were compared to thickness measurements made using an alternative AFM tip-scratch technique. Correlation to within 15% between the two methods was measured, and the thicknesses of spin-cast films were shown to increase linearly with the concentration of cast solutions. The variation of thickness of the PS films with concentration of cast solutions showed that the film thickness increased almost linearly with concentration. An attempt was made to prepare a carbon sample with a highly oriented structure from poly(acrylonitrile) by using a two-dimensional opening between the lamellae of montmorillonite as the field of carbonization.[342]

4.5.6 Polymerization in Monolayers as LB Films

AFM has been applied to the nanometer-scale investigation of various materials, including hard ceramics and soft polymer samples. *In situ* AFM study on the morphological change of the LB film of cadmium 10,12-pentacosadiynoate during polymerization was investigated.[343]

Polydiacetylenes are conjugated polymers obtained by solid-state polymerization on irradiation of UV light and x-rays.[344] This class of polymers has attracted much attention in terms of their application to nonlinear optical materials. Two spectroscopically distinct forms, blue and red, have been observed. A phase transition from the blue to the red phase of polydiacetylenes was induced by heat treatment[345–347] and irradiation by a laser pulse[348] in single crystals. The difference in the structures of the two phases is that a significant portion of alkyl chains attached to the polymer backbone are disordered in the red phase, whereas the alkyl chains are ordered in the blue phase.[349,350] The LB technique has been used for the fabrication of ultrathin films with well-defined structures at the molecular level. Much work has been done on the fabrication and polymerization of LB films of amphiphilic diacetylene derivatives such as 10,12-pentacosadiynoic acid [DA(12-8)].[351] It has been reported that a drastic color change occurs during the polymerization of the LB films of the cadmium salt of DA(12-8) by the irradiation of UV light. This structural change should be related to the above-mentioned phase transition reported for polydiacetylene crystals, but it has a different feature in that the color change occurs during polymerization in the LB films as well as after the polymerization. There are not many investigations reported in the literature in regard to the structural change accompanied by this color change during polymerization, though the structural characterizations of the LB films during the thermochromic change[352–354] and the color change after polymerization[355] have been reported. In a recent study, data on the structural change of a single-layer DA(12-8) LB film during polymerization by the irradiation of UV light were obtained.[343] The morphological change of the LB film is investigated by *in situ* AFM.

It was found that the color change is accompanied by the development of three-dimensional structures from the film surface. DA(12-8) was synthesized as reported previously.[356] The measurements of surface pressure-area isotherms and the deposition of DA(12-8) monolayer were carried out on a film balance at 17°C. DA(12-8) was spread onto the subphase with aqueous Cd2+ buffer at pH = 6.0. DA(12-8) formed its cadmium salt on this subphase. The monolayers were transferred at 25 mN m^{-1} onto solid substrates using the vertical dipping method. A low-pressure mercury lamp was used as the source of UV light. Images of a single-layer DA(12-8) LB film on mica were studied in noncontact mode. The AFM was operated at line scan rates of 1 Hz, silicon cantilevers with a resonance frequency of 28 kHz, and a spring constant of 1.9 Nm^{-1}. In the case of *in situ* AFM measurements, the sample was not moved from the scanner in order to follow the image changes on photoirradiation in the same zone of the sample. The change in absorption spectrum was also investigated of a single-layer LB film of DA(12-8). Before irradiation, no appreciable band was seen in the visible region. The polymerization process seemed to take place in three steps. In the first step of the polymerization (irradiation time of 0 to 5 min), a band at around 650 nm was observed. The second step (irradiation time of 5 to 50 min), showed a shoulder band situated at shorter wavelengths, with a concomitant decrease of the 650 nm band. The blue-to-red color change occurred in this region. The third step in the polymerization process (irradiation time of 50 to 90 min), showed a band at 535 nm until the system reached a saturated state. Further UV irradiation gave rise to a decrease in the intensity of the absorption band due to the decomposition of polydiacetylene. *In situ* AFM images of a single-layer DA(12-8) LB film during the polymerization have also been studied.

Before the irradiation of UV light, several three-dimensional structures with a height of 5 nm above the planar surface were observed. The morphology of the DA(12-8) LB film remained unchanged in the initial regime, where the film is in the blue phase. In the second regime, a number of three-dimensional domains began to appear from the planar surface. *Ex situ* AFM gave similar results. With increasing irradiation time, the number of three-dimensional domains increased slightly, and each domain was found to grow. In the final regime, the number of three-dimensional domains did not change significantly, and each domain increased its size slightly.

The structural change of the LB films during polymerization was further investigated using FTIR spectroscopy. These data showed the transmission FTIR spectrum of a 20-layer LB film in the CH stretching region as a function of the irradiation time. Before irradiation of UV light, two strong bands were seen at ca. 2920 and 2850 cm^{-1}, which were assigned to CH2 antisymmetric [a(CH2)] and symmetric [s(CH2)] stretching bands, respectively. These bands do not change significantly. In the second and the final regimes, however, both absorption bands became broader, with an increase in the irradiation time of UV light, though the band position did not change appreciably. Similar results were obtained using IR reflection–absorption spectroscopy. This indicates that the packing of the alkyl chains attached to the polymer backbone becomes disordered at the point of color change, which is consistent with the phenomena observed on going from blue to red phase in polydiacetylene crystals and LB films.[357,358]

From these investigations, a plausible mechanism was proposed of the morphological change of the LB film during polymerization. In the initial regime of the polymerization, the absorption band at around 650 nm increased. The morphology of the film did not change significantly. In this regime, the polymerization was assumed to be a topochemical reaction without causing any significant change in the LB film structure. In the second regime, where the blue-to-red color change occurred, a number of three-dimensional structures appeared from the film surface. The blue-to-red spectral change of the polymers was considered to be due to a decrease in the delocalization of electrons along the polymer backbone, which was accompanied by an order–disorder transition of alkyl chains. This was supported by the results of IR measurements. Similar structural changes were observed in the color change of blue to red forms upon heating the LB films of DA(12-8) polymers.[352,353] The morphological change should be related to the change in the absorption spectrum. If assuming the following processes:

1. When the polymerization proceeds, the propagation of the polymer backbone will impose a significant stress on the structure of the LB films.
2. When this stress exceeds a certain threshold, the color change will occur to release part of the stress accumulated in the film.
3. This color change will show a decrease in the effective conjugation length of the polydiacetylene backbone, which could be caused by the disruption of the conjugation due to the displacement of some of the atoms in the polymer backbone.
4. The color change caused by the stress in the film will lead to the modification of the structure of the LB film. This should be responsible for the

order–disorder transition of the alkyl chains and for the morphological change of the film.

5. In the final regime, the 535 nm band will develop, and the morphology of the film will change only slightly until a saturated state is reached.

These studies showed that the morphology of the DA(12-8) LB films changed with the blue-to-red color change during polymerization. The results suggest that the stress accumulated in the film by the polymerization, which is usually considered a topochemical reaction, is released by the order–disorder transition of alkyl chains and also by the modification of the two-dimensional LB film structures. This morphological change is related to those of the LB films induced by the photoisomerization of azobenzene in that two-dimensional LB film structures are modified into three-dimensional structures by the photoreaction.[359–363] The results further indicate that care should be taken in the study of chemical reactions in LB films.

4.5.7 Single-Molecule Force Spectroscopy

As mentioned above, the development of AFM opened vast possibilities for measuring local physical and chemical properties of materials with molecular and even atomic resolutions.[9] The cantilever of the atomic force microscope is a nanoscopic force sensor that is so sensitive it can detect a force around 10 pN. Combining the high lateral resolution with extreme force sensitivity, AFM evolved into a versatile platform for experiments with single molecules (as depicted below).[34,266,267,364]

MOLECULAR FORCE SPECTROSCOPY.......
SINGLE MACROMOLECULE........................
SUBSTRATE..

AFM has been used to quantitatively measure the specific interaction between ligand and receptor, such as avidin/biotin and conjugated DNA strands. This has provided new insights into intermolecular and intramolecular forces and even the underlying molecular mechanism. Recently, a new technique based on AFM-single-molecule force spectroscopy was implemented and opened opportunities to study the mechanical properties of single polymer filaments varying from coiled polymers to polymers bearing superstructures.[365,366] For example, dextran, xanthan, poly(ethylene glycol), and poly(vinyl alcohol) have been stretched, and detailed nanomechanical properties have been obtained. Also, this technique has been used to measure the unfolding forces of individual Ig domains of the protein.[365] Their force spectra provided valuable nanomechanical fingerprint information of these polymers which may deepen our understanding of the mechanical properties of polymers.

The mechanical properties of poly(acrylic acid) (PAA) were investigated on the molecular scale by using AFM.[367] PAA was adsorbed onto a glass substrate and then picked up at one point of the chain with the AFM tip and stretched. The deformation of a single PAA chain under tension was measured and modeled by a modified freely-joint-chain model. The length distribution of loops and tails of adsorbed PAA on the glass substrate was also studied by analyzing the extension of the filaments.

These force curves showed two different distinct characteristics, which corresponded with where the tip touched the hard substrate. As is well known, PAA is an important polymer for a variety of industrial applications.[368] PAA has been widely used to stabilize the colloids, which is closely related to the adsorption. Adsorption of flexible neutral polymers at surfaces has been studied extensively and is now a well-understood phenomenon.[369] The steric and bridging forces between surfaces bearing adsorbed PAA have been studied recently by using AFM.[369] Here, PAA was adsorbed onto a hydrophilic glass substrate, and an AFM tip was used to stretch PAA polymer chains and measure the deformation of a single PAA chain under tension. The experimental details of single-molecule force spectroscopy by AFM have been described elsewhere.[370,371] In general, the polymer was chemically or physically adsorbed onto a substrate. When the AFM tip was brought into contact with the sample, some molecules adsorbed onto the tip due to the nonspecific interaction and made a connective bridge between the sample and the tip. While separating the tip and the sample, these molecules were stretched. The deflection–extension curves were recorded and converted into force–extension curves. The force spectrum of dextran showed a force plateau at 750 pN, which corresponded to a force-induced conformation transition of the glucose ring in the dextran polymeric backbone.

Force–extension curves were measured of PAA under different conditions: (a) force curves of PAA of different lengths, and (b) superposition of normalized force curves of part a. Commercially available PAA was used for this study (molecular weight = $4.5\ 10^5$). PAA was dissolved in water to a concentration of 1 mg/mL; the pH value was between 3 and 4. A known amount of PAA solution was deposited onto a clean glass coverslip and dried in air to form a thin layer. The substrate was rinsed with water to remove the loosely adsorbed PAA and was used for force measurements. AFM was investigated in a 10^{-3} M KNO_3 solution. A silicon nitride cantilever (180 m long and 0.6 m thick) was used. The thickness of the adsorbed PAA layer on the glass was estimated from the approaching trace of the force curve. When the probe tip was brought into contact with a hard nondeformed surface, the deflection of the cantilever would be proportional to the piezo movement (curve a); however, if the sample is soft, the soft sample would deform with increasing cantilever deflection, resulting in a nonlinear contact region of the force curve. If the soft layer is thin, the layer may be fully compressed, and the tip may touch the hard substrate beneath the soft layer at one point; the linearity of the cantilever deflection and piezomovement will appear again. The distance between contact point A and contact point B (the point where the tip touches the hard substrate) is approximately equal to the thickness of the soft sample layer. The difficulty in determining the real contact points A and B may result in the error of estimating the thickness. The contribution of electrostatics to the thickness was neglected, and one can only qualitatively estimate the thickness of the soft layer from the approaching trace of the force curve. In this study, the thickness of the PAA adsorbed layer was around 50 to 90 nm.

As already mentioned, the measurement and interpretation of atomic force microscope data require careful consideration of instrumentation calibration as well as environmental effects due to the appropriateness of theoretical models. Thus, over a decade since the invention of the AFM, the progress about this assessment has been made regarding these important issues and areas of potential future development have been pinpointed. Calibration of the spring constant of the cantilever is important for force measurements. The most widely accepted techniques are the spring resonance method[372] and the thermal noise method.[373,374] A new method for cantilever calibration was described, which was suggested earlier.[370] In this method, dextran was used as an internal standard sample for calibration. It was found that a force-induced conformation transition of the glucose ring in the polymer backbone for dextran existed at 750 pN, which corresponded to a force plateau in the force spectrum.The cantilever was calibrated from force measurements on dextran in phosphonate-buffered saline. The value of 750 pN was divided by the value of the cantilever deflection at the plateau obtained from the force curve, and the spring constant was obtained. The value obtained was 0.069 to 0.003 N/m (force constant was 0.05 N/m). Compared with other calibration methods, this procedure is much easier and more convenient for AFM applications.

Force–extension curves of PAA have been measured. Because the polymer is polydisperse and the point at which the polymer chain is picked up with the tip cannot be controlled, the contour length of the polymer stretched between the tip and sample varies. Despite the different contour lengths of polymers chains, all of the force curves show similar characteristics: the force rises monotonically with the extension of the polymer chain until a rupture point is reached, and then the force drops to zero suddenly. The rupture force can be as high as 1.8 nN. All of the force curves are normalized according to their contour lengths, superimposed. The superposition of these curves clearly shows that the elastic properties of PAA chains scale linearly with their contour lengths. Thus, one can deduce that interchain interaction does not contribute strongly to the elastic properties of PAA, because interchain interactions would scale nonlinearly with the length.

These curves showed the subsequent deformation of a single PAA chain. It clearly showed that the deformation of a single PAA chain is fully reversible, indicating that the experiment is carried out at an equilibrium state. As described in the literature, the force law of polymer chains under tension can be derived from the freely-joint-chain (FJC) model.[376] The FJC model treats a polymer as a chain of statistically independent segments of length Ik (Kuhn length). The force law of the FJC model was shown to be given by Langevin function. The elasticity of a freely-joint-chain purely comes from the entropic contribution. The curve fit with the FJC model was plotted together with the measured force–extension curve. The fit gave a Kuhn length of 0.64 nm. It was found that the freely-joint-chain model is able to describe the elastic behaviors of PAA at low forces, but a great discrepancy between data predicted by the FJC model and the experimental data can be found at high forces.

The failure of the FJC model in the high-force regime may be due to the coarse description of the polymer in terms of discrete segments. The FJC model assumes that the segments of a polymer are inextensible. The data in this study was fitted to a modified FJC model.[375] This modified FJC model was constructed from a fixed number of segments with lengths that increase upon application of stress. From this extended Langevin function analysis, at low forces, the elasticity of the polymer was found to be entropy dominated, which means that most of the work done on the chain would orient these segments' end-to-end vectors (decreasing entropy), whereas at higher forces, additional work would be required to straighten the segments (increasing enthalpy). From these analyses, it was found that the magnitude of Kuhn length was 0.64 nm, and a segment elasticity was of 1300 pN/Å. It was shown that the extended Langevin function could provide a useful model for the elastic properties of PAA in the low- and high-force regimes. This indicates that the elasticity of PAA at low forces is dominated by the entropic contribution; in the high-force regime, the elasticity of PAA is governed by entropy and enthalpy. However, more studies are needed on other polymers in order to make a general conclusion. From data of 30 experimental force–extension curves of different contour lengths, it was found that all of the PAA filaments possessed an almost identical Kuhn length of 0.64 nm and a segment elasticity of 1300 pN. The lengths of these filaments varied from 60 nm to more than 1 μm. From these findings, it was concluded that the linear scaling of the elastic properties with the contour lengths and the identical segment elasticities and Kuhn lengths for all polymer chains, confirmed that predominately individual PAA chains were stretched, and the deformation of a single chain under tension was measured. Because PAA adsorbs strongly onto hydrophilic glass substrates, the resultant adsorbed PAA layer consisted mainly of PAA loops and tails of various lengths. It was reported that the binding of molecules to the tip is a rare event that occurs during a short contact period. Pull ups were not observed in more than 70% of the force measurements. Under these conditions, one may expect that only few molecules can build a bridge between the tip and sample during this short contact period. Further, because the adsorbed PAA layer is continuous, the entanglement of polymer chains sometimes has great influence on the experiment. In some cases, a bundle of polymer chains was stretched, instead of a single polymer chain. As mentioned above, the rupture force in the force–extension curve can be as high as 1.8 nN. This high rupture force could be caused by the scission of a C–C bond in the polymer chain or by the desorption of polymer filaments from the tip surface or the substrate surface. Considering the C–C bond energy of 345.6 kJ/mol (70 k_B T), the first possibility could be discounted. Thus, the rapture force reflects the interaction between PAA and the tip or the substrate.[376] It is likely that several hydrogen bonds between the surface silanols and carboxyl groups and van der Waals interactions are responsible for this high rupture force.

Single-molecule force spectroscopy is not only a powerful method in studying the nanomechanical properties of a single-polymer chain, but it is also useful in studying the length distribution of loops in the adsorbed polymer layer. The adsorption characteristics of a polymer are mainly dependent on configuration of the polymer. As predicted by the FJC model, the polymer chain will be stretched to 95% of its contour length at 102 pN; thus, the length distribution of PAA loops and tails on the surface could be measured.

From these analyses, it was concluded that most of the loops were shorter than 200 nm and that some tails exhibited lengths of more than 1 μm. These data showed that the deformation of single PAA chains under tension can be measured by using AFM and can be modeled using a modified FJC model. The deformation behaviors of single PAA chains may represent some common deformation characteristics of polymers, provided that there does not exist a suprastructure (for example, PEG)[366] and great force-induced conformation changes (for example, dextran).[370] It also shows that single-molecule force spectroscopy is useful for studying the polymer adsorption at solid surfaces.

4.5.8 Mechanical Deformation Studies by AFM of Synthetic Polymers

It is obvious that when a macromolecule is placed on a substrate and subjected to the tip of an AFM, it may lead to some degree of deformation, depending on the experimental conditions. AFM has been used to study the effect of mechanical stress on polymer structures.[377,378] In a recent study, the impact of the molecular structure on the network formation in ferroelectric liquid crystalline polymers, two copolymers, was studied.[378] Two distinct network structures were determined.

The interfacial properties of AB block copolymers are of particular interest because of their ability to act as adhesives, interfacial compatibilizers, and surfacatnts. The LB films of AB diblock copolymers (polystyrene, PS; polyvinylpyridine, PVP) were recently studied.[379] The aim of this study was to investigate the location of the polyelectrolyte chain and the *in situ* structure. The AFM images showed micelle-like structures. The height of these micelles was 60 Å, core diameter was 340 Å, and core distance was 1250 Å.

4.5.8.1 The Detachment of a Polymer Chain from a Weakly Adsorbing Surface Using an AFM Tip

Recently, AFM has been used to investigate the detachment of single polymer chains from surfaces and to measure the pico-Newton (pN) forces required to extend the chain orthogonal to the surface. Such experiments have shown that the force–extension profiles provide interesting signatures that might be related to the progressive detachment of the chain from a surface. Using equilibrium scaling analysis, activation kinetics, and exactly solvable partition functions, one can predict force versus extension profiles for various extension rates. It was shown that the variation in the extension rate can distinguish heterogeneous monomer-surface contacts. The qualitative features presented, such as sawtooth force profiles with detachment forces that decrease with extension, maximal yielding forces at high extension rates, and featureless force profiles at large extension. The statics and dynamics of single polymer chains at surfaces have received considerable attention in recent years. Much of this has been spurred by new experimental techniques such as AFM[380–382] and optical/magnetic tweezers which allow one to manipulate single polymer chains. Some of these studies measure the force required to detach a chain from an adsorbing surface, and, in many of these, the force versus extension profile exhibits sharp discontinuities that have been interpreted in terms of unadsorbed loops of the chain. A number of different single-chain systems have been studied: methylated dextran,[383] end-adsorbed polystyrene,[381] poly(dimethylsiloxane) (PDMS),[380,385] a polyacrylamide copolymer chain, and a telechelic poly(ethylene oxide) (PEO) having chemically modified chain ends.[386] Scaling analysis and self-consistent field theory have been used to explain the force profile for pull-off of polyelectrolyte chains; however, these theoretical treatments focused on large extensions and did not detail the features of loop detachment at shorter extensions, and they did not investigate the effects of different rates of extension.[387] The prediction of force–extension profiles for pulling an isolated polymer from a weakly adsorbing surface with different rates of pulling or extension was not carried out. Theoretical predictions were made using the simplest model: an ideal or Gaussian chain of N statistical monomers of size a, where a fraction of the monomers was pinned to the uniformly adsorbing surface with a contact energy $k_B T$. These monomer-surface contacts separated the adsorbed chain into a series of loops and tails. One may consider "grabbing" a loop or tail and extending this pulled tether a distance dw orthogonal to the adsorbing surface, while simultaneously measuring the force needed to extend the pulled tether. This process can be depicted as follows:

ADSORBED POLYMER CHAIN ON SURFACE------------
MONOMER UNITS PULLED OFF- - - - -

In this analysis, one considers a single monomer-surface contact and considers the pulled tether as simply a "handle" by which one applies a tension to this surface-bound monomer. An energy path of detachment is $U(z)$, where U is a particular monomer-surface potential energy, and z is a distance between monomer and surface. One can estimate an exact form of this potential, and a form assuming general features of the potential, such as barrier height, barrier width, and well depth.

It was assumed to be safe to follow the generalized treatment, extending to higher orders in force if needed, by adopting a specific surface-monomer potential energy function, $U(z)$. Alternatively, one may assume a simple generalized form, a simple cubic potential dependence on z. Further, a surface-bound monomer corresponds to residence in the well or a minimum of the potential energy. For detachment to occur, the system must overcome the potential barrier, ψ, and it does so at a rate dependent on the magnitude of ψ and temperature. Furthermore, an applied tension on the surface-bound monomer will alter the energy pathway, effectively lowering the barrier from 0 to ψ and facilitating detachment of the monomer from the surface. The tension on the surface-bound monomer is the tension of the pulled tether of n statistical monomers and is measured as the force on the AFM cantilever displaced a distance dw from the surface.

In an AFM experiment, the control of the rate of extension of the tether is measured in units of statistical monomer size. Thus, one is applying an increasing tension to the surface-bound monomer according to the prevailing force law of the pulled tether. This may be characterized as the rate of extension according to the activation kinetics of detachment. A fast rate of extension is one where the surface-bound monomer has insufficient time to escape the barrier to detachment, even though the barrier is being continually tension-reduced. In this limit, detachment occurs instantaneously at a large extensional force that reduces the barrier to zero. This yielding force field is the maximum force sustainable to the monomer-surface contact, and no detachment force can exceed this value. On the other hand, a slow extension provides ample time for the surface-bound monomer to escape the contact without appreciable tension reduction of the barrier. It was found that in these two limits, the force profiles (i.e., force versus extension curves) were dramatically different. Fast extensions provide a sawtooth pattern that details each monomer-surface detachment, while the slow extension provides a flat force profile and provides no signature of the detachments. Intermediate extension rates, where detachment occurs with escape over a tension-reduced barrier, provide an interesting combination of these two patterns: a sawtooth force profile at short extension which diminishes into a characterless force at larger extensions. One can summarize these analyses as follows. The scaling analysis was used to describe an equilibrium adsorbed chain and the force required to pull the chain slowly from the surface. The extension rates are sufficiently fast so that individual detachments of monomer from the surface can be probed, but they are slow enough that these detachments are activated events, i.e., they proceed with a nonzero, tension-reduced activation barrier. One can construct profiles for chains that are homogeneously pinned to the surface, as well as chains pinned to the surface with different energies and characterized by the relative rates of monomer-surface detachment and tether pulling. The rate-independent force profiles were described for fast extensions, where the detachment is an instantaneous yielding process, rather than an activation process. Further, it is well known that the unadsorbed monomers form a series of loops and tails, or an adsorbing layer, of height of some definite value. This equilibrium height can be determined from scaling analysis and the chain's free energy.[388]

The chain energy that determines this system consists of two terms. The first term is the energy associated with the reduction in conformational entropy upon confinement of the chain from solution to an adsorbed layer of average height. The second term represents the favorable contact energy resulting from monomer-surface contacts. The equilibrium height of the adsorbed ideal chain is that which minimizes the chain energy and is independent of (large) chain length. As the surface energy decreases, there are fewer monomer-surface contacts, larger loops and trains, and the adsorbed height increases.

When the extension of a pulled tether occurs at a sufficiently slow rate such that the monomer-surface contacts can detach and reform many times over the time scale of the pulling experiment, the force measured provides only averaged information about the strength of the surface-monomer contacts and not about the detachment process. In terms of each monomer-surface contact, the rate of application of tension is so slow that the monomer is able to escape the natural barrier, without the aid of the barrier-reducing tension. In this case, contacts can be lost and reformed many times over an incremental increase in extension or applied force, and the chain and surface can be described in equilibrium terms using scaling analysis. The free energy (in units of k_BT) of an adsorbed chain with a tether of n monomers, extended at a distance from the adsorbing surface, was derived based on the stretching penalty associated with extending a Gaussian loop of n monomers to a distance where the height of the absorbed train is independent of the pulled loop. The rapid reformation of monomer-surface contacts effectively exchanges monomers across contact or adsorption points, and the number of monomers in the extended loop, n, may increase to reduce the stretching penalty of the extended loop. Thus, for slow extension of the tether, the force will be constant, much like pulling a chain through a viscous medium, and will not signify the loss of individual monomer-surface contacts.

It was found that at higher rates of extension, the system does not have ample time to trespass over the full detachment barrier. Detachment took place as an "escape" over a tension-reduced barrier of height, expressed generally to first order in tension. Upon detachment, the m monomers in the loop adjacent to the contact are added to the n monomers in the pulled tether, and the force discontinuously decreases. These analyses were found to be valid for detachment forces (f) between the equilibrium force needed to slowly detach the chain and the maximum force at which the barrier to detachment completely disappeared, i.e., $f_{slow} < f_{det} < f_{yield}$.

It was expected that there will be different detachment systems. One can be a multiple detachment system from a homogeneous surface, which relates to the pulling-off of a chain adsorbed with multiple monomer-surface contacts. It may be assumed that the chain is initially adsorbed onto the surface with contact points evenly spaced along the chain contour, providing loops of a fixed number of monomers n.

It is safe to assume that n is simply the average loop size of the adsorbed chain. The average number of monomers per loop of an adsorbed chain of N monomers with Na/H surface contacts is $n = H/a\ 2/$. Let k be an index that advances by one with each surface-monomer contact lost. Initially, $k = 1$ and the extension and force at the first detachment. For subsequent detachments, $k = 2$, the extension w/a [estimated from the relations: $w/a = (w/a + t/a)$] to be with and $a_0 = 1$. In these analyses, the ratio of barrier width to monomer size was assumed to be constant and of order unity. After detachment, this force is reduced discontinuously by a factor of $k/(k + 1)$. These expressions are valid for pulling forces between f_{slow} and f_{yield}. These correspond to bounds on the dimensionless pulling rates. The profile was constructed from a sequence of detachment events, each detachment being described in terms of its average or expected lifetime. An experimental force profile, in contrast, is comprised of stochastic detachment events, and as such, it would retain the discontinuous forces, but the magnitudes of force and extension at detachment will vary from the averaged values of our predictions. It was found that the average detachment force decreases with successive loss of monomer-surface contacts. This general decrease in consecutive detachment forces becomes more pronounced at higher pulling speeds. The spring constant, or slope (df/dw) between detachments, diminishes discontinuously with the loss of adjacent contacts because of the increase in the number of monomers in the pulled tether. This decrease in spring constant is most dramatic with the loss of the first few surface contacts, with the force attaining a constant value plateau at large extension of the pulled tether. This results in replacement of the linear force profiles between detachment points with a pulling force which grows more strongly with extension and an increase in the detachment force. Irrespective of the particular model, the tether becomes increasingly compliant with the increase in the number of monomers in the tether that occurs with each lost contact, and the detachment force is discontinuous. Similar force profiles for a PDMS chain adsorbed to a silica surface were observed experimentally, as they were for the case of using methylated dextran adsorbed to a chemically modified gold surface.[380]

It was found that the spring constant decreases with consecutive discontinuities, and that the maximum forces in the sawtooth profiles diminish with extension, ending in a force that is constant over a larger extension. The discontinuities in the profile correspond to detachment of individual contact points that separate the pulled tether from an adjacent loop, and the maximum forces in each sawtooth correspond to the detachment force. These contact points are lost or sacrificed according to an activated process with a tension-reduced barrier. The force required for detachment decreases with loss of successive contact points, which means that the force required to pull off becomes less at each contact point. It was found that between the detachments of contact points, the force was linear with extension. From these data, it was found that the slope or spring constant, df/dw, is inversely proportional to the number of monomers in the pulled tether, and this diminishes with loss of successive contact points. In these studies, the effect of extension rate upon the average force profile for a chain of equi-sized loops was reported. With increased rate of extension, the detachment forces increase from $f_{slow} = k_B T/a$, the equilibrium value, to yield, which was found be larger than the forces at these extension rates. Moreover, each k_{th} detachment occurs at larger extensions when the rate of pulling is increased. It was found that as the pulling rate increases, the force increases, and larger extensions are required to detach each monomer-surface contact. Moreover, with larger pulling rates, the consecutive detachment forces decrease with each contact lost. The pulling rates are not sufficiently large that tension on the contact reaches the yielding force field.

These experiments showed that the detachment forces may sometimes increase with extension rather than decrease.[380,381] In some instances, decreases and increases in the detachment forces were observed.[386] These observations were attributed to the stochastic nature of the detachment event or to the likelihood of loops with different sizes. Experimentally, it would be difficult to distinguish whether the size and breadth of a "tooth" was attributable to loop size or to the stochastic nature of the detachment.

One can envision an experiment where the adsorbed chain is comprised of surface "sticky" monomers at known intervals. In this case, an ensemble of AFM force profiles for the ripping-off of the chain might be used to discriminate loop size in the stochastic, irreversible process. On the computational side, a stochastic simulation that mimics the stochastic barrier escape to detachment may be constructed. The prediction of individual detachment events was limited as averaged events. However, it was shown that the description, cast for loops of different sizes, predicts discontinuous forces at detachments, and that these detachment forces increase and decrease with successive detachments. The dimensionless force, $fa/(k_B T)$, versus dimensionless extension, w/a, plots for the detachment of a chain having unequal loop sizes with a dimensionless pulling rate of $y = 100$ were analyzed. The first loop has two monomers, and each consecutive loop size is double that of the previous loop until the loop size is 32. Loops following an $n = 32$ loop are one half the size of the previous loop until the loop size reaches two. The loop sizes increase and decrease according to $m(k) = rm(k - 1)$, where $m(k)$ is the number of monomers in the kth loop, and $r = 2^{1}/_{2}$. As a result of this geometric series of loop sizes, the detachment forces of consecutive monomer-surface contacts increase and decrease in accordance with the size of the tether, $m(k)$, relative to the adjacent loop,

$m(K + 1)$. Detail in the force profile is more apparent at small extensions, when the spring constant, df/dw, and detachment forces are large. However, at larger extension, spring constant and detachment force diminish and may become indiscernible from experimental noise. A simple ideal chain model scaling analysis, activation kinetics, and exactly solvable partition functions are used to predict force profiles for the detachment of chains from adsorbing surfaces by pulling a loose tether from the surface. Based on these assumptions, useful AFM force profiles were presented. It was assumed that the time scale of equilibration of the monomer-surface contacts is both shorter and longer than the time scale of the pulling experiment and where surfaces provide homogeneous or heterogeneous contacts. Thus, one would expect that when the extension rate is slow, the monomer-surface contact has ample time to exchange monomers between the pulled tether and adjacent loop, and the force is constant as the chain is being ripped slowly from the surface. However, if the extension rate is made faster and is commensurate with the kinetic rate of detachment, then the magnitude of the pulling force details individual detachments of monomers from the surface. It was shown that the force profile will be discontinuous, marking an individual detachment, and that, on average, the magnitude of the detachment force decreases with successive detachments. As the extension rate is increased, the magnitude of the detachment force increases, and larger extensions are required for detachment. At very large extension rates, the applied force is sufficiently large to reduce the barrier to detachment to zero, and the detachment occurs instantaneously at a yielding force field, which characterizes the monomer-surface contact. At these large extension rates, the force profile is sawtooth shaped with detachment forces that are equal for successive detachment events and are independent of extension rate.

The heterogeneous surface with different monomer-surface contact energies was expected to display discontinuous force profiles: strong contacts will give rise to the discontinuous detachment forces, while the disassociation and reassociation of weaker contacts will alter the force–extension curve between these discontinuous detachments. Furthermore, by increasing the extension rate so that it is comparable with the kinetics of weak contact detachment, one may probe the discontinuous detachment events of weak and strong contacts.

These results suggested additional experiments, where the extension rate of the chain, or probe tip velocity, was varied, and the chain was comprised of surface "sticky" monomers spaced at known intervals along the chain backbone, or the adsorbing surface was atomistically patterned. Stochastic simulations are required to construct predictions comparable with individual AFM force profiles. Detail such as monomer–monomer interactions, solvency, and inextensibility would be appropriately included in these stochastic simulations.[386–397]

4.5.9 AFM Studies of Polymers by Force Modulation Methods

It was mentioned above that the nano-Newton force as applied by the tip in AFM on a soft substrate can have some implications. The substrates' elastic properties can be investigated by AFM after some modifications of the instrument. The cantilever can be modulated under controlled conditions if a magnetic cantilever is used.[398] The cantilever was magnetized by attaching SmCo to the tip, and the modulation was controlled by a suitable electromagnet coil:

LASER.......

FEEDBACK
CIRCUIT...............
CANTILEVER (SmCo)................v
PZT....ELECTROMAGNET

The stiffness of polyethylene oxide (POE) of molecular weight 100,000 was investigated on mica substrate. In a topographic image, islands of POE were observed of size 0.5 to 2.0 μm and a height of about 10 nm. Similar data were found for POE mixed with SDS (see Figure 4.10, SDS + POE). The POE and mice can be detected by using the direct force modulation. The contrast of the stiffness image will correspond to the difference of the cantilever amplitudes on the two phases (mica or POE). The cantilever amplitudes on POE, x_{POE}, and mica, x_{mica}, can be:

$$
\begin{aligned}
|\Delta x| &= x_{\text{POE}} - x_{\text{mica}} \\
&= F_{\text{mf}} /(k_{\text{sc}} + S_{\text{PEO}}) - F_{\text{mf}} /(k_{\text{sc}} + S_{\text{mica}})
\end{aligned}
\tag{4.5}
$$

where F_{mf}, k_{sc}, S_{PEO}, and S_{mica}, denote the small modulation force applied by the magnetic force, the spring constant of the cantilever, the tip – POE stiffness, and the tip – mica stiffness, respectively. When the spring constant k_{sc} approaches zero, $|\Delta x|$ becomes maximum:

$$
\begin{aligned}
|\Delta x| &= x_{\text{PEO}} - x_{\text{mica}} \\
&= F_{\text{mf}}/S_{\text{PEO}} - F_{\text{mf}}/S_{\text{mica}}
\end{aligned}
\tag{4.6}
$$

The response of the direct force modulaton is given by:

$$
S_{\text{sample}} = (F_{\text{mf}}/x_{\text{sample}}) - k_{\text{sc}} \tag{4.7}
$$

The data from AFM gave values for effective stiffness for PEO film as 5.3 N/m and of mica 82.1 N/m. Further, polymeric films on inorganic surfaces are found to have importance in many everyday systems. The characteristics of such assemblies are not only determined by molecular adhesion forces but also by the kinetics of the growth process.[3,398]

Investigations have been reported on the monolayers of poly(ethyleneglycol) (molecular weight: 1500) on silica surfaces by AFM. The AFM image of a disordered polymer cluster is given in Figure 4.7.[398]

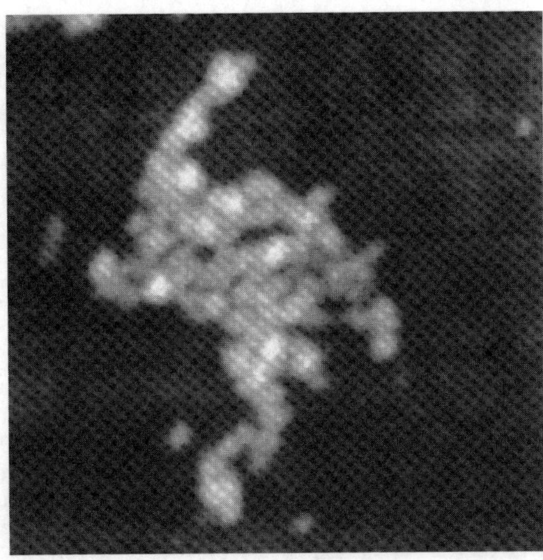

FIGURE 4.7 AFM image of poly(ethylene glycol) (molecular weight 1500). ($1.6 \times 1.6\ \mu m^2$; height range 16 nm). (From Nettesheim et al., *Langmuir*, 14, 3101, 1998. With permission.)

4.6 MIXED MONOLAYERS OF MACROMOLECULES AND LIPIDS

Mixed monolayers of hemoglobin, ovalbumin, xanthan, and virus with Mg-stearate collapsed films were studied as LB films on graphite.[20]

This provided a means of investigating biopolymers as found in their biological environment in the cell lipid-bilayer medium. Furthermore, because the molecular packing of lipids and macromolecules is different, it gives images that can be useful for calibration and other data analyses.

4.6.1 Hemoglobin Molecular Morphology by AFM

Protein molecules have been studied extensively by EM, but not in their natural state, because EM is carried out under vacuum, etc. For example, in membranes, most proteins are surrounded by lipid molecules.[20] The AFM method provides an unique possibility to study these systems as close to natural state as possible. The hemoglobin protein molecule (a tetramer with molecular weight of 68,000) and ovalbumin (molecular weight of 40,000) were found to remain as small clusters on the collapsed lipid film (Figure 4.8).[20]

The presence of protein molecules is easily observed from the light shaded spots. The step heights were found to be ca. 50 Å. This corresponds to the diameters of these molecules. The size of cluster varied, and under higher magnification, one could barely see the outline of each molecule. A more detailed image analysis is under preparation. However, preliminary data as shown here suggest that some 10

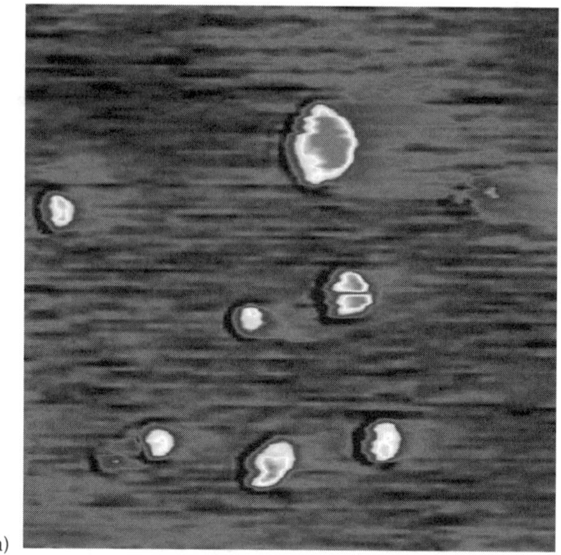
(a)

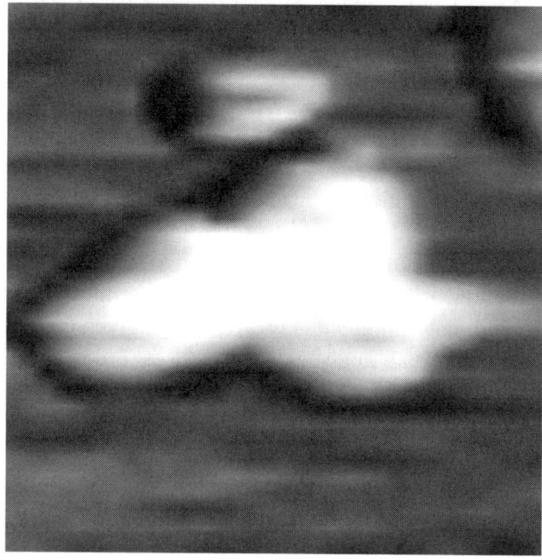
(b)

FIGURE 4.8 AFM images of hemoglobin on HOPG. A 10 μL solution of hemoglobin (0.01 g/L) was placed on HOPG. After the evaporation of water, AFM images were obtained. (a) 20,000 Å × 20,000 Å image; (b) enlarged view as 5000 Å × 5000 Å; (c) three-dimensional image).

molecules are involved in each cluster. At this stage, the method is not sensitive enough to provide better resolution.

It was concluded that the size of a cluster is indicative of protein–protein interactions in the monolayer. There seems to be some kind of higher-order aggregation in these two-dimensional structures. As far as we know, there has been no previous report of the cluster formation in the literature.

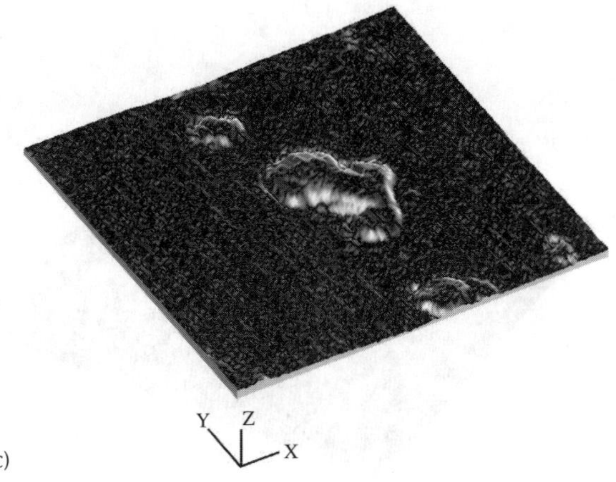

(c)

FIGURE 4.8 *Continued.*

4.6.2 POE + SDS

Mixed systems of POE and SDS were investigated by AFM (Figure 4.9). The AFM images of POE with molecular weights 4000 and 50,000 showed differences that could be related to their molecular weights. The larger macromolecules formed distinct long strips on HOPG as substrate, as expected. These images should provide useful information about POE and SDS interactions, if using high-resolution image analyses.

4.6.3 Mixed SDS + Gelatin on HOPG

The AFM image of mixed SDS + gelatin (1:1 g/g) is shown in Figure 4.11. It is seen that gelatin has aggregated into large globules with height of 1000 A. These data indicate that SDS and gelatin do not form even distribution surfaces.

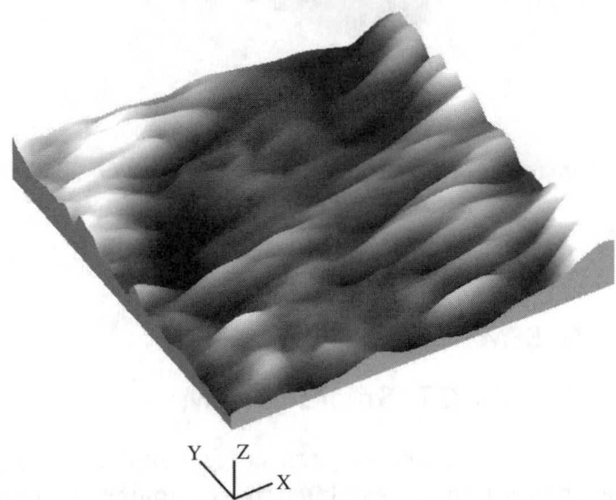

FIGURE 4.9 AFM image of POE, molecular weight 50,000 (9 μm × 9 μm). AFM images were taken of samples where an aqueous solution of concentration 1mg/mL was evaporated.

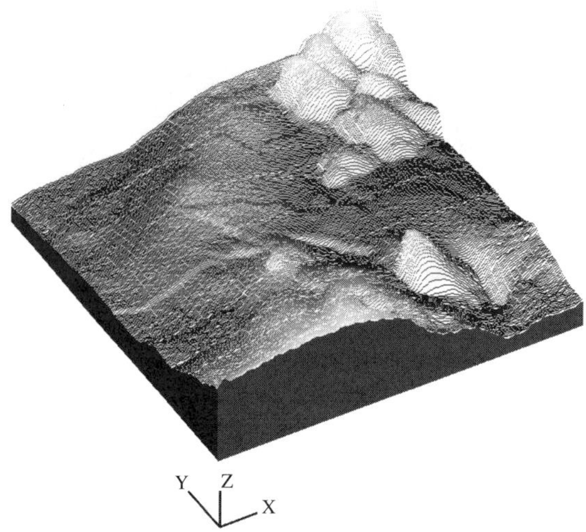

FIGURE 4.10 Mixed POE + SDS AFM image: three-dimensional (30,000 Å × 30,000 Å).

FIGURE 4.11 AFM image of SDS + gelatin (1:1 g/g) on HOPG. (45,000 Å × 45,000 Å × 1214 Å).

4.7 DIVERSE MACROMOLECULAR PROPERTIES AS STUDIED BY SPMS

4.7.1 Electron Transfer (ET) Studies by AFM

In many macromolecular systems, electron charge transfer (ET) phenomena are encountered. The charge transfer exhibited by an ionophore molecule, such as melittin, has been extensively described in the literature.[20] In fact, ET plays an important role in most of the biological cell membrane function and transport phenomena. It therefore becomes imperative to be able to study it using AFM. Myoglobin is an important protein in regard to biological systems.[241] Its primary role is to bind oxygen in muscular activity. The protein molecule, myoglobin (Mb), was studied as placed on graphite basal plane and on self-assembled didodecyldimethylammonium bromide (DDAB) mono- and multilayers with *in situ* tapping-mode AFM, cyclic voltammetry, and differential capacitance measurements.[399]

On graphite, Mb molecules adsorb and aggregate into chain-like features. This aggregation phenomenon has also been reported for other similar systems (see hemoglobin or gramacidin). The aggregation indicates an attractive interaction between the adsorbed molecules. In contrast, the molecules are randomly distributed on the DDAB layers. The adsorption on DDAB drastically changes the domains and defects in the DDAB layers due to a strong Mb-DDAB interaction. On both bare and DDAB-coated electrodes, the protein undergoes a fast electron-transfer reaction involving $Fe^{3+} + 1e^{-} \rightarrow Fe^{2+}$ in the heme group. The structure of the DDAB film is potentially dependent. At low potentials, the film is in a solid-like phase. When the potential is raised to ~0 V, the film transforms into a liquid-like phase via a first-order phase transition. The liquid-like phase may be responsible for the fast diffusion of Mb through the DDAB layers. The charged protein molecules are an important class of macromolecules.[20] Understanding the interactions between proteins and surfaces is critical in many fields of biomedical science, from biosensors to biocompatible materials.[368] In the development of biosensors, for instance, a critical step is to immobilize proteins with intact functions onto the surface of a transducer.[400–403]

Although many proteins spontaneously adsorb onto solid surfaces, the adsorbed proteins often denature or adapt undesirable orientations on the surfaces.[366,404–407]

A widely studied method is to coat an electrode with a layer of organic molecules.[366–369] If appropriate molecules are chosen, the layer can serve as a cushion that prevents adsorbed proteins from denaturation and as a guide for the proteins to adapt a desirable orientation. A novel extension to this method uses alkylthiol monolayers on Au electrodes.[408–415] The alkylthiols are known to self-assemble into ordered monolayers, with the thiol end anchored onto Au or Ag electrodes.[416]

An appropriate functional group on the opposite end can bind a protein, thus immobilizing the protein to the electrode. More recently, membranes of surfactants[417–423] and natural lipids[424–426] have been used to coat electrodes by self-assembly or LB methods. Because these systems mimic the physiological environment of many membrane-associated proteins, they provide an effective way to immobilize proteins onto electrodes for biosensor applications and provide nice model systems for studying the biological functions of these proteins. The latter point is important, because directly studying proteins in live cells is rather difficult, and the model systems allow a variety of powerful techniques, such as AFM, surface plasmon resonance (SPR) spectroscopy, and quartz microbalance, to be used. In this text, data were reported on a combined *in situ* tapping-mode AFM and electrochemical study of the structural and electron-transfer properties of myoglobin (Mb) on bare graphite electrode and on DDAB layers self-assembled on the electrode from DDAB vesicle solution. Mb is a heme protein that functions in transport and short-term storage of oxygen. The electron-transfer reaction of Mb on electrodes has been studied by several groups.[422,427,428] Also studied was the reaction on bare and on DDAB-modified pyrolitic graphite electrodes.[422] The DDAB film was cast onto the electrode by placing a drop of DDAB solution onto an abraded graphite and then drying it overnight. The film had an estimated thickness of ~20 m and was described as a liquid crystalline-like multilayer film with a large amount of water trapped between the layers. From these studies, it was found that the protein on the DDAB-coated electrode exhibited a fast electron transfer. The fast electron transfer was attributed to the prevention of impurities from adsorbing onto the electrode by the DDAB film, and to desirable orientation of the protein in the DDAB film.

It was shown that a combined *in situ* AFM and electrochemical measurement of Mb on well-defined DDAB films prepared by self-assembly could provide useful information about the adsorption of Mb onto the electrodes. The electrode used consisted of ZYH-grade HOPG. The electrode was cleaved and covered immediately with solution to minimize contamination. Myoglobin from horse skeletal muscle was dissolved in 30 mM tris(hydroxymethyl)aminomethane buffer (Tris-HCl, Fluka) to make a 0.2 mM solution (50 mM NaBr, pH 7.4). DDAB was dissolved in chloroform to a concentration of 100 mM. Small unilamellar vesicles of DDAB were prepared as follows. In the first step of the preparation, 500 μL of the DDAB solution was added to 50 mL of the Tris-HCl buffer. Then, chloroform was evaporated from the aqueous solution by bubbling N_2 into the solution for ~30 min, resulting in a 1 mM DDAB solution. To prepare uniform DDAB vesicles, the mixed solution (DDAB/buffer) was sonicated to clarity in an ice bath for ~40 min.

In these AFM experiments, a multimode SPM was used. Commercially sharpened Si_3N_4 tips attached to triangular cantilevers were used. The set point was adjusted to minimize the force between the tip and the sample in each measurement. The potential of the graphite electrode was controlled with Pt and Ag wires as counter and quasi-reference electrodes, respectively. The Ag quasi-reference electrode was calibrated against an Ag/AgCl electrode (3 *M* KCl), and all the potentials here are quoted versus Ag/AgCl reference electrode.

The cyclic voltammetry (CV) was measured with a potentiostat. A Pt wire and an Ag/AgCl electrode were used as counter and reference electrodes, respectively. For the differential capacitance measurement, a lock-in amplifier interfaced with a PC equipped with a data acquisition board were used. CV and differential capacitance measurements were carried out in an environmental chamber continuously flushed with N_2.

Mb adsorption on freshly cleaved HOPG with AFM and cyclic voltammetry was investigated.[434] In these studies, the experiments were carried out by imaging the HOPG in Tris-HCl buffer and following the addition of Mb solution (50 μL) into the cell, under continuous AFM tip scanning. The AFM images were obtained during the adsorption of the protein. The images showed a clean HOPG surface in the buffer initially. Diagonally crossing the image was observed a surface step that is a common feature on HOPG. After introduction of Mb into the buffer, the image was streaky, which indicated the adsorption of Mb. Slowly (~15 min), the adsorbed Mb began to appear as elongated features. The elongated features continued to evolve and formed chain-like features over time. In about 40 min, a Mb layer, consisting of the chain-like features, was completed. A typical chain was about 60 nm long and 7 nm wide. Higher-resolution images showed that each chain consisted of blob-like features of ~6 nm in dimension, which were identified as Mb molecule. This was concluded, because the dimension of the blobs was almost the same as that of Mb determined by x-ray crystallography.[430] Further, the coverage estimated by counting these blobs was about 2 × 1012 cm^{-2}, which agrees with the coverage determined from the cyclic voltammograms. The aggregation of Mb into chains could be ascribed to an attractive interaction between the adsorbed Mb molecules. In contrast to previous AFM study, it was found that cytochrome-c, a similar redox protein, distributed randomly when adsorbed on HOPG surfaces.[431–434] The adsorption of cytochrome was clearly observed as blobs in AFM, and complete protein coverage was observed after 10 min (Figure 4.12).[434]

This difference indicates that the attractive interaction does not exist between adsorbed cytochrome-c molecules. Another difference between Mb and cytochrome-c is that cytochrome-c forms a complete layer on HOPG within a few minutes, much quicker than Mb on HOPG. This time difference is not due to a large difference in the adsorption time between the two proteins. In fact, Mb adsorption begins a few minutes after Mb is introduced into the sample cell, but it is initially too mobile to be clearly resolved by AFM. It may be that the much longer time required for the formation of a Mb layer is because the Mb molecules need extra time to rearrange into the chain-like structure that is immobilized for AFM imaging. These adsorption studies of Mb onto bare HOPG were measured by TM AFM.

The cyclic voltammogram (CV) data of the adsorbed Mb clearly showed a pair of peaks corresponding to the reduction and oxidation of Mb. This was in contrast to literature data[422] that found no redox peaks in the CV of Mb using a bare graphite electrode. One possible explanation for this discrepancy could be that our graphite electrode was a freshly cleaved basal plane of HOPG, while theirs was abraded with sandpaper. The abraded graphite electrode exposes a large fraction of edge plane, which may prevent Mb from forming the stable chain-like features.

Assuming that the electron transfer involves one electron per molecule, the protein coverage from the peak areas in the CV is 1.8 × 1012 cm^2, in agreement with the AFM images as mentioned before. The coverage is similar to 2 × 1012 cm^{-2}, the coverage for cytochrome-c on graphite, which is reasonable because of the similarity in the dimensions of the two proteins. The heights of the anodic and cathodic peaks are proportional to the scan rate, which is consistent with the redox of an adsorbed species. However, the separation between the anodic and cathodic peaks dccreases initially and then increases as the scan rate increases. This behavior is not expected for a simple electron transfer of adsorbed species and is in sharp contrast to the behavior of cytochrome-c on graphite, which shows a monotonic increase in peak separation.[432] This abnormal behavior coincides with the formation of Mb aggregates, but further study is needed to elucidate the origin. After these studies of Mb on bare HOPG electrode, Mb on DDAB layers self-assembled on HOPG. Layers of DDAB were prepared by exposing the HOPG electrode to the DDAB vesicle solution. The thickness of the DDAB layers was controlled by the amount of the vesicle solution to which the electrode was exposed.

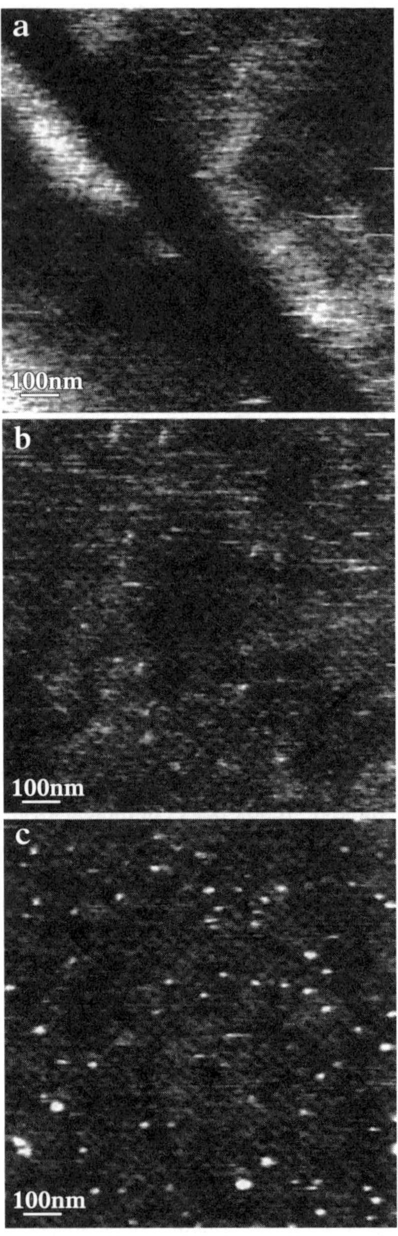

FIGURE 4.12 AFM images obtained after adsorption of cytochrome on lipid monolayers (on graphite). (a) at $t = 0$, (b) after 5 min, and (c) after 10 min. (From Boussaad et al., *Langmuir*, 14, 6215, 1998. With permission.)

Measurements were made by imaging a freshly cleaved HOPG in the buffer solution. After a clean HOPG image was obtained, later one introduced various amounts of DDAB solution into the AFM cell.

AFM image of HOPG was obtained in Tris-HCl buffer ~1 h after injection of 25 μL of DDAB vesicle solution into the cell. The image showed a DDAB layer self-assembled on the surface. The layer was uniform in thickness but consisted of many defects or pinholes. The thickness measured from the height profile of the image was 1.6 nm, corresponding to a monolayer of DDAB. Because the graphite surface is hydrophobic, the hydrophobic tails of the DDAB molecules will be expected to face the surface, while the hydrophilic heads point toward the water phase.

The monolayer of DDAB has been reported to be fragile, and imaging it in contact mode may destroy it. Self-assembled DDAB monolayers on HOPG from a DDAB vesicle solution were investigated by TM AFM. In this study, multilayer DDAB films were obtained by increasing the amount of DDAB vesicle solution in the AFM cell.

In order to estimate the thickness of the film, one can zoom in on a small area and switch to contact mode. After repeatedly scanning the area under a large force in contact mode, and then switching back to tapping mode and zooming out, the scanning tip swept off the DDAB molecules and created a square crater in the area. The depth of the crater was ~4.8 nm, corresponding to a monolayer plus a bilayer of DDAB. This observation is consistent with a recent study of lipid vesicles on various electrodes with quartz crystal microbalance (QCM), which found that the vesicles could spread on a hydrophobic surface and formed monolayer and bilayer films.[433]

The cast films of DDAB were prepared using the same procedure as described in the literature.[422] AFM images of such a cast film showed that the cast films were not as uniform as the self-assembled films. A hole in the middle of the image was created by repeatedly scanning the area in contact mode under a large force. The height profile plot showed that the hole was 120 nm deep. Because the AFM tip may not penetrate all of the layers, the AFM-measured thickness represents only the minimum thickness of the film. The AFM-tip-created hole was rather stable upon repeated scanning of the area in tapping mode, which means that the film is in a solid-like phase. This appears to contradict a previous differential scanning calorimetry study, which found that the cast DDAB film was in the liquid crystalline phase at room temperature and transformed into the solid-like gel phase below ~15°C.[422] It was suggested that the AFM may reveal only the layers adjacent to the electrode surface, which are, indeed, solid-like. Beyond the solid-like layers are many liquid crystalline-like layers that were measured by scanning calorimetry but could not be imaged by AFM. This explanation is supported by the fact that the estimated thickness of the film is 20 m, much thicker than the 120 nm estimated from the AFM images.

Images were obtained of the DDAB layers, and Mb adsorption was studied on these by introducing Mb solution into the AFM cell. Mb adsorbs onto the DDAB mono- and multilayers and forms a stable monolayer on the surfaces within minutes. Figure 4.6 is an AFM image of Mb on the DDAB monolayer, which reveals the individual Mb molecules as blob-like features. This is completely different from Mb on bare HOPG electrode, where the Mb molecules aggregate into chain-like structures. This difference can be understood on the basis of the following considerations. On atomically flat HOPG, the adsorbed Mb can diffuse around on the surface more freely and form aggregates. However, on DDAB surface, the adsorbed Mb molecules cannot diffuse freely because of their strong interaction with the DDAB. The interaction with DDAB immobilizes Mb almost immediately upon adsorption of the protein. This explanation is supported by the following evidence. First, the defects on the DDAB monolayer disappear immediately upon Mb adsorption, which means that the interaction between Mb and DDAB is strong enough to change the packing of the DDAB monolayer. Second, Mb molecules on the DDAB films can be clearly imaged by AFM immediately after adsorption, while they take nearly 40 min on bare HOPG before a stable AFM image can be formed. The AFM images as obtained on DDAB multilayers were found to be similar to those on the monolayer. The CV of the adsorbed Mb on the DDAB monolayer exhibits a pair of peaks corresponding to the redox reaction of the protein, which is similar to that on bare graphite electrode. The coverage of Mb obtained from the integrated redox peak areas is 2.4×10^{12} cm^{-2}, which was accepted to be that of a closely packed monolayer of Mb.

The separation between the anodic and cathodic peaks was found to be larger than that of Mb on bare graphite, which may indicate a slower electron transfer. The separation was also found to increase as the scan rate increased, from which the electron-transfer rate was estimated to be about 80 s^{-1}.[432] While the CV for the Mb on DDAB was similar to that on bare graphite, the redox potential was shifted negatively by ~0.05 V for Mb on DDAB.

On increasing the thickness of the DDAB layer, the CV becomes more like Mb on the cast film.[422] The data showed that the thickness of the film estimated from the amount of DDAB and AFM images corresponded to ~15 bilayers. The peak currents were found to increase more slowly than a linear function of the scan rate, as expected for an adsorbed Mb layer, but faster than the square root of the scan rate, expected for a diffusion-limited process. This behavior can be interpreted in terms of thin-layer electrochemistry. At low scan rates, diffusion is sufficient to bring all the Mb molecules in the thin layer to the electrode surface, and the current is proportional to the scan rate. When the scan rate is fast, the electron transfer becomes diffusion limited, and the current is proportional to the square root of the scan rate. The separation between the cathodic and anodic peaks increases monotonically as the scan rate increases. Using the peak separations at fast scan rates (>10 V/s) and the diffusion coefficient from the literature, the electron-transfer rate estimated on the basis of the published curves is about 6×10^{-3} cm/s. This value is in agreement with the electron-transfer rate reported for Mb in the DDAB film cast on graphite electrodes.[422] The diffusion coefficient estimated from the CV data was in the order of 10^{-6} to 10^{-7} cm^2/s, which agreed with the value obtained for the cast film.[422] This value was found to be almost the same as the diffusion of Mb in bulk solution. It was somewhat unexpected that, near the electrode, the DDAB film is solid-like. The AFM images indicated that Mb interacts strongly with the DDAB film upon adsorption and can penetrate into the film, but the diffusion is still expected to be much slower than that in bulk solution. The DDAB film was also investigated as a function of potential using differential capacitance and AFM. These images of a self-assembled DDAB layer in Tris-HCl buffer were obtained at different potentials –0.4 to 0.4 V. Below ~0 V, the DDAB film was stable and had a thickness of ~4.8 nm, corresponding to a bilayer on a monolayer. Whereas, on increasing the potential to ~0 V, the image suddenly became streaky, indicating a phase transition in the DDAB film. The streakiness was due to the fact that the film in the new phase was unstable under the scanning tip of AFM, so the new phase was more liquid-like. On increasing the potential to 0.2 V, the image became less streaky and revealed some remaining DDAB molecules on the surface. The phase transition was found to be fully reversible. On lowering the potential below 0 V, the DDAB film grew back immediately. Because no DDAB was present in the solution, the recovery of the DDAB layer means no desorption took place during the phase transition.

The reversible phase transition was observed in the differential capacitance data. As the potential increased from negative to positive, the capacitance also increased sharply at ~0.15 V. The capacitance returned to the lower value as the potential was cycled back to negative potentials, but at a more negative potential (~0.05 V). This hysteresis effect was expected for a first-order phase transition, which was observed in the phase transition of many interfacial films. As expected, the interfacial capacitance in the presence of the DDAB film at negative potentials was found to be smaller than that in the blank buffer solution. After the phase transition, the capacitance increased but was lower than the capacitance in the buffer solution. This indicated that the DDAB molecules remain on the electrode surface but with a smaller effective thickness due to, probably, penetration of counterions into the film. This observation was consistent with the AFM images.

In these studies, a stepwise change of the capacitance near 0.1 V was ascribed to the phase transition observed by AFM. The capacitance measurements were made by superimposing a 38 Hz ac modulation with an amplitude of 10 mV to the electrode potential (scanned at 0.1 V/s).

In a recent study of LB film of insoluble surfactants on gold electrode, a potential-induced desorption and adsorption of the film was observed.[435–438] Using electroreflectance spectroscopy and light-scattering measurements, it was found that the desorbed surfactant molecules form micelles trapped near the electrode surface. In the present system, the potential did not induce desorption of the DDAB film within the studied potential range. Instead, the potential induced a phase transition from the solid-like bilayer phase at low potentials to a liquid-like phase at high potentials. This reversible phase transition may have been triggered by the electrostatic interaction between the charged electrode surface and the positively charged head group of DDAB, which changed from attractive at low potentials to repulsive at positive potentials. For a complete understanding of this phase transition, further study is needed. Nevertheless, the observation of the phase transition provided an explanation of the unusually high diffusion coefficient of Mb through the DDAB film based on the following consideration: Mb diffusion in the solid-like layers is slow, but when the potential is cycled to positive potentials, the layers transform into the liquid-like phase, through which Mb diffuses quickly to the electrode surface. Another factor that enhances the transport is the strong interaction between Mb and DDAB, which allows the individual Mb molecules to penetrate into the DDAB film.

Furthermore, Mb spontaneously adsorbs onto freshly cleaved HOPG and forms a monolayer in which the individual Mb molecules aggregate into a chain-like structure. This phenomenon was also observed in the case of hemoglobin (Figure 4.9). In these images, one could clearly see a pattern of aggregation. From surface topographic analyses, steps were found that corresponded to individual hemoglobin molecules.

The aggregation indicated an interaction between the adsorbed Mb molecules. The adsorbed Mb undergoes a fast electron-transfer reaction, corresponding to

Fe^{3+} + 1e – Fe^{2+}. The coverage of the adsorbed Mb determined from the total amount of charge transfer is in agreement with the AFM images. Mb also spontaneously adsorbs onto the DDAB layers self-assembled on the HOPG. The adsorption dramatically changes the defects in the DDAB film, indicating a strong interaction between Mb and DDAB. Instead of forming the chain-like aggregates, Mb molecules are randomly distributed on the DDAB. The adsorbed Mb exhibits a pair of peaks in the CV, corresponding to the electron-transfer reaction. The characteristics of the CV depend on the thickness of the DDAB film. On a thin DDAB film, the CVs can be described in terms of an adsorbed species, similar to that of Mb on bare HOPG. On a thick DDAB film, the CV has the characteristics of a diffusion-limited process, similar to that of cast DDAB film reported in the literature.[422] The DDAB film is not stable at all potentials. At low potentials, the DDAB appears to be solid-like, as judged by the well-defined domain boundaries and defects. Increasing the potential triggers a phase transition, transforming the solid-like phase into a liquid-like phase. This phase transition may explain the unusually fast diffusion of Mb in the DDAB film extracted from the CVs.

4.7.2 Other Diverse Macromolecules

4.7.2.1 Xanthan

Xanthan (biopolymer), produced by the fermentation broth of the bacterium *Xanthomonas campestris*, is a polysaccharide that has been used in different industrial (food, tertiary oil recovery) and other (pharmaceutical) applications. Xanthan molecule consists of a linear cellulose backbone, with links between β-D-glucose residue, with a three-sugar side chain attached to every second glucose. This gives rise to a macromolecule with ionic characteristics and a comb-like polymer. Xanthan structure results show a characteristic morphology.

The xanthan molecules (approximately 10^6 molecular weight) are found to be arranged as linear-shaped structures (while globular proteins, e.g., hemoglobin and ovalbumin, are present as aggregates). This corresponds to clusters of xanthan molecules. The radius of gyration has been reported in the range of 80 to 90 nm. Furthermore, these structures have broken away from the lipid film. It is well known that a xanthan molecule (due to its high molecular weight) leads to high viscosity in aqueous media. This means that xanthan must be aggregated in a different state than other proteins, such as hemoglobin and ovalbumin. These analyses clearly show the differences. Furthermore, xanthan forms some compact gel-like structures (the so-called *fish-eye*). These images might be related to these fish-eye characteristic aggregation phenomena. The three-dimensional image clearly shows the molecular arrangement in such mixed SAMs.

The morphology of polymers has been extensively investigated by STM and AFM. STM topographies of xanthan gum, pipette deposited and spray deposited[440,441] onto HOPG and mica substrate were reported. A solution of xanthan of concentration 10 μg/mL was deposited on HOPG and allowed to dry under ambient conditions. The topographies showed entangled molecules forming a dense network across the substrate surface. The diameter of strands varied from 6 to 16 nm.

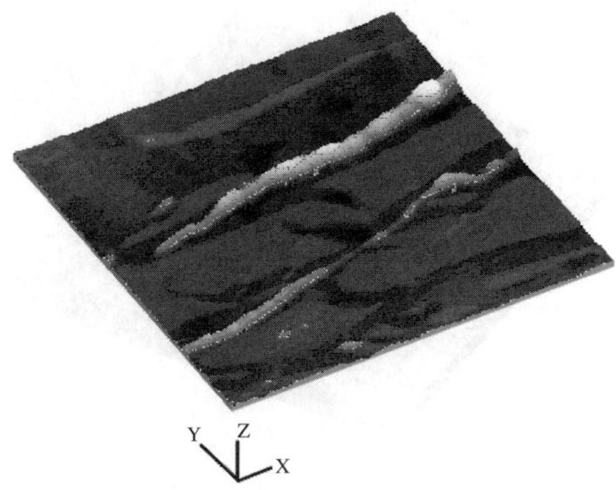

FIGURE 4.13 AFM image of xanthan in mixed LB film with Mg-stearate (30,000 Å × 30,000 Å × 496 Å).

Despite the theoretical problems associated with explaining how STM can image biological molecules deposited onto solid substrates, STM has been used to obtain images of xanthan placed on HOPG. Xanthan sample was deposited by two different methods. (a) Calcium xanthan solution of concentration 1 mg/mL was deposited as a drop of volume 5 μL on HOPG. After drying, the surface was scanned by STM. (b) A xanthan solution was placed on HOPG. STM was used for obtaining the image after the sample was dried overnight. The images clearly showed molecular diameter of 2.8 nm, which agreed with other literature data. Xanthan was added to Mg-stearate monolayers prior to LB film on HOPG. The AFM images showed clearly the long molecular structures of xanthan (Figure 4.13).

4.7.2.2 Immunoglobulin G (IgG)

Immunoglobulin G, IgG, was investigated by STM.[442] IgG is a molecule consisting of two heavy chains (50,000 molecular weight each) and two light chains (25,000 molecular weight each). These chains are known to be arranged in the form of a "Y" shape. The configuration of immunoglobulin G, IgG, was investigated[442] by STM. The length and width of each arm from x-ray data was reported as 8.5 nm × 6 nm. The height of the molecule is reported as 4 nm. STM images of IgG showed dimensions of 7 × 9 × 4 nm^3, in agreement with the x-ray and TEM data. These data convincingly demonstrated that STM can provide images consistent with other microscopy data. No distortion or damage from the tip was reported. An AFM image of mixed IgG (human) + SDS (1 g/1 g) is given in Figure 4.14. It can be seen that IgG molecules are not mixed completely with SDS molecules.

Actually, as experienced by many investigators, the images are mainly of two kinds of resolution: sharp images with no tip–substrate interaction or images with changes shown during each scan that are produced by the tip interacting with the substrate. However, in some cases, this interaction may abruptly stop and sharp images can be obtained. This means that exact analyses are not available at this

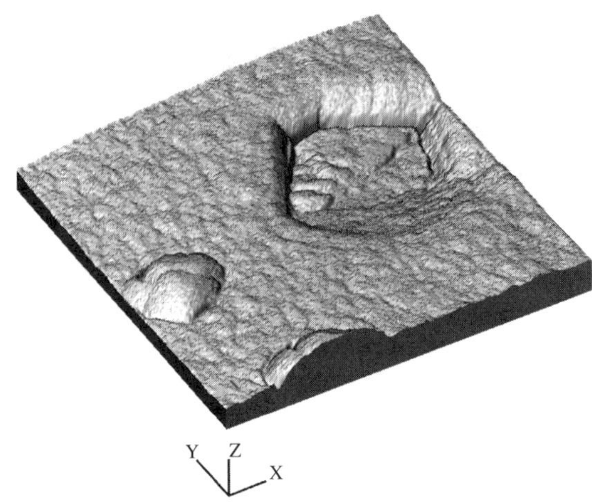

FIGURE 4.14 AFM image of IgG (human) + SDS (1 g/1 g) on HOPG (30,000 Å × 30,000 Å).

stage. These interactions could be due to many reasons. In this book, these aspects will be considered throughout. It has been argued that AFM or STM should principally provide the same kind of morphological features as reported by electron microscopy,[443] which may depend on the radius (or rather the shape) of the tip, as well as on the geometry and the physicochemical state of the biological material (for example, adsorbed water). The less corrugated the objects are, the less important becomes the actual tip shape. This is found for near-atomic resolution images in the case of LB films. In the case of STM, better resolution can be obtained by averaging the signal-to-noise ratio (STNR). However, these arguments are not well described, and further investigations are needed.

In order to understand the physical properties of new materials based on SAMs, the LB method was used in combination with AFM.[149] The LB method is now recognized as the most important procedure for making tailored materials for future molecular engineering products. The LB films will have the same application as the already well-known method of producing thin films of materials on solid surfaces by vapor chemistry. The amphiphile molecules in LB films incorporate disparate chemical functionalities in their hydrophobic tail groups and hydrophilic head groups.[3,149] However, the structure studies of LB films are just beginning to be well understood after the application of STM and AFM to these film assemblies. Prior to this, the structures of LB films were not completely understood, because only x-ray diffraction could be used for such investigations.[149,444]

4.7.2.3 Gramicidin and Other Ionophores

An AFM image of collapsed gramicidin was analyzed.[20] Domains were observed with size consisting of a few hundred molecules. It was also shown that due to cantilever–gramicidin interaction, the image changed. The breakup was fast (< 5 sec) (Figure 4.15). The series of images show how the cantilever–gramacidin interaction led to the rearrangement of the molecules. In most cases, the images stabilized after

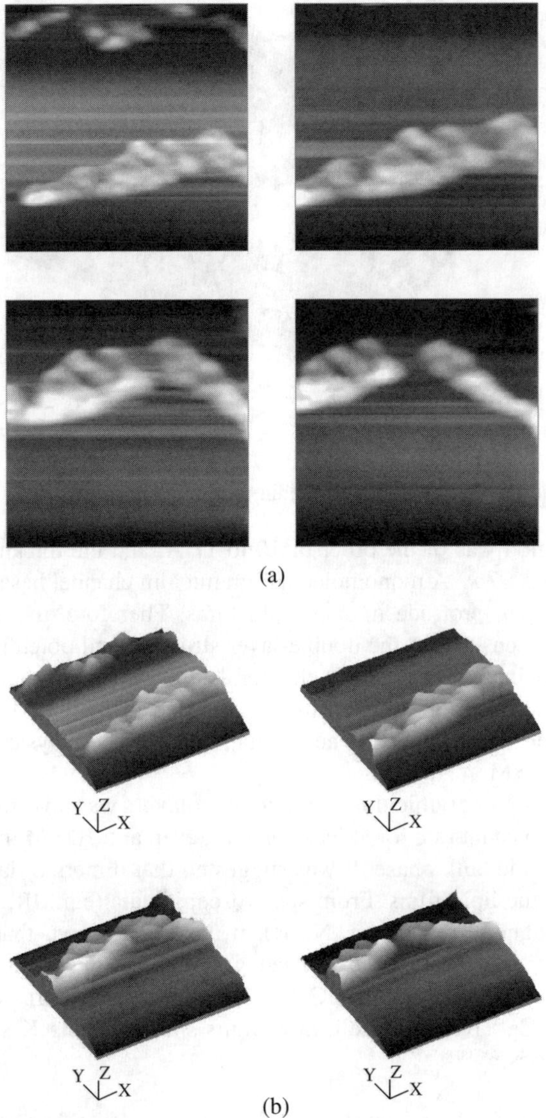

FIGURE 4.15 A series of gramicidin images showing where the tip has moved the molecules: (a) two-dimensional, (b) three-dimensional (45,000 Å × 45,000 Å).

a certain number of scans. In some experiments, the image showed cluster formation of gramicidin molecules (Figure 4.16). It is interesting to note the almost repeat cluster formation. These images show the dynamic AFM observations, where the tip–substrate forces were changing. This observation shows that AFM can be useful in understanding the molecular forces involved in aggregation mechanisms.

Gramicidin channels in lipid bilayers were recently investigated.[445] The transport of Tl^{+} was investigated. Capacitance measurements indicated that the thickness of

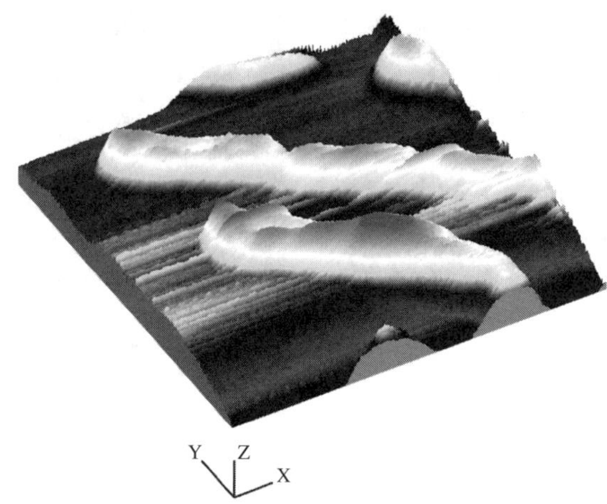

FIGURE 4.16 Cluster formation of gramicidin molecules in AFM (45,000 Å × 45,000 Å).

hydrocarbon region was of the order of 10 to 11 Å, and the thickness of the polar group in lipid was >7 Å. A monomolecular gramicidin channel has a width of 13 Å, which means it will protrude in such lipid films. Therefore, the rate-determining barrier would be sensitive to the double-layer structure and potential profile at this point. The AFM images provide much useful information on these postulates. In contrast to valinomycin, gramicidin in the presence of high KCl in the subphase does not show elevated ΔV or the new collapse state. These systems remain to be investigated by AFM.

The penetration of gramicidin into lipid monolayers was investigated and found to proceed at a constant rate for at least an hour, even at 5 10^{-7} M initial gramicidin concentration in the bulk phase. It was suggested that dimers or larger aggregates could penetrate the lipid films. From spectroscopic data (e.g., IR, circular dichroism; nuclear magnetic resonance, NMR), it has been shown that Ca^{++} interacts with gramicidin and that a head-to-head gramicidin dimer can have two Ca^{++} binding sites located near the –COOH terminal. The transport of Cs^+ and K^+ is blocked by this Ca^{++} binding. Melittin exhibits similar strong K-specific penetration characteristics.[20,100]

4.7.2.4 Mixed Monolayers of Virus Cell or Fusion Peptide Cell

Many viruses have adopted the strategy of membrane fusion (in lipid bilayer) in order to enter their hosts.[447–450] The fusion reaction is triggered by a peptide that can penetrate the host cell membrane. To understand this initial step in fusion reactions, a monolayer has been used as a model system.[20,449,451,452] Before a virus can enter the host cell, it adsorbs to the cell via a plasma membrane molecule, which is recognized by a viral protein.

A large number of investigations are being carried out using AFM, whereby the molecular picture might be elucidated. A virus (*Moloney*) when studied on lipid films shows that the particles prefer to orient at the edges of the collapsed lipid film. A

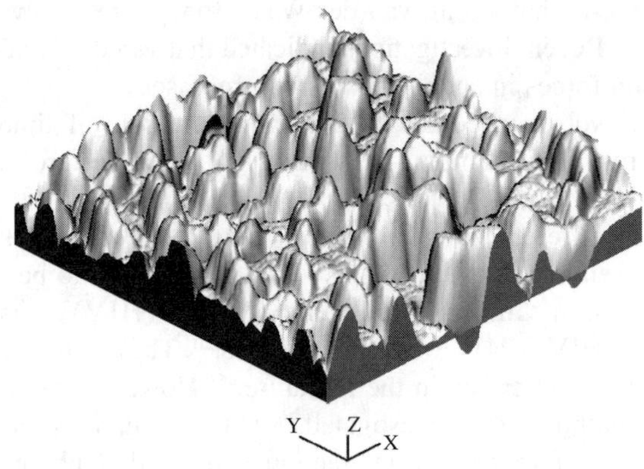

FIGURE 4.17 A three-dimensional AFM image of Moloney virus embedded in Mg-stearate LB film (10,000 Å × 10,000 Å × 207 Å).

three-dimensional image is given in Figure 4.17. It clearly shows the virus embedded in the lipid LB film. The step height analyses shows that the virus has a diameter of 527 Å. This agrees with the electron microscope literature data.[447,448] The geometrical interparticle orientation provides information about the forces of interactions. In other words, different viruses would exhibit different orientation images.

Biological membranes are known to have many functions: they act as a permeability barrier, and must, consequently, transport matter and information across the boundary between the exterior and interior phases; they can be excitable; and they serve to give each cell individuality.

Monolayers of different virus fusion peptides were investigated.[449,451,452] The effects of pH and electrolyte concentration have been studied. Furthermore, the molecular structures in monolayers were investigated using STM and AFM.[20] It was shown for the first time in the literature that van der Waals forces play an important role in fusion (as determined from monolayer studies).

Because the characteristics of fusion are complicated, the monolayer model system has been used in the literature for obtaining information about structure and function. There are basically two important kinds of interactions, e.g., van der Waals forces and charge–charge interactions.

Further, in regard to virus fusion, currently known are two groups of small peptides that are of importance:[453] ionophore peptides and virus fusion peptides. And, some fusion peptides (such as in the influenza virus) exhibit ion channel activity.[454]

The first step in virus-cell fusion is determined by the virus fusion peptide interaction with the lipid bilayer membrane of the cell. The fusion process is also known to be related to the disease, such as inflammation and cell–cell fusion. The fusion peptides are involved in the virus-cell infection.

The different fusion peptides, such as influenza virus, semliki forest virus (SFV), and human immunodeficiency virus (HIV), were described in the literature.[453] However, few studies have been reported in regard to their surface activity as monolayers. All of these peptides are hydrophobic and form stable monolayers. The hydrophobic characteristic suggests that strong van der Waals forces are involved in the virus-cell fusion process. Recent investigations indicated that van der Waals forces are the primary interaction forces in such virus fusion processes.

These studies were carried out to determine the role of lipids in virus-cell fusion mechanisms. For example, it is known that cholesterol is required for infection by SFV.[455]

Monolayers of different virus [influenza virus (INF, which is known to cause the common flu), semliki forest virus (SFV, which is known to be spread by mosquitoes), and human immunodeficiency virus (HIV)] fusion peptides (INF-HA2:SFV-E1:HIV-GP41) were investigated.[20] These virus fusion peptides have been extensively described in the literature.[453] However, the monolayer properties have not been thoroughly investigated.[20,449] These peptides form stable monolayers at the surface of water. The interaction with lipids (sphingomyelin/cholesterol/dioleyl- and dipalmotyl-lecithin) is found to be greater at lower pH (=4) than at higher pH (=7) in the monolayers. This is in accordance with similar studies with other peptides and virus in monolayers (and in vesicle).[20] There is a need for more detailed analyses of these data.

The virus-cell fusion process is an energetically unfavorable process, and the virus coat proteins are responsible for such fusion. While the hydration around the cell surface hinders fusion, the hydrophobic forces promote the process. Furthermore, it was reported that[453,456] fusion peptides are short (16 to 26 amino acids) and relatively hydrophobic (low solubility in water). The reason for this might be that the size and shape of fusion peptides are of the same size as the thickness of the bilipid membrane. Some fusion peptides have prolines near their centers. This is also found in the case of melittin (with 26 amino acids). Another aspect of interest is the conformational segregation of apolar and polar amino acid residues in proteins.[20,100,453]

An AFM image of mixed melittin + SDS (1 g/1 g) is given in Figure 4.18. The melittin molecules are seen to be arranged in clusters.

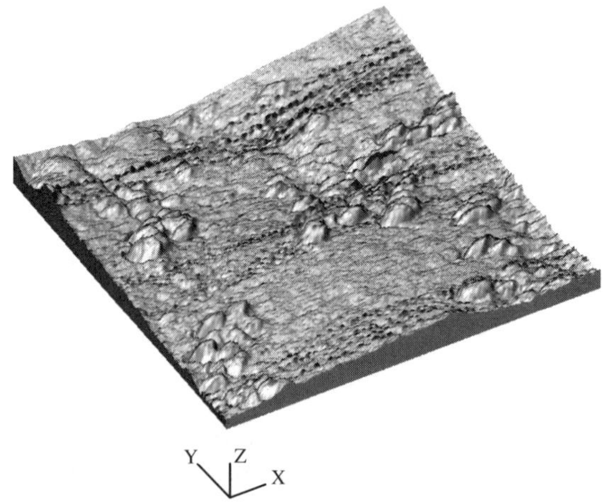

FIGURE 4.18 AFM image of melittin + SDS on HOPG (45,000 Å × 45,000 Å).

X-ray crystallographic studies of various proteins revealed a characteristic feature in protein structure. Many helical segments in some proteins (globular and especially melittin and the virus fusion peptides) contain sequences where apolar residues occur regularly at every third or fourth position. This gives in helical space a polar face and an apolar face. In aqueous media, it is thus found that due to entropic factors, these proteins tended to aggregate (through hydrophobic forces). The hydrophobic residues associate away from water phase. The aggregates generate cylindrical multimers penetrating into or through the thickness of bilipid membranes. For example, alamethicin tends to aggregate to form decamers as the electrolyte concentration increases.[20] In fact, monolayer data of melittin was explained on the basis of these considerations.[20,100] These monolayer studies showed that melittin could be present as tetramer at high concentrations at interfaces.

Three different cell fusion peptides were investigated by AFM.[3] The purpose of these studies was to determine if different peptides form different structures as studied by AFM resolution. The peptides from different viruses were used: HIV-GP41; the protein responsible for influenza virus fusion, hemagglutin HA-peptide; and SFV.

All peptides were dissolved in DMSO (0.01 mg/mL), and a 10 mL sample was placed on HOPG. After evaporation overnight of the solvent, the samples were analyzed. The morphology of the peptide SFV (SFV69) was as follows. The diameter of the molecule was ca. 20 Å, which agrees[458] with the molecular weight of 20,000. The different peptides showed different morphological images (Figure 4.19). The SFV image showed the following dimensional analysis:

Step height = 20 Å
Length = 2000 Å = 2000 Å/step height (20 Å) = 100 peptide molecules

This finding is in agreement with the data for lysozyme, as described below. The length of the aggregate corresponds to 100 peptide molecules. The differences in these images between peptides were shown to be related to amino acid composition and sequence.

The differences in the morphology of these peptides can be ascribed to their hydrophobicity characteristics.[20] These studies also indicate that the effect of solvent on the polymer configuration can be investigated by AFM studies. This observation is important in the studies of cell fusion mechanisms. It is well accepted that virus fusion starts by a step height triggering the integral membrane protein to change conformation, thus exposing an hydrophobic domain that can mix with the lipid bilayer of the cell membrane.

AFM has been used to study the morphology of neurons.[459]

4.8 MONOLAYERS OF SYNTHETIC POLYAMINO ACIDS

Because synthetic poly-α-amino acids have proven to be useful models of protein structure,[241] monolayer studies of the former are of great interest. The

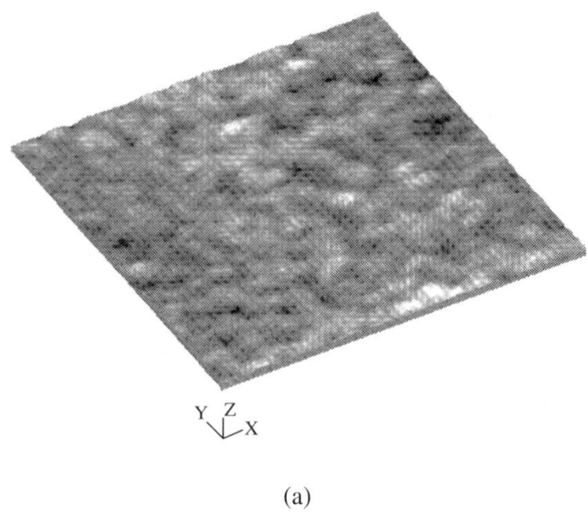

(a)

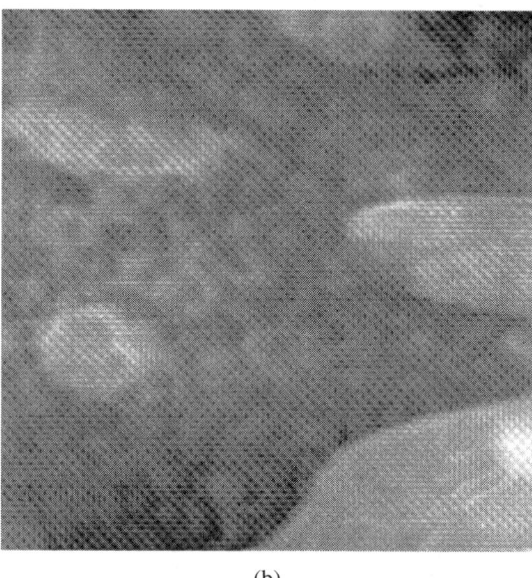

(b)

FIGURE 4.19 AFM images of fusion peptides:[63] (a) HA; (b) HIV-GP41 (45,000 Å × 45,000 Å).

hydrophobic-lipophilic balance (HLB) of some polyamino acids, such as polyalanine, polyphenyl derivatives, are of suitable magnitude to give stable monolayer assemblies at air–water interface.[3]

Surface pressure (π) versus area (A) isotherms of monolayers of different synthetic polyamino acids were reported.[3,100] The π versus A isotherms indicated a close-packed α-helical arrangement, because a steep rise was observed, and the inflection or plateau arose from the collapse of the monolayer to form a bilayer.[20] However, later studies indicated that through the analyses of collapse state, α-helical polymers could be clearly distinguished from β-polymer.[20,100] The surface potential (ΔV) versus A isotherms indicated that the surface potential approximately followed the relation ΔVA = constant = 300(mV) 30 (mN/m). These data should be better explained by considering the contribution to ΔV arising from the various parts of the polymer molecule. AFM analyses of LB films of PBLG were performed, indicating helical structures (Birdi, unpublished). AFM images of polyglycine spread on HOPG were investigated. The effect of solvent was investigated by comparing images from solutions in water or dimethyl sulfoxide (DMSO). The image from water as solvent showed a helical structure, while from DMSO solution, a more compact structure was found (Figure 4.20a,b). This showed that in good solvent (i.e., water), the polymer is more extended than in a poor solvent (DMSO). It is important to notice that the analysis shows that this has the same value (ca. 35 nm) in both cases. As pointed out by some investigators,[20,241] an important element that was neglected in most studies was consideration of the conformation of the polypeptides and their monolayer properties. Investigated were π versus A isotherms of well-defined polymers, as regards molecular weight and conformation.[20,460]

A lot of useful information is given by π versus C_s isotherms of α-helical and β-form polymers. The isotherms for the α-helical forms were reported to give characteristic plateaus at the π_{col} state.

The isotherms for the β-forms, on the other hand, showed an inflexion at the collapse state. Furthermore, the relationship between the magnitude of π_{col} and conformation stability was described in detail.[460] It was reported that π_{col} for the β-form polymers was related to the stabilizing forces in the bulk phase. However, no such relation was found for α-helical polymers.

The greater stability of poly-(γ-benzyl-L-glutamate) (PBLG) as compared to poly(methyl-L-glutamate) (PMLG) in solution has been attributed to a difference in side-chain interactions.[241,460] The efficient shielding of the hydrogen-bonding helical backbone from the solvent by the bulkier benzyl group is undoubtedly one of the factors responsible for the greater stability of PBLG. AFM studies of PBLG on graphite show helical conformation.

Another contribution to stability would be the nonbonded (hydrophobic) interaction between the phenyl moieties, which would be absent in the PMLG and poly(ethylglutamate) (PELG). Consequently, because of the weak hydrogen bonding at the interface (i.e., water–backbone interaction), when PBLG is in the α-helical conformation, the magnitude of π_{col} would be lower than that for PMLG or PELG. Thus, it is safe to conclude that these polymers retain their α-helical structures at the interface. This was recently confirmed from the AFM studies of PBLG.

It is also possible, that in the α-helical conformation, the intramolecular side-chain–side-chain interactions in PBLG, in bulk, contribute to the higher stability, while these interactions would not be expected to contribute to the magnitude of π_{col}. However, intermolecular side-chain–side-chain interactions between helical polymers would be expected to contribute to the magnitude of π_{col}. Thus, PMLG and PELG show higher π_{col} values, because the intermolecular side-chain packing of the methyl and ethyl groups is efficient, while the bulky benzyl groups hinder compact packing. It was concluded valuable physical information can be obtained from studies on π_{col} for polymer monolayers.

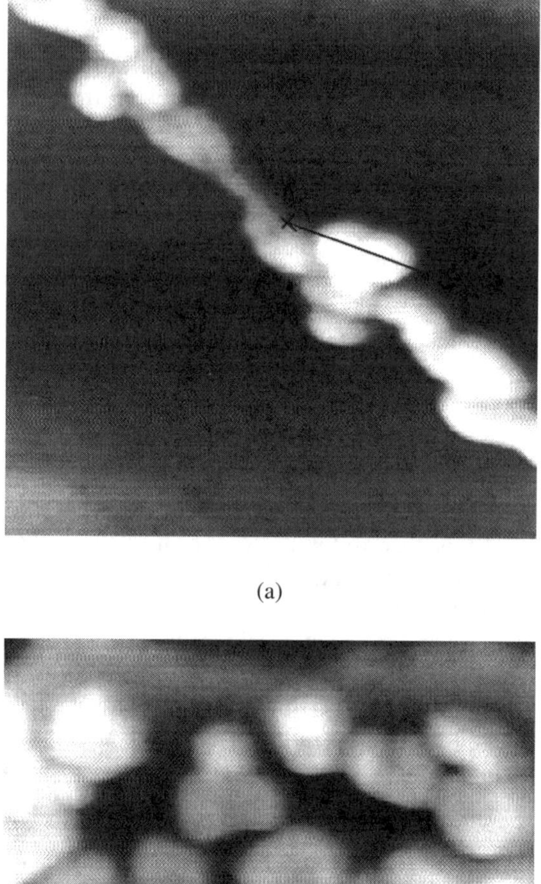

(a)

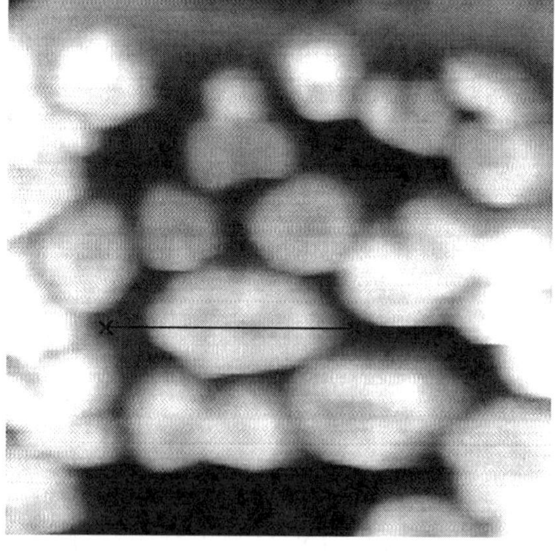

(b)

FIGURE 4.20 AFM images of polyglycine spread on HOPG from different solvents: (a) water (2000 nm × 2000 nm); (b) DMSO (3000 nm × 3000 nm).

The effects of subphase composition and temperature on α-helical PBLG were investigated.[20] The effect on α-helix was studied by adding dioxane-dichloroacetic acid (DCA) in subphase. The data were analyzed by the virial equation of state.

The effects of dimethyl formamide (DMF) (a helix-inducing solvent), $CHCl_2COOH$ and CF_3 COOH (helix-breaking solvents), and urea (a protein denaturant) on the conformation of poly(β-benzyl-l-aspartate) (PBLA) in monolayer, at air–water, was investigated by using IR spectroscopy of the collapsed films.[20] No direct effects of DMF, DCA, or trifluoroacetic acid (TFA) were observed, while urea had some effect on the structural changes of PBLA. The PBLA molecules were probably in the right-handed α-helical conformation, irrespective of the subphase composition, while in the solid state, they were in left-handed β-structure.

The AFM image of PBLG + cholesterol as LB film on HOPG is given in Figure 4.21. This image clearly shows that PBLG is present as helical, and it also shows no mixing with cholesterol LB-film phase. A more detailed analysis of PBLG LB-films was reported in the literature, and images were compared with molecular models (Figure 4.21).[264] The α-helical structure in poly-benzyl-aspartate is clearly seen in Figure 4.21c.

It is currently believed that the structure of water at the biosurface is significantly influenced by constituent macromolecules.[461] In other words, water is considered to affect the properties and structure of biopolymers at the interfaces. Recent evidence suggests that the water adjacent to interfaces differs significantly from that in the bulk phase.[20]

The mechanisms of the collapse process of bidimensional mixtures of poly-γ-methyl-L-glutamate (PMLG) (in α conformation) and cholesterol were investigated.[20] It was shown that the collapse mechanism can be ascribed to a process of nucleation. These monolayers were also subjected to microscopic investigations. The AFM analysis of mixed PBLG and cholesterol showed some evidence for molecular interaction.[3] It can also be seen that PBLG is helical in the monolayer. This is an important observation, because it confirms the analysis as delineated in earlier literature.

In Figure 4.22a,b, the results of poly(γ-benzyl-glutamate) by STM were given.[264] These images revealed details of the molecular structures with the α-helical conformation as expected. Images by STM were compared with macromodel, and these agreed with the images of STM.

4.9 BIOPOLYMER SAM STRUCTURES AT INTERFACES BY STM AND AFM

The biological activity of proteins and enzymes is strongly dependent on the molecular configuration, i.e., the native structure. Such macromolecular structures as α-*helical,* β-*sheet* and *random-coil* are well-defined three-dimensional structures. Although electron microscopy has been useful in determining such structures, the advent of STM and AFM has added much information in recent years.[462] In fact, the first picture of DNA by STM was a highly acclaimed result. The conformation of polymers at liquid interfaces was investigated after removal on solids. However, adsorbed biopolymers on solids under liquids have provided much useful information. However, AFM has some limitations in regard to the biological systems considered.[463] The development of specimen preparatory techniques and instrumentation will eventually reveal high-resolution surface structures. The greater challenge of

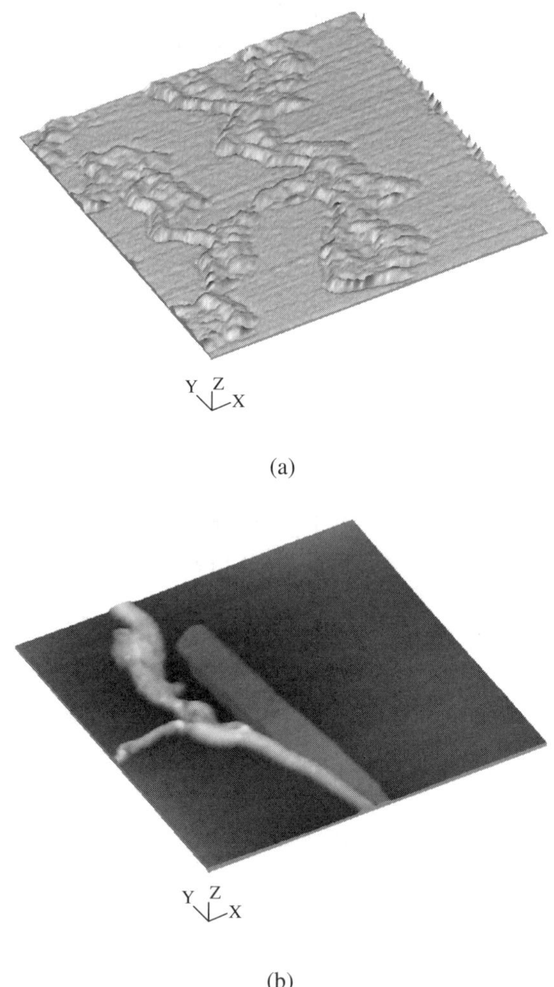

FIGURE 4.21 AFM image of (a) PBLG (30,000 Å × 30,000 Å); (b) mixed PBLG + cholesterol LB film on HOPG (60,000 Å × 60,000 Å); (c) poly-benzyl-aspartate (45,000 Å × 45,000 Å × 711 Å)

AFM is for it to show images where conformational changes in macromolecules can be studied *in situ*. An important step toward realizing developments in biotechnology and nanotechnology is gaining information on the spatial distribution of proteins at interfaces at the nanoscale.

Adenine and thymine molecules, when placed on graphite, were investigated by STM, and one could easily distinguish[243] the two molecules (thymine contains a single hexagonal ring, and an adenine molecule has a double-ring structure). These efforts are for using STM and AFM for the possibility of obtaining the sequence of DNA. This has been suggested to be possible by fixing DNA molecules on a substrate and distinguishing individual bases.[16]

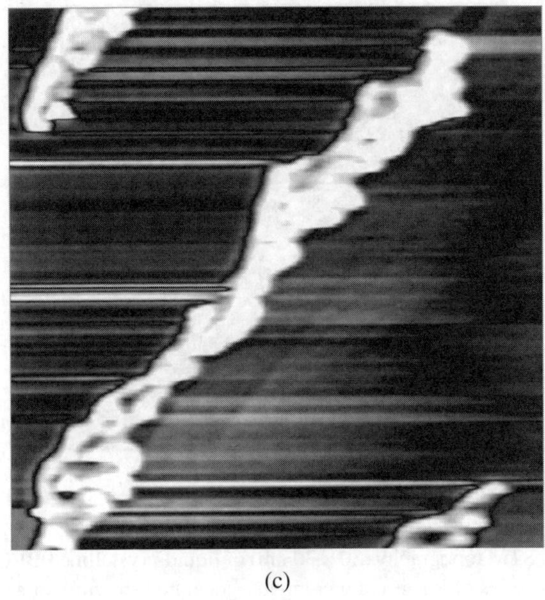

(c)

FIGURE 4.21 *Continued.*

Interactions between polyelectrolytes and globular proteins can be found in any biological system. The classical example is the synthesis of proteins in cellular systems, where every major step in the synthesis process (DNA transcription, RNA translation, and protein folding) is controlled by electrostatic and hydrophobic interactions between the macromolecular constituents.

Furthermore, in all biological systems, the need for three-dimensional protein structures becomes urgent, as is the present state with the bacterial proteins (cholera toxin), which are known to produce the disease cholera. These investigations have been going on since 1977 using protein crystals[260] and x-ray diffraction methods. This knowledge about the three-dimensional protein structure is essential in order to produce new kinds of vaccines.

The AFM has become a well-established and valuable tool in the three-dimensional topographical analysis of biopolymers.[264] The three-dimensional structure of most of the biopolymers is within a range of 4 to 10 nm (40 to 100 Å) in lateral dimensions by 1 to 5 nm (10 to 50 Å) in height. Unfortunately, in some cases, in this size range, the finite size of the probe tip leads to relatively large systematic enlargements. Therefore, this observation needs more extended investigation.

Antibody–membrane interactions have been studied by AFM.[464] This is a particular system where information relating to the nature of the attachment points and the bond strengths can be estimated.

The shape of the AFM tip is intimately related to this lateral enlargement phenomenon on both, because of simple, hard-surface, geometric considerations and because sources of interactive forces are distributed over the tip surfaces as well as

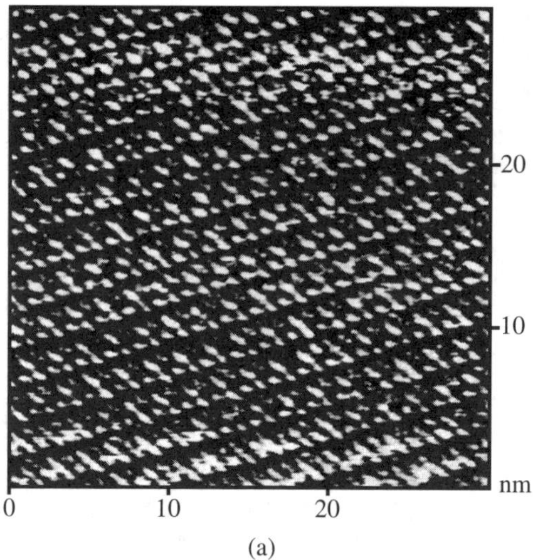

FIGURE 4.22 (a) STM topography (30 × 30 nm) of liquid-crystalline PBLG on HOPG (1268 mV/0.2 nA). (b) A series of images of a computer-generated structure of a 40-residue section of PBLG (macromodel). (From Breen, J. J. and Flynn, G. W., *J. Phys. Chem.*, 96, 6825, 1992. With permission.)

the sample surfaces. Furthermore, some effects of the asymmetry of the tip on images[465] have been reported.

The antibody–membrane interactions have been studied by STM/AFM.[57]

This is a particular system where such information as the nature of the attachment points and the bond strengths can be determined.

Mixed monolayers of hemoglobin, ovalbumin, xanthan, and virus with Mg-stearate collapsed films were studied.[3,112,113] as LB films on graphite. This provides a means of investigating biopolymers as found in their biological environment[100] in the cell lipid–bilayer medium. Height analysis additionally provides calibration controls.

Hemoglobin protein molecule (a tetramer with molecular weight of 68,000) and ovalbumin (molecular weight of 40,000) were found to remain as small clusters on the collapsed lipid film (Figure 4.22).[113] The presence of protein molecules is easily observed from the light-shaded spots. The step heights were found to be ca. 50 Å. This corresponds to the diameters of these polymers. The size of cluster varied, and under higher magnification, one could barely see the outline of each molecule. A more detailed image analysis is under preparation. However, preliminary data as shown here suggest that approximately 10 molecules are involved in each cluster. At this stage, the method is not sensitive enough to provide better resolution. However, the size of cluster is indicative of protein–protein interactions in the monolayer. There seems to be some kind of higher-order aggregation in these two-dimensional structures.

Casein (as found in milk) *molecules* are known to form aggregates. The AFM image of casein[3,113,466] placed on HOPG is given in Figure 4.23. The image shows

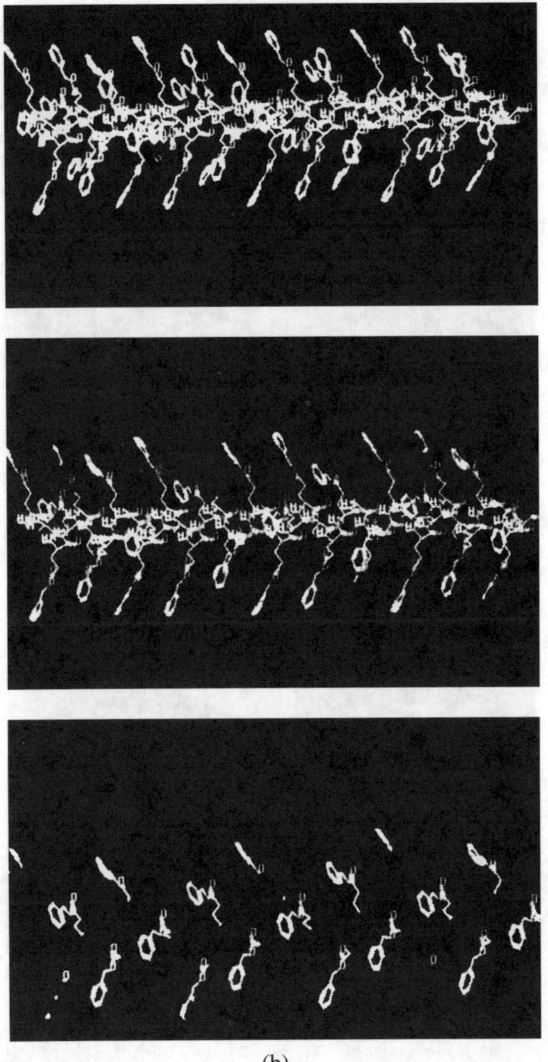

(b)

FIGURE 4.22 *Continued.*

highly monodisperse aggregate sizes of the micelles, which shows the capability of AFM application to such aggregate equilibria studies. The regular size and shape as found in AFM images are in agreement with other literature data (such as from light scattering).[20] The size analyses is as follows:

Diameter = 1980 Å
Maximum height = 200 Å
Step height = 60 Å (= diameter of casein)

FIGURE 4.23 AFM image of casein placed on HOPG. Scan sizes: (a) 30,000 Å × 30,000 Å; (b) 10,000 Å × 10,000 Å. Step height = 60 Å.[3]

The data needs further analyses at higher resolution. However, the steps as observed in these images indicate that casein aggregates (micelles) are present under these conditions. The step height is almost equal to the known diameter of the casein molecule (= ca. 50 Å). This means that the aggregate consists of ca. 900 casein molecules. These images also show that the tip has apparently no effect on the resolution. This is noticed from the sharpness of the edges in the images. A more detailed analysis will be given later. The magnitude of maximum height corresponds to three molecules. In other words, under high resolution, one should be able to image molecular details. This subject is being pursued by many investigators and is mentioned wherever success has been reached.

These observations clearly show that the molecular dimensions as found for casein confirm the valid dimensions by AFM. This conclusion has been contested in the literature, and therefore, needs further analysis before it can be resolved satisfactorily.

4.9.1 Determination of the Surface Potential of Crystals of Biopolymers by AFM

The electrostatic surface forces near the charged protein molecules are not easily investigated, because no direct experimental method is available.[20] However, AFM technique can be used for such investigations, as shown for a typical protein. For example, studies have been made of bacteriorhodopsin (bR), a purple membrane protein, which functions as a light-driven proton pump in Halobacterium halobium.[467] The protein has seven major α-helical components with a retinal chromophore in a lipid membrane. The protein can be crystallized as large, uniform, two-dimensional crystals up to 10 m in diameter.[468] These studies have described the major parts of its three-dimensional structure with a resolution of 0.35 nm using electron crystallography.[470]

A more precise resolution has recently been obtained, and the structure was analyzed with a precision of 0.30 nm.[471] The positions of the major surface amino acid residues responsible for proton transduction were identified. While knowledge of the bR crystal structure is quite advanced, measurements of the bR surface properties are not satisfactory and often contradictory. It is well known that it is the double layer at the protein–water interface that mediates processes such as ion adsorption and proton transfer at the surface.[20,472] The protein surface charge density and surface potential are therefore important characteristics of the protein.[20]

The surface potential as a function of pH also provides insight into the identity of the chemical functional groups present on the surface of the protein. Until now, there were no techniques available for measuring the surface potential of proteins directly, apart from electrophoresis. It was reported that the silicon nitride tip of the AFM is repelled by the surface of the purple membranes.[473]

However, it was not possible to quantify this interaction, and it could only be concluded that the surface possessed a negative charge. In a recent study, the use of a silica probe to measure quantitatively the surface potential of bR crystals under a range of solution conditions was described. The potentials of these silica spheres have been measured in literature under a wide range of pH and electrolyte conditions.[473]

By replacing the complicated pyramidal tip with a sphere of known electric potential, it becomes possible to analyze force data and extract the surface potential of an unknown surface. While this technique has been applied to alumina surfaces and polymer-coated mica, to date, there have been no attempts to use such a calibrated probe to measure the properties of more complicated surfaces, such as biological materials.[474] It is well known that measurement of the surface forces relies on three prerequisites. First, large two-dimensional crystals are required to ensure that the Derjaguin approximation can be applied to the interaction measurements.[3]

Second, the crystal must be attached nondestructively to a suitable substrate without ripping or wrinkling the two-dimensional crystal surface. Finally, the two-dimensional crystal must be located by nondestructive AFM imaging in such a way that a calibrated probe sphere can be brought into reproducible extension–retraction cycles with the protein surface. It is, however, exceedingly difficult to image with a conventional tip and then to move a second silica probe tip into the exact location of the surface feature to be analyzed. The alternative used in these studies was to image the surface and locate protein islands with a conventional cantilever using tapping mode to minimize sample damage. A sphere mounted on the neighboring cantilever is then brought into close proximity to the bR crystal, and the crystal is then reimaged with the silica sphere. By using this procedure, it was possible to obtain good quality submicron scale images with spherical tips, without contamination of the tip and without damaging the surface of interest prior to collection of force–separation data. The interaction forces between a silica sphere and two-dimensional bacteriorhodopsin crystals (bR) were measured in $NaNO_3$ solutions at different pH values and at two electrolyte concentrations.[474]

Using the known potential for the silica sphere as a function of pH, the surface potential of the bR crystals could be determined. The isoelectric point (IEP) was found to be pH 4.3. The crystals had a uniform thickness on substrates of 5.2×0.5 nm, and the Hamaker constant[3] for the bR-H_2O-SiO_2 system was estimated to be 0.5 to 1.0×10^{-20} J at pH values near the IEP. Analysis of the force curves suggested that the surface charge on the protein surface remained constant during approach of the silica colloid probe. In a recent study, the bacteriorhodopsin crystals were prepared according to procedures from the literature.[471]

The bR crystals (3 mg/mL), ranging in size from 0.5 to 5.0 m in diameter, were stored at room temperature in a solution of 0.1 M potassium phosphate buffer at pH 5.1 containing 6 mM octylglucoside, 0.2 mM dodecyltrimethylammonium bromide, and traces of NaN_3. The bR crystals were stable and could be kept without denaturing for several years in this solution. Prior to use, the mother liquor was diluted to 3 g/mL with water, and the crystals were dialyzed for 10 days with a membrane. In ordinary water, the crystals were negatively charged due to ionization of surface carboxyl groups. To ensure adequate adhesion to the negatively charged mica surface used as a substrate, the following procedure was used to reverse the charge on the mica surface. An amount of 50 L cm^{-2} of 0.1 mM quaternized poly(2-vinyl pyridine) (MW 30,000, pH 5.6) was placed onto freshly cleaved mica using a micropipet, and was allowed to stand for 2 h in the presence of saturated water vapor. Then, the mica plate was rinsed in a stream of distilled water to remove any weakly adsorbed polymer. Immediately, 50 L cm^{-2} of the dialyzed (pH 5.6) bR solution was added by pipette to the modified mica surface and allowed to equilibrate. The adsorption of bR onto cationically modified mica was found to be slow and difficult to quantify. Optimal adsorption was obtained when the bR solution was washed from the mica with a stream of water after 5 to 15 min adsorption. The mica was then used directly for imaging and force curve acquisition in the fluid cell without drying. Under these preparation conditions, the bR crystals were consistently found to be in the form of isolated, scattered islands with a surface coverage of 2 to 5%.

The sample and probe alignments were carried out as follows. The silica spheres used as probes were of diameters of 4.8 μm. The silica sphere was attached to AFM cantilevers using a 24 h curing epoxy resin. The spring constants for the cantilevers were measured by the method described in literature, with the necessary corrections.[475] A value of 0.2 N m^{-1} was used in all calculations. AFM in multimode was used for the measurements. In a typical experiment, one of the two cantilevers on the end of each substrate was modified by attachment of the colloid probe. The unmodified cantilever was used to image the bR bearing mica substrate in the fluid cell using tapping mode with extremely small drive amplitudes and surface damping. A 400 mesh TEM grid was sandwiched between the magnetic mounting disk and the transparent mica surface so that when suitable crystals were found during imaging, the position of the colloid probe cantilever could be located to within 5 m of the two-dimensional crystal using the mesh position as a marker. The crystal was then re-imaged with the silica sphere. The best images were obtained when the tip was allowed to engage the surface with no x–y rastering (i.e., zero scan size) and was then immediately withdrawn from the surface by increasing the tapping mode set point signal. Following this step, the x–y rastering was then switched on and the set point relocated until an image of the surface appeared. This procedure minimized the force exerted on the surface by the tip. Fresh electrolyte was flushed through the cell as required, and the surfaces were equilibrated for 5 to 10 min at each new pH prior to force measurements.

The data manipulations were carried out as follows. The resultant force data were converted from deflection voltage versus piezo travel to force versus separation.[476,477] The resultant force curves were normalized using the sphere radius, which was determined by optical microscopy and SEM to within 2%. The combined error in the spring constant and sphere radius was estimated to be 5 to 10%. The analysis program allowed computation of the theoretical force–separation curve for a variety of symmetric and asymmetric electrostatic interactions. In principle, interactions between dissimilar surfaces may obey one of four boundary conditions, depending on whether the interaction occurs under conditions of constant surface charge or potential on each surface. However, at low electrolyte concentrations, the force profiles at separations greater than one Debye length are similar, regardless of the boundary conditions assumed, and the fitted potentials are based on the repulsion at separations of one to two Debye lengths.[3] The bR crystals used in this study were previously characterized by electron microscopy and electron diffraction. The image was computationally reconstructed with a resolution of 0.3 nm from images taken by cryogenic electron microscopy at liquid He temperatures.[471]

The lattice spacing was estimated to be 6.24 nm. It was concluded that the quality of the crystals was ideal, and could extend up to several micrometers in scale, with only minor dislocations and lattice defects. This is essential, because the electrostatic interaction with the silica probe is an average of the surface ionization properties of the crystals over a length scale comparable to the silica probe radius.

Computer-reconstructed micrographs of a single bR two-dimensional crystal were explained in relation to AFM images. Images were obtained of a bR crystal adsorbed on cationic mica surfaces. The first image was taken in tapping mode with a standard silicon nitride cantilever with $k = 0.2\ Nm^{-1}$. The crystal was found to be remarkably smooth, and the bR crystal was 5.2 nm high and >3 μm in diameter. It was not possible to resolve individual protein units in tapping or contact modes in electrolyte solutions.[478,479]

The best tapping-mode images were consistently obtained with weaker cantilevers. The image of the same crystal obtained with a silica sphere attached to the AFM cantilever gave almost the same result. The image quality was, however, not as sharp, but it was concluded that one can use a calibrated probe sphere over small surface areas. In some cases, it was also observed that the bR crystals were folded on the mica surface.

The images of a bR crystal on a cationic mica surface were obtained in 0.1 mM $NaNO_3$ using a silicon nitride cantilever $k = 0.2\ N\ m^{-1}$ (in tapping mode). Images of the derivatized mica surface revealed small strands of polymer, with an average thickness in water of 0.5 nm. The apparent surface coverage of the strands was <5%. The same crystal was also analyzed with a 2.42 m silica sphere attached to the cantilever. The images could easily distinguish between the monolayer bR crystals with sphere or pyramidal tips.

It is known that bR crystal surfaces are structurally asymmetric. One of the membrane surfaces contacts the extracellular solution; the other contacts the cytoplasm. The former surface has a cracked morphology, and the latter evinces a more pitted topography. Both of these surfaces were investigated previously using SEM, and the effects of surface charge on adsorption were determined.[480] For positively charged surfaces, the cytoplasmic side was consistently found to be exposed to solution after adsorption. This distinction is important, because both surfaces have different amino acid sequences; consequently, their surface charge densities are not the same, and their isoelectric points will differ. One of the advantages of using the cationic mica as the sample substrate was that one can readily distinguish whether the colloid probe is over the bR crystals prior to an experiment, even in the absence of a suitable image. At pH 5.6, the silica probe is strongly negatively charged. The bR crystal is also weakly negative, but the PVP-coated mica is weakly positively charged.

This characteristic allows one to determine that the sphere is correctly located. It was found that the force curves obtained with a silica probe were located directly over a bR crystal and from the polymer-coated mica surface 5 μm distant. At pH 6, the negatively charged silica sphere probe was repelled by the protein crystal, but it was electrostatically attracted to the surrounding cationic mica substrate.

Different force curves data were measured using the procedure where AFM cantilever toward the surface at various pH values in 0.1 mM $NaNO_3$. In general, the best fits were obtained when the silica sphere and the bR surface were assumed to interact at constant surface charge. It was reported that silica–silica interactions measured with the same spheres used here were best fit with constant surface charge boundary conditions at pH > 6, whereas at low potentials, a constant surface potential gave better agreement.[474] At pH values above 7, there was usually no apparent jump in, but at lower potentials, a jump-in was observed. For the data with bR surfaces, the region of constant compliance showed no evidence of softness at the low surface loads applied. In general, attractive curves were obtained between pH 3.5 and 4.3. Above pH 6.8, the jump into contact disappeared. The data were analyzed by assuming a Hamaker constant of 1×10^{-20} J and a Debye length of 29.2 nm. The surface charge density on each surface was fixed during the approach of the surfaces.

The potentials extracted via the fitting procedure were denoted "surface" or "diffuse layer" potentials, because potentials refer to values obtained from electrophoresis of proteins or particles under the application of an electric field. However, it was demonstrated that the electrophoretically derived values and those found by AFM force curve analysis were numerically equal for inorganic metal oxides.[473] The diffuse layer potentials for the silica sphere, obtained from silica–silica force data reported elsewhere, and the deduced values of the bR surface potential from fits using the dissimilar surfaces double-layer program were plotted as a function of the solution pH.[474] These studies showed that at high silica surface potential, the interaction force was almost independent of silica potential but quite sensitive to the chosen value for the bR surface. At lower pH values, the surface potentials of both surfaces decrease, and the relative error increases.

The magnitude of bR IEP is about pH 4 to 4.5, where the surface potential of the silica is only –30 mV. A more precise value is difficult to ascertain without a more accurate value for the Hamaker constant. The best fit to the data over all pH values leads to a composite Hamaker constant between 5×10^{-21} and 1×10^{-20} J. Upon increasing the ionic strength to 1 mM $NaNO_3$, there was a decrease in the Debye length to 9.8 nm, but no significant change to the short-range repulsion. The surface potentials found for the bR crystals were generally low and negative. These data were consistent with the presence of a low density of acid carboxyl groups on the cytoplasmic surface of bR, and the result agreed with the structural analysis.[471] In these studies, a small excess of negative aspartic acid residues on the cytoplasmic surface was found. These residues were associated with lateral proton transduction to the entrance of the bR channel. The value of pH 4 for the IEP may have been an indication that the major functional moieties bestowing charge on the cytoplasmic surface of the bR crystals were the carboxyl groups associated with the aspartic acid residues.

In earlier data, a value of pH 4.5 to 4.7 in 100 mM NaCl was reported.[481] The addition of Ca^{2+} did not shift the PZC.[481,482] In 0.1 M NaCl at pH 7.5, a surface potential of –23 mV was obtained.[483] If the surface charge was assumed to remain constant at lower ionic strength, the Gouy–Chapman equation would yield a value at 0.1 mM of –189 mV, which is too high to be realistic. A novel technique was reported as based on shifts to the Raman band from bR crystals, from which a value of –30 to –35 mV was deduced in water at pH 6.[484] However, this method involved the addition of 0.06 mM cationic dye to the bR solution, which may have altered the surface charge on the crystal. More recently,[485] a titration method was used from which a value of –60 mV at zero salt concentration was calculated; the pH was not clearly stated but was apparently close to pH 6. The accumulation and interpretation of the AFM data continue, but bR crystals were found to be sensitive to light, and the scattered laser light from the AFM may cause a photopotential to develop on the particle surface. It has been shown that electrophoretic mobility is altered by various light adaptation procedures.[486] The literature values suggest low negative surface potentials as found in this work, but the solution conditions employed in the various studies make a more quantitative comparison difficult. It is thus clear that even at low salt, and in alkaline solution, where ionization of the carboxyl groups should be quantitative, the observed potentials do not exceed –100 mV.

These analyses showed that there are 9 bR molecules per (hexagonal) unit cell. On the basis of the computer reconstructed charge distributions at the cytoplasmic surface,[471] it was estimated that there are four effective negative charges at the surfaces of each bR molecule. This corresponds to a maximum surface charge density of 36 e/79.8 nm^2 or –7.23 C cm^{-2}. The measured potential of –75 mV corresponds to a surface charge density of –0.24 C cm^{-2}. This low net charge density may be related to strong counterion binding or by assuming the presence of a similar surface density of cationic amine residues near the cytoplasmic surface. However, the EM analysis appears to rule out a sufficient density of such amines on the surface. Conversely, counterion binding is often invoked to explain low potentials on carboxylate latex particles in alkaline solution. More recently, AFM measurements of double-layer forces between mercaptoundecanoic acid monolayers on gold were rationalized in terms of strong sodium ion binding.[487] It is, therefore, useful to consider whether similar binding constants can explain the low surface potentials found here. The surface concentrations of protons and sodium ions are linked to their respective bulk concentrations. Data was analyzed that gave rise to two explicit relations between the surface charge density and the measured diffuse layer potential in the AFM for any value of the solution pH. The solution can be readily found with a root-finder program.

As described in the literature, sodium ions and protons may equally be considered the potential determining ion.[487] Regardless of how high the pH value, the background electrolyte will limit the maximum surface dissociation, and consequently the magnitude of the surface charge density. (The equilibrium constant, K is related to the expression: $[H^+]_{surf}/Ka[Na^+]_{surf}/KNa$.) The maximum surface potential at pH 9 will be determined by the magnitude of pK(Na), while the onset of ionization at pH ~4 is primarily determined by the value used for pKa. Because the environment at the protein surface may be hydrophobic, depending on the nature of the amino acid residues around each aspartic acid site, the pKa of the carboxyl group will probably be substantially higher than the typical values of alkane carboxylic acid molecules in bulk solution, which lie around 4.8. In fact, each of the nine acid residues may have its own unique pKa value.

These analyses gave reasonably good agreement with literature data.[3] The values of acid and sodium ion dissociation are not much different from those found for the self-assembled monolayers on gold, and the pKa is clearly much higher than that found for free carboxylic acid molecules in aqueous solution. The proton binding constant is about 1000 times higher than that for sodium ion, but in alkaline solution, sodium binding will dominate. The various fitted curves indicate that the pK values can be estimated to within about 0.5 by the fitting procedure.

The adhesion values obtained with the AFM were not reproducible enough to yield quantitative data. However, there was a consistent decrease in adhesion strength as the pH was increased from the IEP to pH 9, as would be expected. The pull-offs were clean, and there was no indication of multiple contacts between sphere and surface during extension or retraction of the cantilever. In the pH range 2 to 4, the surface potentials on the bR and silica surfaces obtained were low. Under these conditions, attractive interactions were observed. In principle, if the surface potential is low enough, the attractive interaction between the surfaces can be attributed to van der Waals interactions between the surfaces. For a sphere–plate interaction, the force at small separations (<10 nm) is given by: $[(A_{123}R)/(6D)]$ where A_{123} is the Hamaker constant for the asymmetric silica–water–bR interaction, D the surface separation, and R the sphere radius. Reasonably good fits to a nonretarded van der Waals equation were hard to obtain, but the composite Hamaker constant was consistently found to lie in the range $A_{123} \sim (0.5\text{–}1.0) \times 10^{-20}$ J. This value was consistent with the jump-in distances obtained. The value was likewise consistent with values obtained in SFA measurements of other proteins attached to mica and with the geometric equations. The values, as reported in literature, which are around two orders of magnitude less at 2.5×10^{-22} J appear to be too small.[487] They were obtained indirectly from light scattering measurements. Values of order 10^{-21} to 10^{-20} J must be expected from the density of proteins, and in order to explain their refractive indices, which are usually on the order of 1.45 to 1.5.[487] The investigations thus showed that by using colloid probes to measure the surface potential of proteins, other complex surfaces can be measured. If the colloid probe is small and smooth, it can be used to image and locate the surface of interest. IEP was determined for bR of 4.3 and a composite Hamaker constant of $(0.5\text{–}5) \times 10^{-20}$ J. Furthermore, a maximum potential of about –80 mV is attained at pH values between 7 and 9, in 0.1 mM $NaNO_3$.

4.9.2 Molecular Recognition of Biomolecules by AFM

In many complicated biological systems, the phenomenon of molecular recognition is the most important aspect of these reactions. In AFM, nanoforce is used to monitor the surfaces. The range of forces that measured between the molecules varies over at least six orders of magnitude. In a recent study, a new approach was used to study the molecular recognition of biomolecules by using AFM.[488] In this report, direct assessment of the mechanical limit beyond which biomolecules are rendered inactive was investigated. The methods used were AFM, optical microscope, and colorimetric techniques. The system studied was streptavidin and DNA. In simple terms, the procedure used was as follows:

1. A monolayer of biomolecules was covalently immobilized on a glass substrate.
2. A load was applied to the surface with the AFM tip while imaging by the microscope.
3. Incubation with the HRP (horseradish peroxidase) molecular recognition conjugate and development resulted in staining of the biologically active areas.

This study is relevant when considering that with AFM imaging, there may be some effects on the biological molecules due to the applied stresses. It was found there was no effect when the applied force was less than 1.5 nN. The activity of streptavidin was lost at pressures larger than 1.5 nN.

The analysis of the stress present under a laterally translating tip is a complicated four-dimensional vector field. The data were analyzed by a simple model. It was assumed that at some applied threshold pressure, p_{th}, the stress experienced by the molecules will produce a strain necessary to inactivate the molecule. The corresponding threshold radius, r_{th}, can be given as:[488]

$$r_{th} = ((3R_{probe}F_{load})/(4E^*))^{1/3}(1 - p_{th}((\pi^3 R_{probe}2)/(6F_{load}E^{*2}))^{2/3})^{1/2} \quad (4.8)$$

where E^* is the reduced elastic modulus, F_{load} is the net applied load, and R_{probe} is the probe radius. It can also be accepted that under imaging, the AFM probe apex will be rastered with respect to the sample surface in the x–y plane by two orthogonal triangular functions. These studies showed the possibility of assessing the effect of nano-Newton loads on molecular activity in biology.

In a recent study,[489] chelator lipids were studied by AFM. This study relates to a novel polymerizable lipid containing a chelator head group for the reversible binding of proteins. The polymerizable diacetylene lipids were spread as monolayers on buffer solutions. AFM images as a function of time for systems with DNA complexes were studied.

4.9.3 Applications of Transverse Dynamic Force Microscopy (TDFM) and AFM to Membranes Microscopy

In a recent report, the origin of phase contrast in TM AFM was investigated using two complementary SPM techniques, AFM and shear force microscopy, also called transverse dynamic force microscopy (TDFM).[490] The sample chosen for this study was Nafion, and specifically, the membrane in different hydration states by virtue of its cation form. Differences in probe–sample adhesion throughout a sample, caused by an inhomogeneous distribution of surface water, were an important phase-contrast mechanism. A new variant in three-dimensional force imaging, phase-volume imaging has been a useful tool in the interpretation of phase contrast. With the use of TDFM, approach curves were obtained, while the frequency spectrum around resonance was measured. This enabled the damping of the probe oscillation amplitude and the shift in its resonant frequency to be decoupled. Knowing the true oscillation amplitude of the probe, it was also possible to quantitatively determine the elastic and dissipative parts of the probe–sample interaction. Distinct regimes were found at different probe–sample separations.

AFM is known to provide varying degrees of image resolution depending on the procedures used as delineated above. TM AFM is more suitable than contact mode for imaging delicate samples because of the lower lateral forces. It has been applied to many polymer systems.[490–494] It also has the added advantage of being able to obtain phase images and topographical data. Tapping-mode phase imaging is a relatively new AFM technique. It can differentiate between areas with different properties, regardless of their topographical nature.[495–497] The phase angle is defined as the phase lag of the cantilever oscillation relative to the signal sent to the piezo driving the cantilever.

TDFM is a dynamic probe microscopy in which the detected force is perpendicular to the probe, hence "transverse." The first use of this technique was with the *shear force microscope* (ShFM) as a distance control mechanism in SNOM.[498,499] In the past few years, this particular experimental setup was applied in the study of different samples in which the term "shear force" was not appropriate. (It recalled the idea of shear between surfaces that is not true in all cases.) For this reason, a more general description was required. In TDFM, the cantilever is oriented perpendicularly to the sample and oscillates parallel to its surface. The interaction between the tip and the sample can be measured at different distances of separation by observing the change in amplitude and the relative phase of the cantilever oscillation. The shear force is often used in TDFM to obtain topographic images of the surface. The oscillation amplitude of the probe decreases monotonically when approaching the surface; by using the amplitude signal in a feedback loop, it is possible to scan the surface at constant height. If the system is monitoring amplitude and phase at the same time, it is also possible to record phase information while keeping the amplitude constant.

TDFM was also used for force spectroscopy. In this case, the probe was held over one point of the sample surface, and its amplitude and phase were recorded in a series of approach and retract cycles. The experimental quantity that characterizes the cantilever and its interaction with the specimen is the frequency spectrum across the resonance peak. An original technique that records this information at different tip–sample distances (real-time frequency spectra) was described in a recent study.[490]

The evaluation of the forces from these measurements was made by using the dynamics of the vibrating probe that has to be modeled and assumptions made on the actual interaction force. In these analyses, the force is considered as a combination of dissipative and elastic restoring components. This assumption is based on the experimental evidence of a viscoelastic interaction between the probe and the specimen in normal humidity conditions.[500,501] Nafion is a commercially available perfluorosulfonate cation-exchange membrane. It is generally used as a perm-selective separator in chlor-alkali electrolyzers and as the electrolyte in solid polymer fuel cells.[502,503] Perfluorosulfonate cation-exchange membranes are used in these applications because of their high ionic conductivity and their high mechanical, thermal, and chemical stabilities. The industrial applications of Nafion have prompted considerable research.[504,505] Structurally, Nafion consists of a hydrophobic tetrafluoroethylene (TFE) backbone with pendant side chains of perfluorinated vinyl ethers terminated by hydrophilic ion-exchange groups. The difference in probe–specimen adhesion between the hydrophobic backbone and hydrophilic side-group regions of the polymer allows the spatial distribution of these two regions to be observed using tapping-mode phase imaging.[494]

The two complementary SPM techniques of AFM and TDFM have been used to investigate the difference in phase contrast exhibited by two Nafion samples differing only in cation form (H+ and Cs+). For each ion form of Nafion, the same probes were used under identical imaging conditions for AFM and TDFM. Initially, standard AFM techniques were applied to Nafion, and a test sample was obtained before investigating the membrane further with many more novel techniques, such as phase–volume imaging and the collection of a real-time frequency spectrum.

To understand fully the reasoning behind the experiments performed, it is essential that the nature of the probe–sample interaction for these two complementary techniques be understood. In standard TM SFM, the tip intermittently contacts the surface, resulting in a minimization of the destructive lateral forces. This allows the study of soft surfaces and weakly adsorbed molecules on a substrate. Alternatively, the measurement of the phase lag of the cantilever oscillation with respect to the excitation force contains information about the interaction between the tip and the sample, allowing compositional contrast on heterogeneous surfaces.[497] The origin and nature of the phase contrast have been subjects of debate and discussion for the past few years.[507–523] During each oscillation cycle beginning with the tip furthest from the specimen surface, the tip feels a negligible force, then a long-range attractive interaction, and finally a repulsive force as it approaches and contacts the sample. Despite the complexity of the interaction and its effect on the cantilever dynamics, theoretical simulations and experiments of the cantilever dynamics in air have shown that phase contrast arises from differences in the energy dissipation between the tip and the sample.[524] These considerations were ascribed to the surprisingly simple harmonic cantilever response. In fact, calculations and experiments show a sinusoidal movement of the cantilever for the usual cantilever parameters in air, that is, spring constant and quality factor on the order of 10 N/m and 100, respectively. This allows the phase shift to be related analytically to the energy dissipated in the tip–sample interaction.[524]

It is known that when the cantilever is far enough from the sample, the tip oscillates freely, and the phase shift is 90. As the cantilever oscillates in the proximity of the sample, the oscillation is damped as a consequence of the interaction between the tip and the sample, and a linear decrease of the damped amplitude is produced as the probe approaches the specimen.

As the cantilever approaches the sample, the phase shifts to 180. This corresponds to the noncontact regime in which an attractive interaction is responsible for the damping of the cantilever oscillation. This interaction shifts the cantilever resonance to lower frequencies, producing a phase shift lower than 90. In the other case associated with the intermittent-contact regime, the repulsive force produced during the tip–sample contact displaces the resonance to higher frequencies.

The effect of a tip–sample interaction, which involves energy dissipation, is the displacement of the noncontact solution to higher phase shifts and the intermittent-contact solution to lower phase shift values. More dissipative features will appear lighter in the noncontact regime, whereas they will appear darker in the intermittent-contact regime. An experimental curve is a combination of both solutions. As the cantilever approaches, the attractive force is responsible for the damping of the oscillation, and the cantilever oscillates in the noncontact regime. As the cantilever approaches further, the tip contacts the surface intermittently, and a sudden change in the phase shift is observed as a consequence of the transition of the cantilever oscillation from noncontact to the intermittent-contact regime.

The two possible mechanisms for a decrease in the oscillation amplitude of a probe as it is brought into contact with the sample include a damping or a change in resonance frequency. These two mechanisms are indistinguishable using conventional feedback methods. The experimental setup required to perform a real-time frequency spectrum enables the two components of the probe–sample interaction to be separated.

Images of different cation forms of Nafion were obtained using the same cantilever under identical imaging conditions. Images produced included the following: (a) A 1 μm tapping-mode AFM topography image of Nafion 115 H+ imaged under ambient conditions; (b) a phase image corresponding to part a (Z-scale, 25 nm and 10, respectively); (c) a 1 μm TM AFM topography image of Nafion 115 Cs+ imaged under ambient conditions; and (d) a phase image corresponding to part c (Z-scale, 25 nm and 60, respectively). The dynamics of an oscillating cylindrical probe can be modeled using the simple harmonic oscillation theory, but, in this case, the continuum mechanics model for a cylindrical bar was used.[525,526] It was found that the oscillation amplitude and resonance frequency of the probe change when it is interacting with the surface. The experimental data were described with the help of a model, where the interacting force was assumed to be an elastic and a dissipative component. The importance of a liquid film between the sample and the probe for the TDFM contrast mechanism was already reported. When the probe was close to the surface (~10 nm), the liquid film became confined and displayed solid-like behavior.[500,501] This gave rise to a viscoelastic shear force, dependent on the sample surface via the chemical bonding between the first liquid layer and the surface, and on the amount of surface water, which is humidity dependent.[502] A force with an elastic and a dissipative component can simulate a viscoelastic interaction. It is well known that molecules such as the membrane will rehydrate readily if exposed to an environment of high relative humidity (RH).[503]

Careless handling can result in the membrane ion-exchanging from the acid (H+) to a salt form (e.g., Na+ or K+). All Nafion samples were routinely prepared by refluxing with concentrated nitric acid and deionized water (50/50 v:v), then deionized water alone, to ensure that the membrane was in the H+ form and free from any chemical impurities. Strips of Nafion H+ were then converted to the Cs+ form by immersion in a 0.1 M $CsNO_3$ solution for a week.

In AFM of membranes, H+ and Cs+ samples were mounted on magnetic stainless steel sample and placed inside a multimode AFM. The samples were imaged using tapping-mode phase imaging and a standard silicon cantilever (~40 N/m). The samples were imaged using the same cantilever under identical imaging conditions. All studies were made under controlled relative humidity. The properties of the two ion forms of Nafion were also investigated by attenuated total reflection FTIR spectra. The effect of surface water on tip–sample adhesion was investigated by using a specimen surface prepared to have hydrophobic and hydrophilic domains. The test samples were cleaned properly (with detergent, rinsing with water, dipping in a mixture of chromic and sulfuric acid, rinsing with water, dipping in 5 M NaOH, rinsing with water, drying). After drying, the slide was placed on a small beaker of aminopropyltriethoxysilane, which was allowed to evaporate onto the slide for 5 min. The adhesive properties of the test sample were studied by obtaining multiple-force distance curves over the two regions, and force volume imaging allowed an image to be built up of the tip–sample interaction at each pixel from an array of force curves.[504] It was applied primarily to the study of elastic and adhesive properties of nonhomogeneous substrates.[505–508] Each pixel in the image allows one to analyze the phase contrast for any tip–sample separation, hence, the word volume.[509] This is reported to be a novel AFM technique. Phase–volume (PV) images of the different cation forms of Nafion (64×64 pixels with 64 data points per approach and retract curve), were obtained for analysis.

A Nafion 115 H+ sample was imaged under ambient conditions using phase–volume imaging at several different free amplitudes and ratios of set point to free amplitude. A scan size of 500×500 nm^2 coupled with an array of 64×64 pixels was used to ensure that features comparable with the cluster size could be detected. The images could then be analyzed by taking slices through the images at specific tip–sample distances. Phase images of Nafion 115 Cs+ were then obtained using the same cantilever under the same conditions with a variety of different free amplitudes and ratios of set points to free amplitude to determine the optimum conditions. These were then used to obtain the PV image.

In TDFM studies, the probe was mounted on a piezoelectric actuator that drove it at one of its resonant modes. The oscillation amplitude was detected using the laser reflection detection system (LRDS) that provided the true measurement of the vibration amplitude necessary to quantify the tip–sample interaction. The magnitude for the typical values for the oscillation amplitude were found to be around 10 nm.[511] An uncoated optical fiber probe was used rather than a metallic probe owing to the similarity of its surface characteristics to those of a silicon tapping-mode cantilever. The probe was prepared using the same method as that described for SNOM probes.[94,510]

To test the importance of the relative humidity in the TDFM contrast mechanism, the microscope was placed inside an in-house built environmental chamber. A Nafion H+ sample was mounted onto a 1.5-cm-diameter magnetic stainless steel sample stub and placed in the TDFM. The relative humidity could be reduced using nitrogen gas passed through molecular sieve material or increased by bubbling nitrogen gas through water and into the chamber. Once the desired humidity was reached, the sample was allowed to equilibrate. Images were obtained as a function of RH [high (~50%) and low (~10%)]. In addition to the images, amplitude–distance curves were obtained at each of the humidities. This process was then repeated for the Cs+ ion form of the membrane.

In TM AFM and TDFM, the feedback mechanism uses the probe oscillation amplitude signal to control the probe–sample separation. However, under conventional feedback procedures, it is not possible to differentiate between a damping of the amplitude and a shift in resonance frequency. It was reported that the liquid confined between the probe and the surface may be responsible for a decrease in the oscillation amplitude and a resonance frequency shift. These effects can be detected by measuring the frequency spectrum of the probe while it is approaching the surface. In this way, the two components of the interaction are separated and measured as a function of tip–sample distance. Using the TDFM, it is possible to perform a real-time frequency spectrum by simultaneously exciting two modes of the probe. The procedure used was as follows. The first frequency was kept constant at the first (or second) resonant peak and was used to monitor the oscillation amplitude during the approach and retract cycle. The second frequency was swept from just below to just above the second (or first) resonance frequency of the free probe, by using the sweep mode of a signal generator. This mode of operation will sweep the driving frequency up and down continuously, and the sweep time can be as low as 50 ms. Amplitude and phase spectra were monitored continuously as the probe approached the sample surface.

AFM was carried out by using the tapping mode, and topography and corresponding phase images of Nafion H+ were obtained under ambient conditions. The images of Nafion Cs+ were obtained using the same cantilever under identical imaging conditions. The phase range was significantly larger in the Cs+ ion, 60 as opposed to 10 for the H+ form. There was no significant difference in the topography images, therefore, topographic coupling is not responsible for the change in phase contrast. Attenuated total reflection (ATR) data for the two substances clearly showed differences due to the presence of greater water content of the H+ form. This showed that an electrostatic force could be responsible for the difference in phase contrast. An optical image of the mixed hydrophobic/hydrophilic test sample was obtained. Water droplets were clearly visible on the right-hand side of the interface between the silanized glass and cleaned glass, running diagonally from top right to bottom left. The interface was not particularly sharp because of the way in which the silane was evaporated onto the glass. Force curves were obtained (as described above) with respect to the left and right of the interface. The energy required for the cantilever to break free from the surface was considerably higher to the right of the interface, where the water droplets were clearly visible. The energy could be calculated from the area under the curve, if it is assumed that Hooke's law was obeyed and that the cantilever spring constant was ~0.12 N/m.

The pull-off energy was ~1.2 f J and ~7.6 f J for areas left and right of the interface, respectively. It was clear from these results that any surface water dramatically increases the adhesive force between the tip and sample. Although these features are over an order of magnitude larger than the proposed cluster size in Nafion, this contrast mechanism would still apply when imaging at higher resolutions. The elastic response of Nafion increased with frequency, at the frequencies at which the cantilever was being driven ~250 kHz, and so the viscoelastic energy loss would be negligible.[513,516]

Therefore, any energy dissipation in the tapping interaction is primarily the result of tip–sample adhesion rather than a viscoelastic energy loss. Differences in surface adhesion over a sample caused by an inhomogeneous distribution of surface water are probably an important phase-contrast mechanism. Normal tapping-mode topography and phase images of Nafion were obtained in a bistable regime. In a bistable regime, the cantilever can oscillate in noncontact and intermittent-contact regimes.[514–517]

From the mathematical point of view, the state of the oscillation depends on the initial conditions. In practice, the cantilever oscillates randomly in noncontact and intermittent contact during the scanning, producing a flipping of the phase shift between negative and positive values. The sudden transitions from noncontact to intermittent-contact regime are accompanied by artifacts in the form of depressions in the topography image. Line profiles across the images showed four transitions between the noncontact and intermittent-contact regimes. At each of these transitions, corresponding to the intermittent-contact regime, a depression of up to ~5 nm (25% of the Z-scale) and a phase shift of more than 90 was seen in the image. From this observation, one could conclude the importance of avoiding a bistable regime in phase imaging of a surface. The procedure used was to take 14 slices through a phase–volume image of Nafion Cs+ at intervals of 3.7 nm. The oscillation amplitude of the cantilever when free was in excess of 50 nm. These data showed that each of the slices had the phase contrast that would have been obtained in traditional tapping-mode phase imaging at different set points. At cantilever–sample distances slightly larger than the free amplitude (cantilever position 37 nm), the interaction may be too weak to produce any meaningful phase contrast. On the other hand, a significantly higher phase shift >60 was observed in certain regions for a cantilever position from 22 to 33 nm. In these regions, the cantilever oscillated in the intermittent-contact regime. Once below a tip–sample separation of about 15 nm, the images were predominantly in the intermittent-contact regime, and the phase contrast was reduced to about 20. However, a few points, which showed up as white squares in the dark regions, were still in the noncontact regime, even for a damped amplitude lower than 10%. If slices taken in the noncontact (cantilever position, 44.3 nm) and intermittent-contact regimes are compared, the same features are clearly visible in both images — there has been a contrast reversal. These data showed that the system conforms to energy dissipation in the noncontact regime. Further, topography and corresponding phase image and the three types of amplitude–distance curves, obtained during phase volume imaging of a Nafion Cs+ sample using a free amplitude in excess of 5 V and a set point of 2.5 V were obtained. The location from which the phase–distance curves were taken was clearly seen in the phase image. It was also observed that the phase–distance curve may start off in the noncontact regime before swiftly moving into the intermittent contact regime, where the phase angle tends to ~90, indicating that there is little energy dissipation. In the second type, the data curve takes noticeably more force to move from the noncontact to intermittent-contact regimes because of damping caused by an attractive force, which is probably electrostatic. Once in the intermittent contact regime, the phase angle is considerably lower, ~60, indicating that more energy is being dissipated. Inside regions were made up of the second type of curve. A third type of data curve will be expected where, despite the high free amplitude and low set point, the cantilever never actually made contact with the surface because of particularly strong damping. The corresponding points in the topography image were found to appear high. This arises from the fact that these points are in the noncontact regime, whereas the remainder of the image could be in the intermittent-contact regime.

Phase–volume images (500 nm) of Nafion Cs+ imaged under ambient conditions were studied. The phase–distance data curves could be divided into three types:

1. A curve that moves quickly from the noncontact to intermittent-contact regime.

2. A curve that requires more force to move into the intermittent-contact regime.
3. A third type that remains in the noncontact regime can be found within regions made up of the second type of curve.

The behavior of the first type of curve could be attributed to the hydrophobic backbone of the membrane, whereas that of the second and third types of curves could be attributed to the ion-rich regions of the membrane. The inability to image the Cs+ ion form completely in the intermittent-contact regime, using the same cantilever as that used for the H+ form, despite using double the free amplitude, can again be explained by the greater charge density associated with the Cs+ ion, which is screened less effectively because of the lower water content. This was suggested to produce a strong long-range attractive interaction that shifts the resonance frequency of the cantilever, therefore reducing its oscillation amplitude until the set point is reached. The second type of curve corresponded to the hydrophilic region that surrounds the Cs+ ions. These regions show a lower phase shift with respect to the hydrophobic regions in the intermittent-contact regime, associated with higher energy being dissipated between the tip and the sample. The energy dissipation would be due to preferential water adsorption to the hydrophilic regions. The unbalance between the lower adhesion when the water neck forms and the needed force to break the meniscus should be the responsible factor of energy dissipation. The contrast reversal observed in the phase volume in the noncontact regime supports this energy dissipation model. Only the rupture of a water meniscus between the tip and the sample could produce energy dissipation in the noncontact regime. Although terminology such as soft and hard tapping is useful, what is really important is the tapping regime in which the experiment is performed, that is, whether it is noncontact or intermittent contact.[518] It is particularly important not to work in a bistable regime because of the height artifacts that can be produced. If the phase contrast observed is greater than 90, it is certain that this is the case. Several articles cite phase ranges of 90 or higher; it is likely that the images were obtained in a bistable regime.[513] The effects as reported in the current literature include contrast reversal in the topography, and phase images have been attributed to working in a bistable regime.[519,520]

The procedure for phase imaging is reported to establish the tip–sample interaction regime by obtaining a phase–distance curve and adjusting the free amplitude and set point accordingly. During these investigations, it was observed that different cantilevers required different amounts of force to move from the noncontact to intermittent-contact regimes. This is consistent with a recent study on the effects of tip sharpness on the contrast in phase imaging.[318] The transition occurred more easily with a sharper tip. A higher attractive force appears with blunter tips, as a consequence of the larger effective contact area for interaction.

The TM AFM topography images and corresponding phase image and TDFM topography, and phase images of Nafion were analyzed. It was found that the resolution of the TDFM was somewhat lower than that of the AFM. This could be related to the following: the size of the end of the particular TDFM probe and thermal drift, because scan times are about 30 min, rather than the 4 min for the AFM. The nature of the sample may also have a bearing on the ultimate resolution, because the cluster–network model of Nafion postulates a large-scale organization of clusters with transient connective tubes in constant flux.[521] Comparisons of AFM and TDFM images were made as follows: (a) a 1 m TM AFM topography image of a Nafion H+ sample imaged under ambient conditions (Z-scale, 15 nm); (b) a phase image corresponding to part a (Z-scale, 30); (c) a 1 m TDFM topography image of Nafion H+ imaged under ambient conditions (Z-scale, 400 nm). The resolution of the TDFM was found to be lower than that of the AFM owing to the size of the probe and thermal drift, which is more of a problem for the TDFM because of the longer scan times, 30 min compared with 4 min for the AFM; (d) corresponding TDFM phase image (Z-scale, 30). Four amplitude–distance curves obtained using Nafion H+ and Cs+ samples across a range of humidities from 10 to 52 were also measured. The effect of humidity was easily observed, although one can easily expect Nafion H+ to be of higher hydrophilicity. The difference between the points at which the amplitude starts to be damped and drops to zero indicates the thickness of the water layer over the surface (~6 nm for the H+ form at 46% RH). The hysteresis in the retract curve is probably caused by the presence of a capillary neck. At lower humidities, the approach curves become steeper, the neck formation is not clearly detectable, and the hysteresis in the curves decreases, which indicates a decrease in the thickness of the water layer. Comparison of the approach and retract curves for the Cs+ and H+ samples at 10% humidity shows a steeper approach curve, which indicates that the surface is drier. This would be consistent with the infrared data, which clearly demonstrated the differences in water content of the ion forms of the membrane. In Nafion Cs+, the noncontinuous nature of the surface liquid film causes differences in the force curves. The effect of relative humidity on surface water thickness was measured. Four different TDFM amplitude–distance curves were obtained using Nafion H+ and Cs+ samples at range of humidities: (a) Nafion H+ at 46% RH, (b) Nafion H+ at 10% RH, (c) Nafion Cs+ at 52% RH, and (d) Nafion Cs+ at 10% RH. These studies showed a correlation between humidity and the AFM data. It was found that as the humidity decreased, the slope of the curves became steeper. Also, hysteresis between the approach and retract curves decreased. The depth of the surface water layer was lower for the Cs+ ion form than for the H+ form. Topography and corresponding phase images of Nafion H+ and Cs+ were obtained during the investigation. The phase range obtained using TDFM was significantly lower than that observed with TM AFM because of the different contrast mechanism involved. In TM AFM, the phase contrast was caused by the tip going in and out of the water layer, whereas the TDFM probe is always inside the water layer moving from side to side.

Differences in the hydrophilicity of the sample, resulting in preferential water adsorption to some areas, would still be detectable because of the increased drag on the probe in these areas. A change in phase contrast with humidity was observed; the phase contrast was lower at the lower humidity, which is consistent with a recent AFM study.[494] The phase contrast for the Cs+ form was lower than that of the H+, again indicating that there is less surface water for this ion form. A real-time frequency spectrum surface was measured. The surface was obtained from the frequency spectra taken at different tip–sample distances. These data showed that when the probe approached the surface, the resonance frequency (dotted line) did not change in a monotonic way on a nanometric scale, as usually expected.[522] From the approach curve data, one could distinguish four different regions.

The data were described as follows. A real-time frequency spectrum was obtained using the TDFM, and frequency spectra were taken at different tip–sample distances. Using the mathematical model as delineated above, it was possible to evaluate the elastic and dissipative components of the tip probe from these data. The probe is free, until it reaches a distance of 16 nm from the surface, when the oscillation amplitude of the probe drops as a consequence of the formation of a capillary condensation neck. A small shift of the frequency peak produces a significant change in the amplitude at resonance. In the third region, the dissipative component dominates the elastic component and is responsible for the large decrease in the oscillation amplitude. In the fourth region, the slope of the elastic force is steeper than the dissipative force, and the elastic component will eventually become predominant for z smaller than 4 nm. These results clearly showed that the information provided by the approach curve alone is insufficient to determine the kind of probe–sample interaction. The real-time frequency spectrum is therefore a unique tool to use to decouple and quantify the two components, giving better insight into the nature of the interaction. The origin of phase contrast in Nafion was investigated using the two complementary SPM techniques of AFM and TDFM. A variety of standard and new techniques, namely, phase–volume imaging and a real-time frequency spectrum were used. Force curves obtained over a mixed hydrophobic/hydrophilic test sample showed a much larger adhesive force over the water-rich regions. An increase in relative humidity resulted in an increase in the thickness of the surface water layer and the phase contrast observed with both SPM techniques. Therefore, differences in probe–sample adhesion, caused by an inhomogeneous distribution of surface water, form an important phase-contrast mechanism. Phase–volume imaging has been a useful tool in the interpretation of phase contrast. It has clearly demonstrated the wide range of phase contrasts that can be observed on the same sample. Moving from the noncontact to intermittent-contact regime resulted in a contrast inversion. The most dissipative features were light in the noncontact and dark in intermittent-contact regimes. Further, in a bistable regime, data of height artifacts were produced in the topographic images at the points where the transition occurred. The anomalous results for the different cation forms of Nafion and those in many published studies can be attributed to working in a bistable regime or moving from one regime to another. The importance of phase–distance curves has been highlighted as a prerequisite to imaging to ensure that imaging takes place in any one regime, rather than relying on a standard set of operating conditions. The sharpness of the tip influenced the phase contrast observed, altering the force required to move from one regime to another. It is therefore necessary to obtain a phase–distance curve if a cantilever is damaged or changed to ensure that imaging continues in the same regime. The phase–volume images of Nafion consisted of two main types of phase–distance curves. The first curve moved quickly from the noncontact to the intermittent-contact regime, and once there, the phase angle of ~90 indicated little energy dissipation. These regions were attributed to the hydrophobic backbone. The second type of curve required considerably more force to enter the intermittent-contact regime, and once there, the phase angle of ~60 indicated more energy dissipation. These regions were attributed to the ion-rich regions that would damp the cantilever oscillation with an attractive electrostatic force at longer distances, then, once in contact, dissipate more energy owing to their greater affinity for water. A greater force was required to image the Cs+ ion form in the intermittent-contact regime, compared with the H+ form because of the lower water content and therefore reduced screening of the Cs+ ions charge. When compared with AFM (dynamic mode), it is clear that TDFM differs in two main aspects: the shape and the orientation of the probe with respect to the specimen surface. In TDFM, the cantilevers have a cylindrical tapered shape and are mounted perpendicular to the specimen surface, which allows accurate control of the tip–sample distance, because the probe is not extensible in the vertical direction. This characteristic makes the TDFM a more suitable tool for force spectroscopy: the probe does not jump to contact during the approach, and a constant load rate can be applied by keeping the approach or retract speed constant. Both of these aspects are problematic in AFM. Thus far, efforts have been concentrated mainly on determining the energy loss mechanism in the tip–sample interaction and have neglected the effect of resonance frequency shifts. A real-time frequency spectrum was obtained to decouple the two effects of change in resonance frequency and damping of the oscillation. It was also possible to determine quantitatively the elastic and dissipative parts of the interaction by accurately modeling the dynamic of the TDFM probes. Distinct regimes were found at different probe–sample separations. Although TM phase imaging remains a useful tool for identifying and mapping regions of different properties regardless of their topographical nature, the interpretation is not always trivial. There have been several pitfalls for the unwary, namely, phase inversion and height artifacts.

5 Crystal Structures by STM and AFM

A characteristic of many solids is their crystalline state, i.e., a crystal is a solid bounded by faces meeting in definite angles. Except in the so-called regular system of crystals (cube, octahedron, etc.), the properties of a crystalline solid, such as elasticity, thermal conductivity, refractive index, etc., are different in different directions. In many systems, for example, photography, the microcrystals of silver are of much importance. The application of STM and AFM to the studies of crystal structures has been of much interest. This arises from the fact that the images obtained are three-dimensional as compared to any other microscopic analysis. Because the amount of sample needed is very small, less than μg, it enables studies of a much wider range of substances. Further, even though it may not be obvious, crystal formation is also determined by interfacial forces.[20] In the following a few examples of such investigations will be described. Due to various physical forces, the crystal formation of small molecules is much different than those of macromolecules. Therefore, these two systems will be described under separate sections. The nanocrystals applications in industry and biology have become common during the past decade. The characterization of such particles has been described in current literature.[3] However, as mentioned below, the nanocrystals of various substances are only becoming apparent after the application of STM and AFM to such systems.

5.1 CRYSTAL STRUCTURES OF SMALL MOLECULES

In current literature, studies on nanocrystal structures have been reported on a variety of organic and inorganic substances. It is obvious that macrocrystals may show different behavior and structures than nanocrystals. At this stage, not many studies have been reported on these systems. However, only a few systematic studies on nanocrystal structures of related series of compounds have been reported.

- *Anthracene crystals* — The smallest particles (atoms, ions, molecules) that make up a crystal, and thus also delineate the surface, have specific electric, magnetic, and mechanical properties that can be registered by STM and AFM. Using this procedure, an anthracene crystal surface was investigated. The images showed the 001 face of anthracene crystal, and the molecular size of 4.7 Å was clearly seen. The x-ray data gave a size of 5 Å. Further, the influence of the functional modification (CH_2, CO, CN_2) of alkylated anthrone derivatives of the structures of their two-dimensional crystals were investigated by STM.[523]

- *Iron oxychloride* — Iron oxychloride is a member of a class of metal oxyhalides that crystallize with a layered structure.[524] AFM was used to study the surface structure of single crystals of FeOCl. The FeOCl have a purple/black luster and cleave easily. X-ray/powder diffraction of a crystal are well known in the literature. The AFM measurements of FeOCl crystals compared well with the literature values of the a, b, c lattice constants.
- The material that can be cleaved in air yields micrometer-sized atomically flat regions that can be imaged easily with atomic resolution. The intercalation of aniline and trimethyl phosphite into the FeOCl crystals is evident.
- The STM images of thermally induced dewetting of an iron oxide scale from an Fe surface were investigated.[525] The influence of the presence of S at the metal/oxide interface on substrate topography was also investigated. The oxidation of S-free and S-covered Fe surfaces at oxygen partial pressures of 10^{-7} Torr resulted in the growth of oxide islands. The island sizes ranged from 10 to 20 Å, and the height variation was invariable at the monoatomic level. Furthermore, heating the oxide formed on S-free and S-covered Fe substrates in UHV resulted in dewetting.
- *Alkylated anthrone* — Physisorbed monolayers of alkylated molecular species were investigated as SAMs by STM and AFM.[526] However, little systematic work is available on the effect of a small chemical modification of a molecule on its two-dimensional assembly structure. Further, the influence of functional modification (CH_2, CO, CN_2) of the alkylated anthrone derivatives of the structures of their two-dimensional crystals was investigated by STM.
- The data of the unit cell of 2-hexadecylanthrone structure were as follows:
  - a 740 pm
  - b 3500 pm
  - Gamma 135°
  - Molecules/unit cell 2
  - Interchain distance 470 pm
  - Interlamella distance (ΔL) 2500 pm
- In this study, the two-dimensional crystalline structures of three functionally modified alkylated anthrone derivatives were investigated by STM at the solution–HOPG interface. 2-Hexadeylanthrone can adopt different two-dimensional crystalline modifications that have been observed as coexisting crystallites. STM was found to add much information on such crystalline structures.
- *Galena* — STM of galena and surface oxidation and sorption of aqueous gold were reported in a recent study.[527]

- *Calcite* — The calcite precipitation mechanisms have been investigated by AFM.[528]
- *Superconductors* — The bismuth-based copper oxide superconductors Bi_2Sr_2C and Cu_n O_2n+4 (n = 1–3), comprise a structurally and chemically rich class of materials that exhibit a wide range af superconducting properties.[529] STM has been used to characterize the surface structure of cleaved single crystals. STM images exhibit a one-dimensional superstructure with a modulation period of 25 Å.

5.1.1 Morphology of Crystals of Different Amino Acids by AFM Studies

The three-dimensional structure analyses of proteins requires the crystals for the x-ray analyses. The crystals of protein molecules are difficult to obtain, which is the prerequisite for any kind of electron microscopic analysis. Currently, no clear knowledge exists about which procedure to use in order to obtain a protein crystal for x-ray analysis. In all cases, one is forced to employ brute force and try various procedures and hope for success. However, based on theoretical considerations, one may expect a correlation between the crystal formation and the amino acid composition of the protein molecule. Because all protein molecules are composed of primarily 25 different amino acids, it was useful to determine the crystal structure of each amino acid by AFM, in order to understand the protein crystal phenomena. A solution of amino acid of concentration 0.01 mg/mL was used. A drop of volume 10 μL was placed on HOPG, and after evaporation overnight, AFM measurements were carried out.[3,20,530] Because membrane proteins have been found to be difficult to crystallize, this procedure could be useful in studying the three-dimensional structures of such biopolymers. These data also convincingly show the application of the drop evaporation method for such AFM studies. An AFM image (three-dimensional) of glycine is shown in Figure 5.1.

There are observed differences between the various amino acids. As is well known, amino acids in an aqueous environment are classified among other properties as being charged, hydrophobic, or neutral, as determined by the side-chain group. In other words, the solubility of different amino acids in water would be different, as is also found. The evaporation of water molecules leads to the formation of crystals. The shape of crystals would be expected to be related to the water plus amino acid interaction properties.

It was concluded that basically, there are two kinds of crystal forms:

1. *Sharp-edged* crystal form (proline, phenylalanine, threonine, valine, lysine)
2. *Rough-edged* form (ornithine, dl-serine, tyrosine, tryptophan)

It also seems that the crystals of the hydrophobic amino acids, which also means less hydration (e.g., phenylalanine, proline, alanine, valine), are more sharp edged than the polar amino acids (glycine, cysteine, ornithine). This may suggest that as water evaporates, the hydrophobic nature of the amino acid determines crystal

FIGURE 5.1 AFM image of amino acid glycine crystals deposited on HOPG from their aqueous solutions (three-dimensional: 20,000 Å × 20,000 Å × 1915 Å): (a) diffuse; (b) sharpened.

formation. This further suggests that the protein crystal formation would be expected to be related to the polar–apolar nature and may provide useful information on protein crystal formation.

It was clearly shown that AFM provides much useful information regarding the crystal formation of amino acids from aqueous solutions. It is worth mentioning that macroscopic analyses would not have provided this detail differently. The evaporation mechanisms of water in these systems need to be considered. As water evaporates, the concentration of amino acid increases. The contact between the liquid phase and the solid substrate, HPOG, changes as regards the contact angle. This means that before the initiation of crystal formation, the effect of substrate becomes appreciable. Differences are observed between the various amino acids' side-chain groups,[107,531] in other words, the solubility of different amino acids in water formation of crystals. The shape of crystals would be expected to be related to the water plus amino acid interaction properties.

It also seems that the crystals of the hydrophobic (less hydrating) amino acids (e.g., phenylalanine, proline, alanine, valine) are more sharp edged than the polar amino acids (glycine, cysteine, ornithine). This may suggest that as water evaporates, the hydrophobic nature of the amino acid determines the crystal formation. This further suggests that the protein crystal formation would be expected to be related to the polar–apolar nature and may provide useful information on protein crystal formation. The self-assembly characteristics are obviously different in these amino acids, as expected. It is of importance to recognize that the water–amino acid interaction, as found from heat of solution data, determines the overall property of a given protein from its sequence. These crystal images would help in understanding these water-binding mechanisms. The nucleation growth and molecular-packing structures of guanine and adenine on graphite have been investigated by STM and AFM.[532]

5.2 SURFACE ADSORPTION STUDIES BY SPMS

The chemisorption of molecules on solids has been of interest in many fields.[3,107] In a particular case, the chemisorption of sulfur was extensively investigated on metal surfaces using LEED and STM under UHV.[534,535] It was found that at low surface coverages, adsorbed sulfur formed open lattices on surfaces of metals. Furthermore, different types of S atom lattices were observed by STM in UHV.[536] In a recent study, the early stages of sulfur deposit growth on HOPG caused by HS electrooxidation in neutral buffered solution were investigated by STM.[536] STM images showed deposited sulfur as domains with hexagonal packing. The sulfur electrodeposition on HOPG from a 0.001 M Na_2S + 0.05 M H_3BO_3 + 0.1 $Na_2B_4O_7$ (pH 8.0) solution was studied. The tips used for STM were made by cutting 0.11 diameter Pt-Ir wires. Atomic resolution STM images of sulfur deposited on HOPG showed discontinuous islands. The S atom rows with S–S distances of 0.21 nm in the rectangular arrays were observed.

The quantum chemistry analysis for sulfur adsorbate structures on the basal plane of graphite was made using the MNDO method.[537] It was shown that adsorbed sulfur atom formation in top and bridge substrate positions is favored. Single sulfur atom and sulfur dimer or trimer adsorption bonding to hollow positions on the basal plane of graphite leads to unstable adsorbate configuration. This can lead to the fact that a remarkable increase in tunneling current is observed when scanning over sulfur-covered substrates. This has also been observed in the case of mixed SAMs of alcohols and thiols (see Figure 3.5).

The application of AFM to study the fluid–solid interface is useful for different adsorption phenomena (such as dye) at interfaces. For example, it is well known that cyanine dyes form so-called J aggregates, which are characterized by a sharp, intensive absorption band with a bathochromic shift compared to a monomer band.[538–540] Further, the studies of J aggregates attracted considerable attention as molecular assemblies in systems that can bridge the gap between the physics of a single molecule and structurally ordered crystals and in systems for optical communications based on the high nonlinear optical coefficients of these aggregates because of their unique optical properties (e.g., giant oscillator strength, excitonic coherence).[541]

Previous studies show that pseudoisocyanine chloride (PIC-Cl), 1,1′-diethyl-2,2′-cyanine chloride, forms J aggregates with an intense absorption band at about 572 nm in aqueous solutions at room temperature.[538–543] Studies also show that PIC-Cl forms J aggregates at a solid–liquid interface; in an early report, J aggregates of PIC-Cl at a mica–solution interface were observed (from spectra).[539]

The absorption band of the J aggregate was red-shifted by 7 nm compared to that of J aggregates in solution, and a monolayer arrangement of the dye molecules in the aggregates was proposed. On the basis of a quantum mechanical calculation, a brick–stonework arrangement of the dye molecules in J aggregates of PIC-Cl at a mica–solution interface was later proposed.[545,546] These models were based on two-dimensional monolayer adsorption of the dye molecules. It was observed that an aqueous PIC-Cl solution in optical cells made of soda lime glass exhibited a new absorption band of a J aggregate that was red-shifted compared to that of J aggregates in solution.[546]

On the basis of total internal reflection fluorescence spectroscopy, it was suggested that the formation of the J aggregates be confined to the vicinity of the glass–solution interface. These J aggregates at solid–liquid interfaces disappear when dried. Optical properties of J aggregates, such as the absorption line shape and spectrum shift, were linked to the aggregate structure, such as aggregation number, structural dimension, and dye molecular orientation.[547] Therefore, for a better understanding of their optical properties, the structure of the J aggregates at solid–liquid interfaces needs to be clarified. However, it has been difficult to access the structure of aggregates at solid–liquid interfaces (JL aggregate) because of the lack of suitable analytical techniques to observe *in situ* the structure of JL aggregates. Although the advent of AFM now allows for direct access to the real structure of J aggregates, no AFM studies on the JL aggregates in solution exist, probably because of the difficulty of *in situ* AFM imaging of fragile J aggregates at solid–liquid interfaces.[548–552] In previous studies, one could solve this difficulty by using TM AFM, and for the first time, the real structure of the J aggregate of PIC-Cl at a mica–liquid interface in an aqueous PIC-Cl solution was revealed.[553,554] Contrary to previously accepted models in which J aggregates at a mica–solution interface have two-dimensional monolayer structures, AFM images showed that the aggregates have three-dimensional leaf-like island structures. Furthermore, the morphological change revealed by AFM imaging suggests that the J aggregates grow by a Volmer–Weber-type growth process.[544,545]

The J-aggregation process of pseudoisocyanine chloride at a mica–solution interface was investigated by using TM AFM and polarization absorption measurements.[555] At a mica–solution interface, pseudoisocyanine chlorides form J aggregates that have three-dimensional leaf-like island structures. The islands are anisotropically oriented with respect to the lattice of a mica substrate. This anisotropic alignment may result from the epitaxial interaction between the positively charged N atoms of the molecules and the regularly aligned negative holes left by the dissociating K+ ions. On the basis of the orientation of islands to the lattice of a mica and the epitaxial interaction, one can propose that in the internal structure of the J aggregate, the long axis of the dye molecules is parallel to the long axis of the islands. In this study, to further clarify the J-aggregation process, the role of a mica surface in the formation of the J aggregates was of interest. Using TM AFM and polarization absorption measurements, the orientation of J-aggregate islands relative to the surface of mica at an initial growing stage of the J aggregates was investigated. It was found that J-aggregate islands are anisotropically oriented with respect to the lattice of a mica substrate. This anisotropic alignment may result from the epitaxial interaction between the positively charged N atom of the molecules and the regularly aligned negative holes left by the dissociating K+ ions.

AFM images were recorded by operating at the tapping mode in liquid phase. For the AFM measurements, mica was fixed on a magnetic steel plate by an epoxy resin and placed in a liquid cell unit. After the mica was cleaved to expose a fresh surface, 30 μL of an aqueous PIC-Cl solution was placed on the mica. The triangular Si_3N_4 microcantilevers had a spring constant of 0.58 N/m, and the drive frequency was 8 to 10 kHz. The integral and proportional gains ranged between 0.5 and 1.2, with the scan rate between 0.5 and 1.0 Hz. The samples for the absorption measurements were prepared by placing an aliquot of the aqueous PIC-Cl solution between a mica film and a hydrophobic glass plate (optical path length ~25 m). Polarization absorption measurements were taken using linearly polarized light (s- and p-polarized lights) and were taken at the same position of the sample cell by attaching a polarizer film in front of the hydrophobic glass side of the cell. The orientation of the J-aggregate islands relative to the surface structure of mica was analyzed as follows. The atomic structures of the surface of mica in air were imaged. The same surface of mica was imaged in a PIC-Cl solution at a PIC-Cl concentration region where the J-aggregate islands form, and the angle between the long axis of the islands and the line connecting the two holes left by dissociating K+ ions was measured. The AFM images were obtained for three separate specimens at the same PIC-Cl concentration. The data of the measured angles were plotted in a histogram.

Previous reports indicated that PIC-Cl forms J aggregates at a mica–solution interface when the concentration of PIC-Cl exceeds 1.0 10^{-4} M.[553,554] The growth process of these J aggregates was examined in detail. The spectral analyses showed quasi-adsorption behavior. At the atomic scale, the lattice structure of the mica surface in air can be recognized and corresponds to the alignment of holes left by the dissociating K+ ions on the mica surface.[556,557]

When mica was cleaved, interlayer K+ ions were divided onto two mica surfaces, resulting in the formation of negative holes left by K+ ions dissociating. As the ordering of the K+ ions is hexagonal between silicate layers in mica, the ordering of the induced negative holes is also hexagonal.[557] These spots have been interpreted as hexagonal rings of oxygen ions corresponding to the position of K^+ ions.[558] The mica surface remained unchanged (i.e., atomically flat) until the J band appeared. The AFM images showed that at the concentration region between 1.0×10^{-4} and 1.0×10^{-3} M, where the intensity of the J band increased drastically. At the concentration where the J band started to appear (1.0×10^{-4} M), leaf-like islands of the J aggregates were observed. The average size of these islands ranged between about 400 to 600 nm long, about 80 to 100 nm wide, and 3 to 6 nm high. The number density of the leaf-like islands increased with increasing PIC-Cl concentration, and then when the PIC-Cl concentration reached 3.0×10^{-4} M, these islands coalesced into larger islands about 2 to 4 μm^2 in area. In these images, leaf-like three-dimensional islands were observed, which correspond to J aggregates at the mica–solution interface.

These AFM images exhibited concentration dependence growth of the J aggregates. This may indicate that at an adsorption–desorption equilibrium of the dye, molecules may exist on the mica surface. Thus, suggesting that the dye molecules may actually be present on the mica surface at concentrations even below 10^{-4} M, although the AFM images showed no change in this concentration region. These observations could be due to the rate of adsorption–desorption of the dye molecules, which is faster than the scanning rate of the AFM tip, and thus, the dye molecules cannot be imaged. Furthermore, the lack of any J band that is characteristic for the J aggregates means that no J aggregates were present. Thus, the critical nuclei may be controlled by parameters such as adsorption/aggregation and dissolution energy.[554] When adsorption/aggregation energy exceeds the dissolution energy, nuclei are formed, resulting in the island formation. In the concentration region below 10^{-4} M, adsorption/aggregation energy is less than dissolution energy, and therefore, nuclei cannot be formed. On the contrary, at the concentration region above 10^{-4} M, adsorption/aggregation energy exceeds the dissolution energy, resulting in the formation of nuclei followed by their growth to islands. AFM images clearly showed the anisotropic growth of these islands. The anisotropic growth of J aggregates was established by determining the orientation of the long axis of the islands. Although the histogram of between 0 and 180 showed scatter in data, it was found that the long axes of the islands were anisotropically oriented relative to the alignment of the holes formed by dissociating K+ ions. The long axis of each island was found to be oriented to one of three directions, with respective peak maximums at 0, 60, and 120, respectively. These three directions coincided with the directions to neighboring holes. Further, it was found that the orientation of the grown J aggregates possessed triangular symmetry on a mica substrate. These differences in absorption spectra measured with linearly polarized light were as expected.[559]

In the case of PIC solution (2.0×10^{-3} M), in the mica/hydrophobic cell it was obtained by using s- and p-polarized lights. Small differences in absorption at the J band were observed between the spectra measured by the s- and p-polarized lights. Furthermore, the absorbance ratio at the J band was reproducible, ~1.15. These results indicate anisotropy of the transition dipoles of the J aggregate and show the anisotropic growth of the J aggregates. The absorption spectra in the short wavelength region (<550 nm) were similar, suggesting that Davydov-type splitting does not exist in the plane of mica.[559]

It was concluded that if the direction of transition dipoles is parallel to that of the long axis of the J-aggregate islands, then the transition dipoles are oriented to three directions (0, 60, 120 degrees, designated by A, B, C, respectively) in equal numbers. Further, assuming that the axes of s- and p-polarized lights are set to 0 and 90, respectively, then the absorbance of the J band measured by using s polarization corresponds to the sum of dipoles oriented to 0, that is, those of direction A, which is cos(60) of directions B and C. Similarly, for the measurement involving p polarization, the sum of the dipoles oriented to 90 of sin(60) of directions B and C is expected to contribute 1.154 to the J band. The directions of s and p polarizations are arbitrary, and therefore, the results indicate that transition dipoles are oriented to three directions as a result of the anisotropic growth of the J aggregate at a mica/solution interface. It should be noted that the polarization absorption measurements were performed at a higher concentration of PIC, while the angle dependence of the islands by AFM was examined at 0.2 mM. Smaller domains have been reported from the optical properties of the J aggregates included in the leaf-like islands.[554] Thus, the J aggregates observed at 2.0 mM were found to consist of the grown islands. The role of the leaf-like island observed by AFM also corresponded to a "macroaggregate" for J aggregates of PIC-Br.[560–562]

This anisotropic growth suggests the existence of epitaxial interaction between the dye molecules and the lattice of a mica substrate during the aggregation process. The highly probable epitaxial interaction is, namely, that the positively charged N atoms of the dye molecules be placed at the negative holes left by the dissociating K+ ions.[544] It was suggested that the driving force for the formation of J aggregates would be the interactions between regularly aligned negative holes and positively charged N atoms of dye molecules. However, counterions might become important in stacking the dye molecules in three dimensions. According to this epitaxial interaction, there are two possible alignments of the dye molecules in the islands: the long axes of the dye molecules are either parallel or 60° relative to the long axes of the islands. and it was proposed that J aggregates of PIC-Cl in solution possess a one-dimensional thread-like structure.[541] The J aggregates form by combining the dye molecules with polymer chains.[551] All of these studies showed that dye molecules are in a brick–stonework structure and are parallel to the long axis of the thread-like structure or of the polymer chains. In terms of energy, this means that this is the preferred alignment for producing the J aggregates. Thus, dye molecules may grow so that the long axis of dye molecules is parallel to the long axis of the islands. This structure is different from that of bulk crystal and inherent to a mica/solution interface.[564]

The observed anisotropy probably originates from the epitaxial interaction between the dye molecules and the regularly aligned negative holes left by the dissociating K+ ions. For further clarification of the anisotropic J-aggregation process, details are needed on the internal structure of the J aggregates.

5.2.1 Studies of Chiral Compounds by AFM

Chiral compounds are important substances in everyday life. The application of SPMs has added much useful information to these systems in recent decades. The study of chiral compounds, such as tartaric acid, has become very important, which was discovered by Pasteur. The physisorbed monolayers of chiral isopthalic and terephthalic acid derivatives at the liquid–graphite interface by STM were investigated.[565] The two-dimensional monolayer grown from a mixture of pure enantiomers separated into domains. These images exhibit a characteristic orientation on substrate graphite basal plane, depending on the enantiomers assembling the domain. By attaching chiral molecules to the probe tip of an AFM, it was found possible to differentiate between enantiomers of a different chiral molecule on a surface.

AFM tip was coated with an acylated phenylglycine that has been used in chiral high-performance liquid chromatography columns.[566] Measurements were made to study interactions on the coated tip with surfaces to which derivatives of a chiral molecule mandelic acid were attached. The magnitude of the adhesion force between the tip coated with phenylglycine and a surface modified with the enantiomer of mandelic acid was more than twice that of a surface covered with S-mandelic acid. These data showed that AFM force measurements can be used to measure chiral interactions.

In a recent study, the chiral segregation by achiral adsorbates was reported by STM. These studies showed manipulation of SAMs with chiral molecules.[567]

5.3 MACROMOLECULE CRYSTALS BY STM AND AFM

The x-ray diffraction of any material requires that the material exists in its crystal state. This has been of much importance in the understanding of protein three-dimensional structures. This information is important, because the biological activity of biopolymers is related to their three-dimensional conformation in their native form. The biological activity (as in the case of enzymes) is lost if a protein molecule is denatured (i.e., in random coil state). In recent literature, some data were reported on the application of AFM to protein crystal studies. Protein crystal structures were also investigated by STM.[523]

Protein conformation is important in regard to its biological activity and stability. The biological activity of an enzyme is lost if the natural conformation is destroyed by denaturation.[20] Protein crystal structures were also investigated by STM.[523]

Protein crystals are known to be unstable in a dry atmosphere. The water content is ca. 43%. Lysozyme was therefore precipitated from an electrolyte-free aqueous solution (either with ethanol or polyethylene glycol).The dimension found was 37 Å upward and 79 Å sideways. This gave the size of the lysozyme as 37 Å × 79 Å. This is in contrast to the x-ray data; however, the tetragonal unit cell of lysozyme[568] gives for eight molecules the dimensions:

$$\mathbf{a} = \mathbf{b} = 79.1 \text{ Å}$$
$$\mathbf{c} = 39.1 \text{ Å}$$

This STM data agree with the lattice constant for the eight molecules. Similar data for other proteins have been observed.

Lamellar crystals of poly(oxymethylene) (POM; molecular weight = 90,000 g/mol) grown from solvent bromobenzene were found to form hexagonal shapes.[569] The crystal height was 8.5 nm. High-resolution AFM images of polyethylene crystalline lamellae showed a pitch height of 0.26 nm and interchain spacing as 0.50 nm.[570] These data agree with x-ray diffraction data.

In a recent study, simultaneous measurements of polystyrene particles with diameters of 144 and 55 nm were carried out using SEM and AFM.[571] The SEM images showed two-dimensional crystals that were hexagonally arranged. The 144 nm particles were identical according to SEM. The AFM images of 38 and 12 nm particles were somewhat diffuse as compared to SEM. This observation requires further investigation.

High-resolution imaging of ionic domains and crystal morphology in ionomers using AFM techniques was reported.[572] In another study, the high-resolution mapping of functional group distributions at surface-treated polymers was reported by using modified tips.[573]

As already mentioned, STM and AFM were extensively used for studying the morphology of DNA from the early stages. The AFM method was used to study some biopolymers, such as DNA on mica, etc.[73,249,251–254] However, the resolution of AFM is known to be limited by the sharpness of the tip, and interpretation of the image has been related to the geometry of the tips.[50] For example, due to the finite radius of the tip, AFM images of DNA are on average seven times broader than the known 2 nm (20 Å) width of DNA. This kind of result requires the description of the resolution of AFM images, especially when comparing images of AFM with EM. AFM images of some solid surfaces might be subject to artifacts, such as the broadening of structures and ghost images of tips due to the finite size and shape of the contacting probe. This means that the shape or radius of the tip used in AFM measurements should be known. One procedure might be of use if the tip happens to leave a scratch on the soft lipid substrate (Figure 3.7).

The three-dimensional crystals of Ca-adenosine triphosphatase (Ca-ATPase) of the[574] calcium pump from sarcoplasmic reticulum were imaged by AFM.

5.3.1 Crystallization and AFM Investigation of a Polymer Structure

It is well known that some synthetic polymers form crystalline phases. In particular, the metastable phase of isotactic poly(propylene) (iPP) was first observed in 1959.[575] Its crystalline structure has resisted analysis for over 35 years and only resolved in 1994.[576] This unusual delay, for a polymer that is known to adopt the standard threefold helix with a 6.5 Å chain axis repeat distance also characteristic of the phases of iPP, is due in part to the fact that iPP cannot be obtained in fiber form; it is converted to the stable form on stretching.[578,579] As a consequence, the phase structure was solved mainly on the basis of hk0 electron diffraction patterns obtained from single crystals; part of the upper layers of the diffraction pattern was accessed by tilting the single crystals in the EM stage.[576] This possibility is, however, limited by the available 60° range of tilt angles of the microscope stage. Accessibility to the remaining part of the diffraction pattern requires that iPP be epitaxially crystallized on adequate substrates. Being a mild orientation process, epitaxial crystallization can be used (and is actually the only means) to "fiber" orient crystal phases that, like iPP, are mechanically unstable.[578,579]

In epitaxial crystallization, the chain (or helix) axis becomes oriented parallel to the substrate surface and, therefore, allows observation with the electron beam normal to the chain axis direction (as in fiber science) rather than parallel to it (as in single crystals).

The difficulty in obtaining the structure of iPP may also be due to its highly unconventional character in polymer science, because it is known to form a frustrated packing of helices.[577] The threefold isochiral helices that build up the trigonal unit cell (with a = 11.03 Å) were found to have different azimuthal orientations. The helices thus have different environments, which implies that at least one of these environments is less favorable. The concept of *frustration* was described based on the two mutually incompatible requirements of close packing in a hexagonal lattice and of antiparallelism of neighbor magnetic spins.[580] The structure of isotactic poly(2-vinylpyridine) (iP2VP) was reported to have different azimuthal settings of the chains, and the resulting different helical environments have been described in much detail.[581] However, the more widespread applicability of frustration for crystal structures based on isochiral threefold helices of chiral, and also of some achiral polymers such as iPP and iP2VP, was recognized only recently.[582]

Frustration in crystalline polymers and biopolymers is further complicated by the fact that two different packing schemes may be considered. In the first, one side chain of one of the threefold helices is oriented (to a first approximation) north, and the side chains of the two other helices are oriented south (a scheme hereafter referred to as NSS); in the second, these side chains are oriented northeast east (in short NEE) or the equivalent northwest west (NWW). The former packing is observed for iP2VP7 and was suggested as a possible model for iPP3 further supported by a recent reexamination of the structure. The metastable phase of isotactic polypropylene (iPP) is crystallized epitaxially on two specific nucleating agents: -quinacridone and dicyclohexylterephthalamide (DCHT). The resulting thin films are investigated by EM and AFM.[583] Epitaxial crystallization yields a biaxially oriented sample of iPP, an orientation that cannot be achieved by mechanical means due to the phase transformation on stretching. Electron diffraction indicates that the iPP (110) plane is the contact face involved in both epitaxies. AFM investigation with methyl group resolution reveals a lateral periodicity of 19 Å in that (110) contact face, which corresponds to the distance between three chains and is a trademark indicator of the frustrated packing of iPP. The AFM data further indicate some variability of the surface pattern, suggesting that two different frustrated structures may coexist in the surface layer; this, in turn, suggests some type of surface reconstruction. Structural requirements that efficient iPP nucleating agents must meet are analyzed.[583] A threefold helix geometry is characteristic of all crystal polymorphs of iPP and the different azimuthal orientations of the two middle chains relative to the corner one, which creates a 19 Å structural periodicity in the (110) plane (horizontal). These structures can be described in shorthand form by the orientation of one of their side chains (arrows) as north south south (NSS) and northeast east (NEE). Helices were suggested to be right handed, and domains made of left-handed helices also exist on account of the chiral but racemic character of iPP. The investigations of the epitaxially crystallized iPP will thus help to obtain a complete structural investigation. Further, it provides a possibility for determining its frustrated packing of helices in crystalline polymers. In the literature, one finds that the nucleating agents for the phase have been proposed and patented. They make it possible to obtain virtually pure iPP,[584,585] by enhancing the usually low "spontaneous" nucleation rate of the phase well beyond that of the phase, and by taking advantage of the faster growth rate of the phase in the "conventional" crystallization range of iPP (i.e., between 140 and 105C).[585,586] Furthermore, the current interest in iPP nucleating agents arises from the fact that, as a result of its metastability and different spherulite architecture, the phase has a lower melting temperature and improved mechanical properties (such as impact strength) than the more common phase. Analysis of the structural relationships between the nucleators and iPP helps establish, at least in their broad outlines, the structural requirements these nucleators must meet. In a recent report, the investigation of iPP epitaxially crystallized on different nucleating agents was given.[583] Large areas of iPP, suitable for structural investigations by EM and diffraction, and by AFM, have been produced. The electron diffraction data are used in a parallel work dealing with a reevaluation of the crystal structure of iPP. The AFM investigations showed the first images illustrating in real space the concept of frustration in polymer crystallography and suggested possible surface reconstruction of epitaxially crystallized iPP.

The iPP samples were of various origins, e.g., mainly two samples of high isotacticity were used and with molecular weights and polydispersities in the 3×10^5 range and 5, respectively. These were as follows: -quinacridone, of formula C20H12N2O2. One of several thiazine derivatives examined for their iPP nucleating effect,[587] namely, triphenodithiazine C18H10N2S2 (hereafter TPDT) of structure N,N'-dicyclohexyl-terephthalamide, C20H28N2O2 (hereafter DCHT), of structure which, according to a recent patent, induces the phase of iPP with a high yield.[585]

The crystal structures of the two former agents were established.[587] This is most unfortunate, because the latter nucleating additive turned out to be the substrate of choice in the present investigation. Also, another patent described the iPP nucleating action of calcium pimelate, which is produced by an *in situ* reaction of pimelic acid [heptanedioic acid: HOOC(CH2)5COOH] and calcium stearate.[588–590] This procedure is incompatible with the experimental requirements, because relatively large (several square micrometers) substrate crystals are better suited for electron and atomic force microscopy. However, a recent patent uses the same nucleating additive but does not rely on this *in situ* reaction.[591] The epitaxial crystallization was performed by depositing crystals of the substrates on thin polymer films (20 to 50 nm thick) cast on glass cover slides by evaporation of a drop of a 1% solution in p-xylene or chlorocyclopentane.[592] The substrate crystals were grown from semidilute (3 to 5%) solutions in appropriate solvents. Suitable solvents are as follows: for -quinacridone, dimethyl sulfoxide (DMSO) or dimethylformamide (DMF), after dissolution at 150°C; for TPDT, DMF; for DCHT, preferrably dimethylacetamide or a mixture of hexafluoro-2-propanol/toluene (in pure HFIP, growth of DCHT is dendritic, which is less suitable for subsequent AFM investigations).

After melting and recrystallization of the polymer in the crystallization range of the phase (between 115 and 135°C), the substrate crystals (which actually lie on top of the iPP films) are redissolved in their own solvent. The polar nature of the latter ensures that the iPP film remains unaffected in the process. The resulting thin polymer films, in which the face in contact with the nucleating agent is the exposed (top) one, are suitable for AFM work without any further processing and for EM after (optional) shadowing with Pt/C and backing with a carbon support film. EM and diffraction are performed with an instrument equipped with a 60 rotation-tilt stage. To preserve the sample from rapid damage under the electron beam, most preliminary observations are made at low magnification, usually by defocussing the diffraction pattern. Bright field and diffraction patterns were recorded on EM and films, respectively.

AFM was carried out in the contact mode (CM). Images were obtained with an A-type scan head (maximum scan size 1 $\mu^2 m^2$). Si_3N_4 tips attached to a microfabricated cantilever (triangular base, 200 m) with a force constant of 0.06 N/m were used. With the help of an optical microscope, the tip was positioned close to the surface of preselected oriented phase regions, which are easily recognizable due to the imprint left by the nucleating agent in the polymer film. The high-resolution AFM images were recorded in the glass fluid cell, which makes it possible to probe the polymer film while it is immersed in liquid media. It is limited on its lower face by the sample (on a cover slide) and on its top face by the glass of the cantilever holder and laterally by an O-ring. Due to the limited elasticity of the O-ring, the lateral range of the *liquid cell* attachment was significantly smaller than in conventional AFM. In these studies, large-scale images were first recorded without the O-ring to visualize the lamellar structure, prior to addition of the immersion liquid (2-propanol, a nonsolvent of the polymer) and high-resolution investigations.

Scanning line frequencies ranged from 1 to 54 Hz when visualizing the lamellar structure (scanned area: 1 μ^2) for high-resolution work. In order to calibrate the distances indicated by the piezo controller, measurements on mica were performed under the same conditions. The imaging procedure was performed by displaying the deflection signal (incoming signal for the feedback system) and the height signal (output of the feedback signal).

It was found that probing polymers in organic liquids reduces the capillary forces caused by water films on their surface. Contrary to measurements in air, the AFM force curves show less hysteresis. Therefore, images can be recorded over a wide range of separation distances between the tip and polymer surface. The high-resolution pictures could only be obtained when reducing the loading forces to values much smaller than 10^{-10} N, as determined from force curves. No bandpass filters were applied during imaging. The quality of the tip was checked by rotating the sample with respect to the tip. Such rotations are easy to monitor in the present case, because the epitaxially crystallized films display a single orientation of the lamellae and the chains over areas as large as 20×30 μ^2. Multiple tip imaging would result in different moire patterns when rotating the sample relative to the tip; such moire patterns were never observed. Further, several sets of experiments were performed using different tips and with various tip/sample orientations with constant and reproducible results. Molecular modeling was performed using commercially available programs.

Furthermore, EM and electron diffraction investigation were carried out on the various substrates investigated. Quinacridone and DCHT were more easily analyzed than TPDT: DCHT in particular is easily obtained in the form of "large" rectangular single crystals (tens of m lateral dimensions) well suited for EM and AFM work.

The crystals were redissolved but leave their imprint in the iPP film. The iPP lamellae are parallel, suggesting a single chain orientation. Because the lamellae were found as edge-on, it is easy to estimate the lamellar thickness, which is not perfectly uniform: it ranges from 15 to 30 nm. In the case of DCHT, the further growth of iPP lamellae into the thicker film of surrounding melt was seen, which shows its nucleating action. It was found that the thinner polymer film used migrated by capillarity under the -quinacridone crystal while in the molten state.

From the diffraction pattern analyses of a composite DCHT-iPP bilayer after partial and complete dissolution of the DCHT crystals, it was found that the latter are significantly thicker than the iPP film. Besides confirming the frustrated structure of iPP, the patterns indicate that an iPP plane of (110) type is in contact with the DCHT crystal surface [zone axis (100)]. Similar patterns were obtained with -quinacridone and indicated that the (110) face of iPP is also involved in that epitaxy. The electron diffraction pattern of the epitaxially crystallized film showed after only partial dissolution of the substrate DCHT crystal: the sharp spots corresponding to DCHT and the broader ones to iPP. The chain axis was found to have a vertical orientation.

Further, it was found that the cell projection of DCHT has a strong 6.5 Å periodicity that matches the c axis layer line of iPP (6.5 Å). A similar matching exists for the "turn" layer line of iPP and a strong layer line of -quinacridone, the structure of which has been established recently.[586]

Whereas electron diffraction indicates that the dense (110) contact plane of iPP is involved in the epitaxy, it does not provide a complete picture of the polymer–substrate relationship. Indeed, given the various possible frustrated structures, it remains to be established, (a) first, which model of the frustrated packing (NSS or NEE) is present in the contact face and, (b) in case it is the NSS structure, which face of the (110) layer is actually in contact with the substrate. Indeed, the opposite faces of the layer have different structures: when seen from the north, the three helices expose one, two, and two methyl groups (a face hereafter described as 122) and when seen from the south, the helices expose two, one, and one methyl groups (hereafter the 211 face). These data were found to be similar to that already reported with epitaxially crystallized films of iPP, for which two structurally different (010) contact faces are possible. As was the case in that study, the next step in the analysis of the epitaxy required probing of the structure of the contact face in real space, i.e., resorting to AFM investigations.[583,594]

Profile analyses of the exposed faces of (110) planes of iPP were carried out. It was found that the opposite sides of the (110) plane did not exhibit the same topology. These faces were characterized by the number of methyl groups in each helix projecting in that face. For commodity, a large unit cell was created, corresponding to two 19 Å periodicities in the (110) plane. In view of the ambiguity inherent to, and left by the above electron diffraction analysis, an AFM investigation of the (110) contact face of epitaxially crystallized iPP appears warranted. These studies were related to the following systems:

1. To determine the existence of a 19 Å periodicity in the (110) contact face of iPP. This periodicity corresponds to the distance between three helices with 19 Å periodicity.
2. To establish if the exact structure of the contact face is involved in the epitaxy, i.e., to distinguish between the two possible frustrated packing schemes (NSS or NEE).

This was based on the fact that the different structures and faces would be expected to have different "AFM signatures," as illustrated by their different profiles, which reflect the different azimuthal orientations of the chains. Based on this, the NEE structure should be rather uncharacteristic and display three rows of "lone" methyl groups, with little difference in height: this structure would nearly approximate a "simple" one-chain structure.

For NSS structures, on the contrary, the 122 face is the most likely to display the three chains, with 19 Å periodicity. Indeed, the rows of "lone" methyl groups protrude above the plane of the "pairs" of methyl groups by an average of 1.5 Å. Similarly, the 211 face should display two rows of single methyl groups, alternating with a "deeper" helix exposing two methyl groups.

A third objective of the AFM study might be considered, because, in the opposite faces of the NSS structure, one or two helices expose two consecutive methyl groups of the chain. AFM should, in principle, be capable of directly visualizing the helical hand of the helices in the contact plane, as has been done for syndiotactic polypropylene. It is worth noticing that this direct determination of helical hand was not possible for epitaxially crystallized iPP, because the chains in the contact plane expose only one methyl group [the helical hand was nevertheless deduced, indirectly, from the relative shifts of the helices in the (010) contact face].

It has generally been argued that the AFM investigations performed under liquid (in "wet" conditions) give consistently better results when high resolution is to be achieved.[594] Images of a large-scale area of the epitaxially crystallized iPP were obtained. These images were technically remarkable, because they displayed large-scale features, namely, the interdigitation and tapering of lamellae seen edge-on, while at the same time achieving methyl group resolution over the entire imaged area. Lamellar thicknesses were found to be uneven, because they vary from about 30 to 50 nm. More importantly, the various chains building up the contact surface were clearly imaged as striations oriented from the lower right to the upper left of the picture, i.e., normal to the lamellar end surface orientation. The lateral periodicity of these striations is prominent in some parts of the figure, in particular, in the center right part. In this part, but more faintly also in many other areas of the picture, the lateral periodicity is 19 Å.

An AFM image of the contact face of iPP showed rows of methyl groups nearly parallel to the lamellar surface and the chain orientation that appears more prominently as striations normal to that lamellar surface in the center right part but also in most areas. Imaging conditions were in 2-propanol, and the size of the imaged area was 200×200 nm^2. The local structure of the contact face was analyzed. It was concluded that the periodicities along the chain axis direction (oriented at 2 o'clock) were more prominent, but striations parallel to the chain axis were also visible. The Fourier transform data showed layer lines 6.5 Å^{-1} apart, corresponding to the c* repeat distance of the threefold iPP helix; a lateral periodicity of 19 Å^{-1} (and multiples thereof). These Fourier transform and filtered pictures showed that AFM "sees" in real space the frustrated packing of iPP materialized by a 19 Å, three chains periodicity in the (110) plane.

From the high resolution, unfiltered AFM image, the chain axis orientation at 2 o'clock was determined. The imaging conditions were as follows: medium,

2-propanol; and imaged area, 12.5 nm × 12.5 nm. It was concluded that the areas with prominent rows of methyl groups 6.5 Å apart (along the chain axis) and separated by the expected 19 Å suggest a packing scheme of the type NSS, which is most likely to display this periodicity, given the "hilliness" of the exposed (110) face. Further, the existence of a single row of methyl groups for the best resolved helix suggested a contact face of type 122. Other areas were also found to show a much fainter 19 Å periodicity, or even a shorter, one-chain (6.35 Å) periodicity. From these data, one could assume that in these areas, the underlying structure was of the type NEE or NWW.

From these AFM data, it was concluded that the contact face of the epitaxially crystallized iPP films may correspond to different frustrated packings. The origin of this "surface polymorphism" cannot be determined at this stage. The NEE or NWW and NSS structures have small differences in packing energies. The substrate does not appear to favor one form exclusively, because the two structures coexist in the contact plane. Other factors must therefore be considered, such as the different local environments of up and down chains that impose in the first place (i.e., on crystallization) one specific structure, or induce some surface reconstruction into the other form upon dissolution of the substrate. This argument appears all the more likely, that, upon dissolution, the surface helices probed by AFM are surrounded by four helices only instead of their usual six neighbors in the crystalline core. The variations in surface organization of the (110) contact face do not seem to result from mechanical interactions with the AFM tip. This possibility must nevertheless be considered, because the different structures are interconvertible through only small rotations of some of the chains: for the NSS structure, it suffices that one chain out of the sequence of three be rotated on its axis by 60 to transform a 122 contact face into a 112 one. Azimuthal reorientations of 30 would transform NEE into NSS structures and would be at the root of the above-mentioned surface reconstructions. However, such mechanically induced transformations appear rather unlikely, given the weak tip–surface forces involved when working under liquid. As a matter of fact, they have not been observed when performing similar AFM experiments with the closely related iPP phase: the contact faces exhibit a uniform pattern of methyl groups.[594]

AFM images were obtained for repeated scans, which implied that the surface was not harmed by the probing of the tip. This surface "strength" appears to be due to the fact that tip-helical stem contacts are local, because they involve, ideally, only one methyl group of the stem, whereas the helical stem is embedded in and interacts with its crystallographic environment over its entire length, i.e., over 30 to 50 nm. It was found that for helices which expose two methyl groups, the latter have not been resolved, possibly because of the small scale (19 Å) "hilly" nature of the contact surface, in which the rows of "lone" methyl groups are projected 1.5 Å above the surrounding surface. As a consequence, it is not possible to "read" the helical hand of the helices, as was the case for syndiotactic polypropylene: it is not, therefore, possible to determine the local "chirality" of the structure.[595] The frustrated structure is chiral, and the iPP molecule, although chiral, is racemic: domains of both chiralities, separated by antiphase boundaries, should and must exist in iPP. When the resolution of the helix hand is reached, AFM will provide a direct means to observe the chirality in different lamellae and lamellar domains of the frustrated iPP structure and, therefore, assess the size of antiphase domains, admittedly under the rather special conditions of epitaxial crystallization.

Data have shown that two major classes of epitaxies of helical polymers have been observed so far. In the first class, dimensional matching with the crystalline substrate takes place via the interstrand distance, which is a prominent physical feature of the helices. Such epitaxies have been observed for iPP and isotactic poly(1-butene) in its form I.[595] The contact planes are made up of isochiral helices with 31 (or 32) symmetry with identical settings and constant relative c axis shifts; as a consequence, the side chains build up regular arrays or, more precisely, linear gratings.

These gratings are involved in the epitaxy, because they match a substrate periodicity. As a result, the matching is enantioselective, because the gratings of side chains, which roughly follow the helical path, are symmetrically oriented relative to the helix axis for right- and for left-handed helices: layers made of right- and of left-handed helices are, likewise, symmetrically oriented relative to the matching substrate periodicity.[592–595] The second class of epitaxies involves, as a rule, less regular polymer contact faces. This structural irregularity may result from a nonrational helix geometry [as for the 113 helix of form II of poly(1-butene)] or from different settings when more regular helices are involved [41 helix of form III of poly(1-butene)]. In this second class, it is always the chain axis repeat distance that plays a major role in the dimensional match with the substrate. The structure of the (110) contact face of iPP, involved in the epitaxies on -quinacridone and DCHT, is clearly of the second type: as a result of frustration, it combines different azimuthal settings and chain-axis shifts. The nearly perfect match between the iPP helix axis repeat distance (6.5 Å) and a similar substrate periodicity was clearly observed. The structural periodicity created by the end benzene rings aligned with a spacing of c/2 = 6.7 Å was also observed from the molecular model of -quinacridone. The phase of quinacridone has a monoclinic unit cell with parameters a = 13.78 Å, b = 3.90 Å, c = 13.40 Å, and d = 79.5. Å. A similar structural analysis is not possible for DCHT, because its crystal structure is not determined. However, the diffraction spots of the substrate on the first layer line of iPP suggest a similar match. Also, the similar chemical structures of -quinacridone and DCHT (linear, elongated molecules ended by benzyl or cyclohexyl rings) suggest that the contact face of DCHT is an alignment of stacks of such rings.

The images showed rectangular surface pattern and the rows of end benzene rings that generate a 6.7 Å grating matched by the 6.5 Å c axis repeat distance of iPP. A second favorable feature appeared to be the orthogonal geometry of the contact face, which matches that of the (110) plane of iPP. Both -quinacridone and DCHT display diffraction spots that can be indexed on a rectangular lattice. The efficiency of DCHT, in particular, appears to result from an additional matching corresponding to reticular distances close to 9.5 Å: the major 19 Å periodicity in the (110) plane (three interhelix distances) corresponded to two substrate periodicities. However, this analysis supposed a near orthorhombic unit cell for DCHT, because epitaxy rests on surface periodicities and not on reticular distances of planes in diffracting position, normal to the surface. The 6.5 Å chain axis repeat distance of iPP was found not to be significantly different from the interchain distance in the (110) plane

(6.35 Å) or, for that matter, the interchain distance in the (010) plane of iPP (6.55 Å). Substrates suitable for the phase may, therefore, act as *phase nucleating agents*, but in that case, the chains (and therefore lamellae) would be oriented at right angles to the "major" iPP epitaxial relationship. Observed occasionally, in bright field electron micrographs of iPP crystallized on -quinacridone, were small patches of lamellae at right angles. It was not possible to determine, with the help of electron diffraction, whether they were of or were iPP. Finally, nucleating agents have other crystal faces with different lattice parameters and surface structures, which may nucleate other crystal phases. Such a dual nucleating activity was indicated for -quinacridone ("E3B is also a very efficient nucleating agent for the modification") and is suggested by the consistently lower phase contents found for this nucleating agent (usually in the 70% range, as defined by the k factor) compared to, for example, DCHT (k > 95%).[596] It was concluded that the end (010) surface of -quinacridone crystals is consistently decorated with iPP quadrites. And, the main (and minimal) structural characteristics required for nucleating agents of iPP are existence in the contact faces of a 6.5 Å periodicity that matches the iPP chain axis repeat distance and an orthogonal cell geometry in that contact plane. Because these requirements are purely geometric (i.e., involve dimensional matchings only, but not the helix chirality), it was concluded that the nucleating agents of the chiral iPP crystal phase do not permit morphological recognition of the enantiomeric domains. On the other hand, the nucleating agents of the racemic iPP, which interact more tightly with the helical paths of the first deposited layers, permit this discrimination through the different lamellar orientations. Furthermore, the epitaxial crystallization of the metastable polymorph of isotactic polypropylene has been achieved on single crystals of selected nucleating agents. Electron diffraction investigations indicate that the iPP (110) plane is the contact plane. In the absence of the crystal structure of DCHT, the exact epitaxial relationship could not be worked out at the molecular level. A lattice matching between the c axis periodicity of iPP (6.5 Å) and a corresponding distance in the substrate crystal face appear as prominent features in the epitaxy: nucleating agents with this 6.5 Å periodicity and an orthogonal geometry of the contact face are likely to induce the iPP polymorph. However, the similarity of the chain axis repeat distance and the interhelix packing in the (010) plane of the phase may explain that some iPP nucleating agents may also induce the phase.[595] AFM with methyl group resolution of the (110) contact plane reveal a lateral packing of helices with a 19 Å periodicity, which results from different azimuthal settings of the chains and is a trademark of the frustrated packing of helices in iPP. This AFM study was the first "real space" illustration of *frustration* in polymer crystallography. Evidence suggested that different iPP packing schemes exist in the contact plane, which have been defined as NEE or NWW, and NSS. For the latter, out of two possible profiles, that denoted 122 is the actual contact plane, with one "tip" and two "bases" of the threefold helices exposed. The limitations of resolution, possibly linked with the small-scale "hilly" nature of the contact plane, did not allow direct observation of the helical hand to assess the (co)existence of lamellae and lamellar domains with different helix chiralities, a possibility resulting from the crystallization of the chiral but racemic iPP in a chiral, frustrated phase.

6 Studies of Solid Surfaces by SPMs

The surface of solids plays an important role in many processes. In order to understand these processes, one needs to have the knowledge of the surface molecular structures. SPMs provided much useful information as regard to the molecular structures at solid surfaces. The biggest impact of SPMs has been that solid surfaces can be investigated under ambient conditions. Further, because many surface reactions take place under fluids, these experiments could be, for the first time, investigated by SPMs. The molecular picture of solid surfaces has become important knowledge that was poor prior to SPMs. SPMs have also allowed one to study changes in solid surfaces under dynamic conditions at molecular scale. For example, the adsorption of gas molecules on solid surfaces under dynamic conditions has been studied, as described below.

6.1 WETTING PROPERTIES OF SOLID SURFACES

Solid surfaces exhibit interfacial properties in contact with fluids, which are related to surface tension of solids.[3,20] The macroscopic wetting properties are estimated from contact angle[3] analysis at the solid–liquid interface. In a recent study, the wetting properties of silanated surfaces were investigated by AFM.[597] A variety of interfacial forces can be investigated by using AFM. In the noncontact mode, one mainly estimates the van der Waals forces (image resolution ca. 10 nm). On the other hand, the ionic repulsion forces are measured in the contact mode. These studies would provide an understanding of the interaction forces (attractive and repulsive) in much greater detail. It is seen that by AFM, the magnitude of distance of separation at which two bodies jump into contact can be determined. The exponential and power laws have been found to fit experimental force curves. The structure of the reduced $Ti0_2$ (110) surface was determined by STM.[598]

Surface modification of fine particles by various alkylsilanes and alcohols is used industrially to vary the surface wettability from hydrophilic to hydrophobic and to control the properties of particles, such as aggregation, dispersion, and adhesion. Based on these industrial applications, a number of studies have been carried out to clarify the relationship between wettability and surface properties.[599–602]

For example, the wettability of modified glass after long storage times in different media was investigated.[603] However, in most of the cases, much attention had been paid to macroscopic properties, and the contribution of the geometric structure of the modifier chains to wettability was not fully analyzed. The development of AFM has made possible the direct imaging of modified surfaces, and many studies[604–608] have been made with this powerful tool during the last 10 years. The adsorption of octadecyltrichlorosilane (OTS) on fumed silica was examined using AFM and resolved the kinetics of the silylation process.[609] In another example, the results of FTIR analysis and the AFM image of trichlorosilane molecules on silica substrate were compared, and the formation mechanism of the self-assembling monolayer was clarified.[605] However, because of the limits of the applicability of AFM, most studies have been restricted to flat surfaces, and obtaining morphological information of colloidal particles had to overcome the problems of probe-induced particles.[610–613] The relationship between wettability and surface geometric structure of modified silica particles and glass plates was investigated. Surface modification was performed by the autoclave method with 1-dodecanol to control the surface wettability. The preferential dispersion test proved that wettability varied at a surface modification ratio of 20%, which coincided with the changing point of the geometric structure of modifier chains determined by the adsorption method. The geometric structure was also evaluated by AFM, and the hexagonal packing of chains of the modifiers in water and in air could be estimated at a high surface modification ratio. Surface images of the nanosized particles on a glass plate were accomplished in water by taking advantage of the hydrophobic attractive force, which was proved by adhesion force measurements. In a recent study, the surface geometric structure effects on the wettability from the aspect of molecular order using alkoxylated silica particles and glass plates were determined.[614]

The geometric structure of modifier chains was evaluated by adsorption of nitrogen or neopentane, and the wettability was examined by the preferential dispersion test. In addition, AFM was used for imaging the modified surface and the measurement of adhesion forces between tip and modified glass plates. The direct imaging of nanometer-sized particles was obtained in water by the new sample-stabilization techniques that can provide fixation of a particle on a glass plate.

Nonporous amorphous silica particles with a surface area of 195 m^2/g (Aerosil 200) and slide glass were used. The mean diameter of silica particles was 12 nm. Surface treatment (alkoxylation) was achieved by the autoclave method at 235°C and 30 atm for 1.0 h.[613] The surface modification ratios of silica particles were measured by thermogravimetric/differential thermal analysis. The surface modification ratios were calculated by dividing the number of modified groups by the number of silanol groups (2.8 nm^{-2}) on unmodified particles.[614]

The thermogravimeteric (TG) and differential thermal analyses (DTA) were carried out over the range 25 to 400°C, and the combustion of the modified group was determined from the exotherm peaks of DTA at 200 to 300°C.

The degrees of hydrophilicity and lipopholicity balance (HLB)[3] were tested by the preferential dispersion test to evaluate wettability. The preferential dispersion test was performed by dispersing silica particles into distilled water and hexane solution under ultrasonic wave for 5 min, and the dispersion tendency was judged after standing overnight. Silica particles were outgassed at 160°C for 4 h under 10^{-5} Torr to ensure that the surface was free from physisorbed water. Adsorption isotherms of neopentane at 0°C and nitrogen at –196°C were determined by the volumetric method. The usual value of N2 = 0.162 nm^2 was taken as the cross-sectional area of a nitrogen molecule, and the cross-sectional area of the neopentane molecule was calculated as neo = 0.614 nm^2 from the monolayer capacity on unmodified samples.[615,616]

The specific surface areas obtained from nitrogen adsorption were calculated by two methods. One method was the specific surface area (m^2/g) calculated by the Brunauer–Emmett–Teller (BET) equation, and the other method was the surface area (m^2/g) calculated where the work of adhesion is equal to that for the unmodified surface based on the Gibbs adsorption equation.[3]

The equation to estimate the surface area (m^2/g) according to the latter method, is known to be related to Avogadro's constant (mol^{-1}), the adsorbed amount (mL STP g^{-1}), and the relative pressure at which the monolayer is formed on the unmodified sample.

In these studies, the excluding area (nm^2/chain) per modifier was estimated from its relation to the specific surface area for the unmodified samples and the number of modifier chains on the samples. The specific surface area (m^2/g^{-1}) was calculated from neopentane adsorption isotherm by using the Gibbs adsorption equations.[20,617–619] The exclusion area (nm^2/chain) from neopentane adsorption was equal to the area occupied by the modified group at the neopentane adsorption temperature of 0°C. The values of the exclusion area obtained from nitrogen and neopentane adsorptions reflected the geometric structure of modifier chains at –196 and 0°C, respectively.

Surface imaging and force measurement with AFM of glass plates were measured in water and in air with an AFM, in the contact mode. Silicon nitride cantilevers of 100 or 200 m length with pyramidal sharpened tips were used, and the loading force on the cantilevers was set at 0.13 nN in the repulsive mode. The spring constants used were 0.09 and 0.02 N/m, and the radius of curvature of the tip was estimated as 30 nm.[620] The adhesion force between the tip and modified glass plates was measured in water in the "force curve" mode. In this operating mode, the substrate displacement was controlled by the applied piezovoltage, and the interaction force was recorded as the voltage from the split photodiode detector. The photodiode voltage and piezovoltage were converted via calibration standards to the normalized force–separation distance curve. In advance of the measurements, the cleanliness of the surface of the tip was guaranteed by gas etching for 5 sec. The wettability of the samples was first evaluated by the preferential dispersion test. The results of the tests for modified samples showed that the samples exhibited hydrophilicity over the range of surface modification ratio of 0 to 20% and changed completely to hydrophobic at a surface modification ratio of >20%.

It was assumed that the surface state changed from a surface modification ratio of 20% because of the influence of the modified group. The specific surface areas were calculated by the BET method. The specific areas gradually decreased from a surface modification ratio of 20%. It was reported that the areas decreased with the increase in surface modification ratios, because the modified group partially closed the pores.[621] However, this assumption is not valid for the nonporous smooth particles used in this study. A more appropriate explanation was proposed,[622–626] however, that nitrogen molecules were reported to adsorb preferentially to silanols, and weak adsorption was observed on organic surfaces compared with silica bare surfaces.[626] Accordingly, the decrease in specific areas by nitrogen adsorption indicated that the silica bare surfaces were covered with modification groups and varied to organic-like surfaces from a surface modification ratio of 20%. This theory is suitable for the results of wettability that changed to hydrophilic, also at a surface modification ratio of 20%.

The specific surface areas were determined by nitrogen adsorption as a function of the surface modification ratio, and the BET method[3] was used to calculate the specific areas. From these studies, although the change in the surface state was clarified, the mechanism of how the geometric structure of modifier chains reflects the surface wettability remained unsolved. The exclusion area is the occupied area of modifier chains at each adsorption temperature, and it reflects the geometric structure of the chains. Therefore, a larger exclusion area means that the modifier chains lie horizontally to the silica surface when extending the chains, and the decrease in area indicates that the chains are likely to stand perpendicular to the silica surface. Further, the area may be overestimated after the packing of the modifier chains has been established, because the space between the chains would also be added as inaccessible sites. The modifier chains may be assumed as frozen conformation at nitrogen adsorption temperature.[627] The interpenetrating step of the modifier's layer among particles might occur in the case of neopentane adsorption for a high surface modification ratio.[625]. However, neopentane adsorption reflected the structure of the modifier chains more practically, because the properties of the particles were usually evaluated at room temperature.

The magnitudes of the exclusion areas were determined from nitrogen and neopentane adsorptions. The areas were estimated from the work of adhesion for unmodified particles. It was reported that long alkoxy chains conformed to the "Loop-Train-Structure," with anchoring of the terminal methyl groups by the nitrogen adsorption method.[628] Accordingly, it was not the modification but the temperature that determined the chain structure by the nitrogen adsorption method. On the other hand, E_{neo} steeply decreased until reaching a surface modification ratio of 20% and slightly increased from a surface modification ratio of 20%. Because the modifier chains can arrange instead of freeze at neopentane adsorption temperature, the gradual decrease means that the modifier chains came to stand, from lying over the silica surface, with the increase in surface modification ratios. This change occurred because the modifier chains interfered with themselves as the surface modification ratio increased, and the minimum value at a surface modification ratio of 20% indicated that the modifier chains began to arrange packing over the silica surface. This assumption is supported by the value E_{neo} = 0.20 nm^2 at surface modification ratio of 20%, which is close to the area of 0.19 nm^2 occupied by a hydrocarbon chain in an alkane single-crystal structure. The continuing slight decrease from a surface modification ratio of 20% contributed to the overestimation, as the packing of modifier chains got closer and closer. The change in the geometric structure from lying to standing at 20% coincided with the wettability, which also varied to hydrophobic at 20%. The geometric structure of the modifier chains influenced the wettability, as shown from the results of gas adsorption tests. The geometric structure of modifier chains was also evaluated by AFM imaging in air and water. In advance of nanoscaled scanning, the unmodified glass plate was observed in water after removing the contamination by H_2O_2 and thoroughly rinsing with distilled water. The topography image of the unmodified glass plate in water was measured. This showed that the surface consisted of hillocks with an average width of ~100 nm, and the surface roughness was 22.5 nm. The image area was 1000 nm × 1000 nm. The surface exhibited morphology with features of hillocks, with an average width of ~100 nm. However, the surface image in water changed drastically after modification, especially at high surface modification ratios. The images of a modified glass plate showed numerous small grains, with a similar surface roughness of 21.5 nm. To observe these small grains more precisely, the scanning area was narrowed, and from these images, the mean diameter was calculated as 18.6 nm. This value was corrected to 15.1 nm by considering the convolution effect between tip and sample. It was found that the estimated diameter was almost equal to that of silica particles, which were modified in the same batch as the glass plates. It was asserted that the morphology of the glass plate could not have changed with the modification process. Hence, it was concluded that silica particles diffused and attached, spreading uniformly on the glass plate, in the modification process using the autoclave method. A small change in surface roughness from 22.5 nm to 21.5 nm also agreed with this concept. An AFM topography image of modified slide glass was scanned in water. The surface modification ratio was 50.1%, and the image area was 1000 nm × 1000 nm, where many small grains could be observed. Thus, one could observe nanosized particles without any special fixation.

This subject on the long-range strong attractive force between the hydrophobic surfaces in water media was discussed in the literature.[629] Accordingly, the strong interaction should exist between particles and scanning tip in this study, and the interaction between particles and tip was set as the repulsive force mode. Consequently, it was assumed that surface imaging of silica particles was enabled by the right choice of the media for observation and the setting of the interaction force loaded on the tip. According to this sample preparation method, surface images at various surface modification ratios were taken by AFM in water from the large scan area to a smaller area. No significant change in the image could be observed in the large scanning area of 500 nm × 500 nm, but regularity appeared at high surface modification ratios in the small area of 5 nm × 5 nm. In addition, Fourier filtered (FT) image and a transformed image into the autocorrelation pattern were also measured. The FT images showed that some spots were arranged in loose hexagonal packing, and the area occupied by one spot was calculated as 0.31 nm^2 from the autocorrelation image.

AFM images were also obtained of 50.1% modified slide glass imaged in water. The scanning area was 5 nm × 5 nm. The top of the cantilever was made of silicon. Analyses of these data showed that the modifier chains were closely packed at 50.1%, and 0.31 nm^2 was almost the same as the exclusion area of ~0.20 nm^2 over a surface modification ratio of 20%. From these considerations, one could estimate that the spot corresponded to the terminal methyl group of the modifiers packing on the silica particles. In the literature, similar data for the terminal methyl group of octadecyltrimethyl chains was reported (i.e., 0.43 nm^2/chain on the silicon surface).[630] In these studies, after the observation in water, the same sample was scanned in air for ca. 1 h. In the large scanning area, a clear image could not be obtained, and the scanning noise appeared quite strongly. This observation was ascribed to a probe-induced effect, which may cause the tip to be swept over the attached particles over the glass surface in scanning.[610–613] The sweeping of particles may take place because only van der Waals forces are present in air, instead of the hydrophobic attraction force in water, and the van der Waals force is generally too weak to fix the particles. However, in the 8 nm × 8 nm area, clear topography images with high regularity were successfully obtained. The regularity consisting of six sites appeared more clearly compared with the image in water, and one spot was calculated as 0.32 nm^2 from the AC image. This value was almost equal to the occupied area of the spot in the water and supported the theory that these spots correspond to the terminal methyl group of the modifiers. However, the reason for the high regularity in air was not discovered. One may consider two possibilities. The first is that the difference in the surface structure between glass slide and silica particles affected the packing arrangement of the modifier chains. The second possibility could be[605] that the copolymer was imaged in xylene solution or in air and the rigid island image observed consisted of aggregated polymer in air after drying of the xylene solution. The data obtained for the polymer were interpreted as it was being dragged on evaporation of the solution and compressed within the island shape. In the same way, if the obtained image is the top of the modification group islands formed on evaporation, it would be quite valid that the modifier chains arranged themselves with high regularity on these rigid islands.

The adhesion forces between the substrate and tip were measured in water media by AFM. A critical increase was seen from a surface modification ratio of ~40%. As previously described, this critical increase was supposed to be the result of strong hydrophobic interactions. Previous investigations reported the same tendency for the adhesion force between silylated silica plate and glass beads, and also reported the attraction between a hydrophobic surface and a polar surface.[98,606]

Accordingly, the critical increase in adhesion force was supposed to be generated by the establishment of a complete hydrophobic surface and to confirm that particles fixed on a glass surface could be observed by AFM at a surface modification ratio of 50%. From these investigations of the relationship between geometric structure and wettability on the alkoxylated silica particles and glass plate, one can conclude the following:

1. Wettability was determined by the geometric structure of modifier chains, and both showed a change at a surface modification ratio of 20%.
2. Nanosized silica particles attached over the glass plate were observed by AFM in water with the repulsive force loaded on the scanning tip.
3. The image of hexagonal-packed modifier chains was obtained in water and in air. The occupied area was estimated as 0.31 to 0.32 nm^2 at a surface modification ratio of 50.1%.
4. The stabilization of nanosized particles could be explained by a strong hydrophobic attractive force, according to the results of adhesion force measurements.

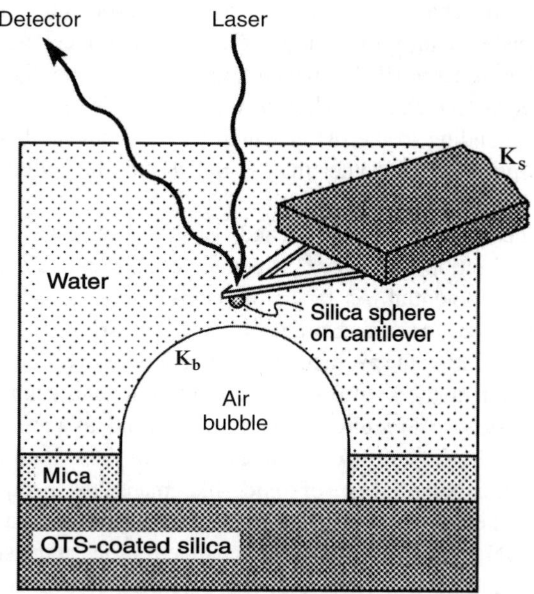

FIGURE 6.1 Schematic drawing of experimental setup used to measure forces between a silica sphere and the interface of an air bubble. (From Drucker et al., *Langmuir*, 10, 3279, 1994. With permission.)

The measurements of hydrophobic and Derjaguin–Laudau–Verwey–Overbeek (DLVO) forces at a silica particle and an air bubble were measured by using AFM.[631] The procedure used is shown in Figure 6.1. An air bubble is constrained on a silica surface coated with octadecyl trichlorosilane (OTS). The bubble could not move in any direction. A silica sphere was attached to the cantilever. The length of the cantilever was 20 or 200 μm, with spring constants of 0.11 and 0.04 N/m, respectively. These studies showed the presence of a strong attractive force between air bubbles and hydrophilic silica particles. The forces in colloidal systems are related to the particle size distribution. In a symmetric system, when the particles are of identical size, the van der Waals force is always attractive, and the double-layer force is always repulsive. However, in the more general case of an asymmetric system, where particles are of dissimilar sizes, the van der Waals and the double-layer forces may be attractive or repulsive. This means that the net DLVO force may be purely attractive or repulsive, or the sign and gradient may vary in a complex way as a function of distance of separation.

6.2 AFM ANALYSES OF SURFACE ACID–BASE PROPERTIES

In the field of surface science and interfaces, it is well known that acid–base interactions play an important role for a large number of phenomena, such as adhesion on polymers, polishing, friction, etc.[3,631,632] The oxide surfaces in an aqueous solution become charged due to amphoteric dissociation of surface M–OH groups. The Bronsted acidity or basicity of an oxide surface can be characterized by the point-of-zero charge (PZC), which corresponds to the pH value required to achieve a zero net surface charge.[3] The surface of an oxide can also be characterized by the IEP corresponding to the pH at which the zeta potential is zero. If there is no specific adsorption of ions other than H or OH, the magnitude of IEP is equal to PZC.

In general, the magnitudes of PZC and IEP are determined by electrophoretic or potentiometric titration methods.[633] In some cases, these data were obtained by using the streaming potential. As far as the acid–base surface properties of oxide surfaces are concerned, force measurements, using a surface force apparatus, have proved useful in evaluating the average interaction between two surfaces as a function of pH.[3,629] The latter method suffers due to limitations and cannot be used for all kinds of surfaces, because it requires smooth surfaces and, as a result, curved surfaces used require areas of ca. 10 μm^2. AFM does not have these limitations.

The electrostatic potential, ϕ, of oxide-like materials in aqueous electrolyte medium is determined by the amphoteric character of the surface charge sites. For example, the Si–ON groups in silica may exchange protons at the surface with water to form Si–OH_2+ and Si–O–. Therefore, two pKs may be defined as follows: pK+ for the site acting as a proton donor, and pK– for the site that acts as a proton acceptor. The PZC is equal to the quantity (pK+ + pK–)/2. It can be seen that $|\phi|$ is an increasing function of pH.[634]

The nature of the electrical double layer is an old problem. It is the key issue in colloid science and soil chemistry (the role of double layers in promoting colloidal stability) and is central to many other fields: electrochemistry (the electrode/electrolyte interface), physiology, and biophysics (electrolyte adjacent to biological membranes).

The magnitude of electrostatic force, according to DLVO theory,[3,629] between two surfaces (similar and dissimilar materials) in an aqueous media is related to the surface potential, ϕ. As an example, the magnitude of double-layer interaction, F_{dl}, is given as:

$$F_{dl}/R_{sphere} = 4\pi\varepsilon\varepsilon_o\phi_1\phi_2 e^{-d/\lambda}/\lambda \tag{6.1}$$

where R_{sphere} is the radius of the sphere, and dl is the distance, where

$$\lambda = [(\varepsilon\varepsilon_o k_B T)/(2000e2N_A I)] \tag{6.2}$$

and

$$I = 0.5\Sigma c_j Z_j^2 \tag{6.3}$$

where φ_1 and φ_2 are surface potentials of the sphere and the plane; I and λ are, respectively, ionic strength and the Debye screening length of the electrolyte; c_i and Z_i are, respectively, the molar concentration (mol/L) and the valency of ionic species

i; d is the distance between the sphere and the plane; and ε, ε_o, k ,T, and N have their usual meanings. AFM force versus distance curves can be reasonably analyzed by this relationship.[631] The magnitude of I is equal to $3/(M\text{–}NaCl)^{1/2}$ Å. Measurements were carried out in KNO_3 0.001 mol/L solutions. The total ionic strength (K + NO_3 + OH^- + H^+) was kept constant at 0.002 mol/L. The magnitude of pH was varied from 3 to 11. Within the DLVO theory, the electrostatic forces between two surfaces (similar or dissimilar materials) in an aqueous electrolyte can be related to the corresponding surface potentials.

The radius of curvature of the Si_3N_3 tip was ca. 300 nm. The spring constant was 0.37 N/m. To gradually change the tip–sample distance, d, the sample was attached to a piezoelectric actuator. Surface force measurements provided information on the charge distribution at a solid–aqueous electrolyte interface. This arose from the fact that the surface potential and charge could be investigated from the interaction between the two double layers. Furthermore, from the variation of these values as a function of pH and ionic strength, the mechanism of surface charging could be determined.

As compared to the surface force apparatus, the AFM method provided the advantage of locally probing the surface, within Debye length. If the AFM tip was crudely defined as a sphere of radius 300 Å, then the effective area of interaction may serve as an estimate of the lateral resolution,

$$\text{Area of resolution} = \sqrt{(2\pi R\lambda)} \tag{6.4}$$

AFM has been used for local characterization of solid surface acid–base properties.[631] In the field of surface science and the physics and chemistry of interfaces, it is well known that the acid–base interactions play an important role for a large number of phenomena, such as adhesion on polymers, polishing, etc.[3,632] The oxide surfaces in an aqueous solution become charged due to amphoteric dissociation of surface M–OH groups, the Bronsted acidity or basicity of an oxide surface can be characterized by the PZC value that corresponds to the pH value required to achieve a zero net surface charge.[3]

6.3 MEASUREMENT OF ATTRACTIVE AND REPULSIVE FORCES BY ATOMIC FORCE MICROSCOPE (AFM)

As two bodies approach with distance of separation of molecular dimensions, short-range attractive forces (van der Waals) and long-range repulsive forces (Coulombic forces) exist (Figure 6.2). Van der Waals forces acting between two different phases have been studied by various investigators. However, the experimental measurement of these forces at very small distances (nm) has neither been extensive nor easy.[3] These forces are measured by the SFA by using direct force. In the literature, one finds studies[7,89–92] using two curved mica surfaces. Recently, AFM[3,635–638] was used to measure these hydrophobic forces. AFM has many advantages over the SFA, because almost any kind or shape of surface can be used.

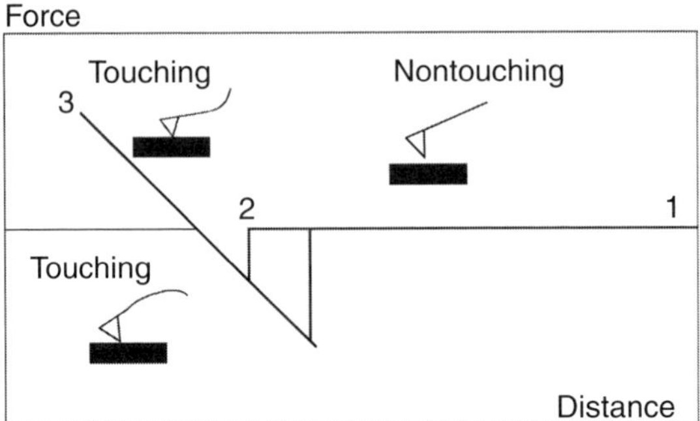

FIGURE 6.2 Force versus distance (between tip and sample) schematic curve. Sample is at a large distance at position 1, and the tip is nontouching, while it has contact (touching) in positions 2, 3, and 4 (schematic).

These forces were measured by the SFA using direct force. By using two curved mica surfaces, in the literature, one finds that some such studies were carried out.[473] These studies will be useful in determining the forces that stabilize self-assembly structures on solids as well as LB films.

Theoretical analysis is based on the following relationships. The relation between the attractive force, F_{att}, between two surfaces of curvature, R_{cur}, is given as:[3,639–641]

$$F_{att}/R_{cur} = C_1 \exp[H_d/D_1] + C_2 \exp[H_d/D_2] \tag{6.5}$$

$$F/R = C_1 \exp(-H/D_1) + C_2 \exp(-H/D_2) \tag{6.6}$$

where H is the closest distance of approach, and C_1 and C_2 are parameters characterizing the magnitudes of short- ($H < 10$ nm) and long-range hydrophobic forces, respectively. However, the hydrophobic forces may also be given as a power law:

$$F/R = -K/(6H^2) \tag{6.7}$$

$$F_{att}/R_{cur} = K_{ham}/[6H_d^2] \tag{6.8}$$

where K_{ham} is the only fitting parameter, as also called the Hamaker constant, Table 6.1.[98,642]

The hydrophobic forces were described to arise from various interactions. For example, these could be due to the entropy increase resulting from the rearrangements in the configuration of the water molecules.

The tip and sample interaction is depicted in Figure 6.2. It is of interest to mention that AFM is a sensitive molecular-force microscope. The forces measured by AFM are of the order of 10,000 to 100,000 times lower than just an insect (ca. 1 mg) sitting on a surface. The hydrophobic forces were described as arising from various interactions, for example, an entropy increase resulting from a rearrangement of the configuration of the water molecules.

TABLE 6.1
Magnitude of Hamaker Constant for Various Substances Immersed in Water (20°C)[3,20]

Solid	Hamaker Constant/J
Polyhexafluoropropylene	$2\ 10^{-22}$
Paraffin wax	$2\ 10^{-22}$
Polyethylene	$2\ 10^{-21}$
Polystyrene	$5\ 10^{-21}$
Copper	$1.4\ 10^{-20}$
Silver	$2.5\ 10^{-20}$
Anastase (TiO_2)	$3.5\ 10^{-20}$
Iron	$4\ 10^{-20}$
Graphite	$5\ 10^{-20}$
Silica	$6\ 10^{-20}$
Hg	$1.3\ 10^{-19}$

At large distances, there is no interaction between the tip and the sample, and the force versus curve is a straight line (as depicted in Figure 6.2). Although electrostatic interactions will have some effect even at large distances, these can be measurable at some intermediate distance. As the distance becomes shorter, the attractive van der Waals forces at position 2 (Figure 6.2) make the tip jump into contact with the sample. This distance would be determined by properties of the sample and the tip and the Hamaker constant. The repulsive forces become active as the distance becomes shorter. During the retraction cycle, the adhesive forces maintain the contact at position 4. The AFM operation can be maintained between positions 3 and 4. The data of such studies[97] are given in Figure 6.3. These measurements show the deflection of the cantilever between the silica plate and the glass sphere coated with octadecyltrichlorosilane.

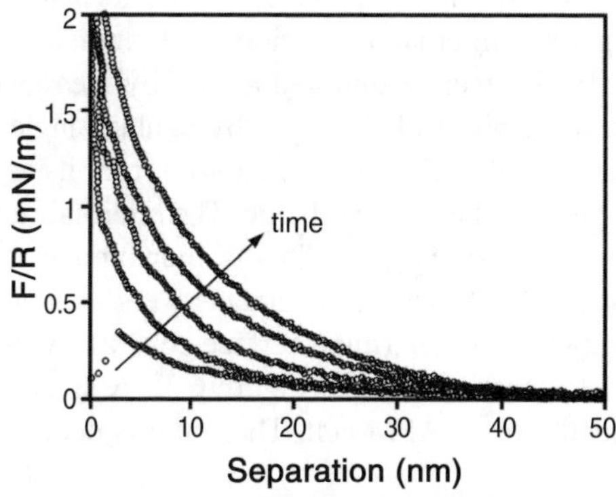

FIGURE 6.3 Force between a silica probe and thiol-adsorbed gold substrate at pH 10.2 as a function of adsorption time (from bottom to top: 0.33, 2.5, 10, 30, and 180 min). (From Hu, K. and Bard, A. J., *Langmuir*, 14, 4790, 1998. With permission.)

In a modified AFM,[98] hyrophobic forces were measured. In usual AFM instrumentation, one measures the force between the tip and the substrate. A glass bead of diameter 10 to 30 μm was mounted on the cantilever. The silica plate (substrate) and bead were sililanated using trimethylchlorosilane (TMCS) and octadecyltrichlorosilane (ODTCS). Contact angles were measured to determine the hydrophobic surfaces. This setup can be shown as follows:

CANTILEVER....SILICA BEAD (20 μm)

....

SUBSTRATE

As will be described later, one can also attach substances other than a silica bead to the cantilever, for example, bacteria or an antibody.

The stiffness constant, k_{st}, of the cantilever is not easily determined. It can be estimated by measuring the resonance frequency with various weights on the end. However, one can also estimate k_{st} by calculation as follows. The expression used was:

$$k_{st} = w_c d_c^3 Y_{si} / 4 l_c^3 \tag{6.9}$$

where w_c, d_c, and l_c were width, thickness, and length of the cantilever, respectively. The value of Y_{si} was 1.1 10^{11} N/m^2, because it was silicone. The magnitude of k_{st} calculated was in the range of 50 to 120 N/m. The magnitude of adhesion force measured for this system was compared with data from other force methods. These analyses gave a value of 430 mN/m, which was considered acceptable.

In the noncontact mode, the net force detected was the attractive force between the tip and the substrate. On the other hand, in the contact mode, the net force measured was the sum of the attractive and the repulsive forces.

Most SAM studies to date have focused on investigating fully covered substrates and their various applications.[20] There have been relatively few studies of the adsorption kinetics of the SAM formation process, so, the formation process of octadecanethiol layers on gold from ethanol solutions was studied.[643]

The adsorption kinetics were monitored *ex situ* by measuring the thickness via ellipsometry and the wettability of the layer by contact-angle measurements. The adsorption of a charged thiol (HSC10COH) on gold from its aqueous solutions was monitored *in situ* by probing the surface charge. The microfabricated AFM cantilever was modified by attaching a silica sphere in a fashion similar to the octadecanethiol in n-hexane procedure.[636,644] Force measurements were conducted between a silica sphere and a silica substrate to determine surface coverage versus time effect in the information of SAM formation kinetics (Figure 6.3). A silica sphere (10 to 20 μm diameter) was mounted into an AFM cell. The setup is depicted as follows:

CANTILEVER WITH SILICA SPHERE.................

THIOL MOLECULES IN SOLUTION...................

ADSORBED THIOL MOLECULES ON SOLID SUBSTRATE....

To assess the thiol adsorption behavior, the Langmuir rate law was employed to fit the adsorption curve obtained here.[3] Typically, if an adsorption process follows this rate law, then the rate of the surface adsorption is governed by the Langmuir isotherm, which is given by:

$$d\theta/dt = k_a(1 - \theta)C_b - k_d\theta \quad (6.10)$$

where θ is the fraction of surface covered, fraction of surface unoccupied; C_b is the bulk thiol concentration; and k_a and k_d are the intrinsic rate constants for adsorption and desorption, respectively. Integration of Equation 6.6 gives the time-dependent surface coverage:

$$\theta(t) = (k_aC_b/(k_aC_b + k_d))(1 - \exp(k_aC_b + k_d)) \quad (6.11)$$

As stated earlier, the thiol desorption process is negligible, i.e., $k_d \ll k_a$. With the substitution of $k_{ohs} = k_aC + k_d$, the simplified relationship becomes:

$$\theta(t) = 1 - \exp(-k_{obs}t) \quad (6.12)$$

Thus, with an assumed k_{obs} value, the surface coverage at any given time can be calculated. Nevertheless, one should realize that the application of the Langmuir rate law to the SAM formation process is a simplification, especially at lower concentrations, where the solution diffusion processes can limit the rate of the adsorption process. The magnitude of k_{obs} was found to be 0.05. The adsorption was found to be of state type.

In a recent study, the rate of oxidation of alkanethiolates in SAMs on gold was investigated by using STM.[645] During the past decade, a number of reports have shown that air oxidation of alkanethiolates takes place. It has been found that the degree of oxidation can be extensive, even in the absence of light, and can be complete within 9 h. From STM images, a correlation between the degree of surface roughness of SAMs and high degree of oxidation was found.

The surface forces between a spherical silica particle and a flat silica surface in aqueous solutions of a nonionic surfactant dodecyl ethoxyethylene (C12E5) were measured by AFM.[646] The silica sphere was attached to the pyramidic silicon nitride tip with a small amount of glue. A SEM image in Figure 6.4 shows a cantilever with silica sphere. The glue was applied using a thin copper wire under an optical microscope. This modification of the cantilever indicated that in the future, many variations of this kind will be applicable to a variety of systems (such as antigen–antibody).

The potential for measuring specific molecular recognition forces between probe-bound ligands and surface-bound proteins using a SFM recently gained much attention. Generally, observed discontinuities in the SFM force–displacement curves are attributed to the breaking of discrete, specific affinity bonds. Study of the molecular recognition system composed of surface-immobilized IgG molecules and SFM probe-bound fluorescein ligands showed the interaction mechanisms.[647] In

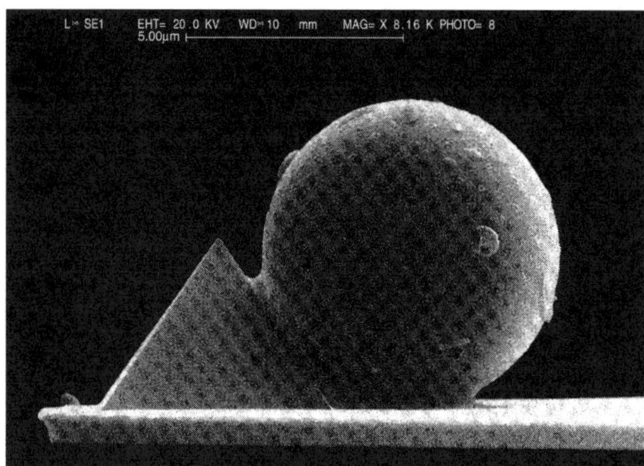

FIGURE 6.4 AFM cantilever with a silica sphere attached to the silicon nitride tip with glue. (From Rutland, M. W. and Senden, T. J., *Langmuir*, 9, 412, 1993. With permission.)

measuring protein interactions with the SFM, it was found that the adhesion curves reflect exaggerated separation distances and a discontinuous stick- and slip-like behavior. The mechanical nature of the experimental system, that of a spherical bead glued to the cantilever, was found to magnify these anomalous effects.

If the probe remains in contact with the sample while it is being withdrawn, relatively large, discrete lateral forces can cause rolling and buckling of the cantilever, leading to a false indication of separation distance. These forces are probably caused by strong areas of localized adhesion due to protein–probe interactions.

Thus, these interactions are the cause of buckling and unbuckling of the cantilever as the probe is translated over a microheterogenous sample. It was also found that this anomalous behavior increased with contact time. This nonspecific protein–probe interaction as measured by SFM was suggested to be useful for studying as an additional technique for developing a better understanding of the general nature of the protein interfacial interactions.

7 Diverse Applications of SPMs (STM and AFM, etc.) and Nanotechnology

Scientists developed specialized instruments in order to control the size and shape of a wide variety of materials at the atomic or molecular levels. The instrumentation as basically used in STM and AFM has been rapidly modified in various directions during the past decade. This has led to instrumentation that can be used in many different systems.[3,20] Therefore, it is necessary to describe these different applications for specific areas in this chapter. It is also apparent that this development will continue in the future with almost the same rapid speed, especially in those areas where the application of STM and AFM has brought about important contributions to science and development. The most diverse area of applications has been the technique of creating patterned SAMs for such diverse application as electronics, biorecognition, and cellular adhesion. The application in biological sensors will be expected to be another major area of development. Especially in applications in diagnostics will one expect many contributions, where small size (nanoscale) is becoming important and necessary. As mentioned above, these applications are only in their beginning stages, and recent inventions indicate that in the near future, much research activity in this area of science is expected.

All advances in science have been based on step-wise knowledge of new analytical techniques. It is obvious that those techniques that are highly sensitive, such as atomic-scale STM and AFM, will require more time to mature and to be accepted in full extent than other less-sensitive methods. In these developments, innovation is based upon molecular scale. The requirement of perfection is the most important criterion. It is therefore only recently that extensive applications have been developed. Also, due to high sensitivity, more time is necessary for such developments. Furthermore, one immediately recognizes that based on these trends, there is almost no limit to future developments.

In recent years, the number of different applications in the areas of STM, SFM, and AFM has increased appreciably. The major areas of these applications are given in the following.

The STM and AFM (FFM and SNOM) allow one to manipulate the sensor (STM = tip, AFM = cantilever or light) at a distance of the order of nm (10 Å) from the substrate under ambient conditions.

In STM, electrons are involved, which makes this system different than that of AFM. In AFM, one can measure forces, which creates some unique possibilities not present with STM.

This means that one can develop so-called nanotechnology products if the movement of the sensor within this range of separation could be controlled, as follows:

- Materials and manufacturing
- Nanoelectronics and computer technology
- Medicine and health
- Aeronautics and space exploration
- Environment and energy
- Biotechnology and agriculture
- National security and other government applications
- Science and education
- Global and trade competitiveness

Nanotechnology is the science of building or manufacturing by controlling the placement of molecules, one by one. These techniques give us thorough, and ultimately inexpensive, control of the structure of matter. Applications of nanotechnology are being studied for manufacturing of products, medicine and health care, space exploration, food production, warfare, and many other uses.

The magnitude of a nanometer (nm) is one billionth of a meter (or a thousand, millionth of a meter). At this magnification scale, any material molecules will become visible. Some three to four atoms fit lined up inside a nanometer. For example, nanotechnology can be considered as a science about building things atom by atom, molecule by molecule. The SPMs allow one to be able to manipulate atoms individually, and place them where you wish on a structure.

In the following, some major developments are mentioned; however, because the possibilities are many, the list is not complete due to space considerations and the rapid rate at which advances are being made. The STM had been used to locally modify surfaces. In the last few years, efforts along these lines have culminated in the ability to manipulate individual atoms and molecules with atomic-scale precision. A goal of STM is to extend our touch to a realm where our hands are simply to manipulate matter on the atomic scale and discuss the physical mechanisms involved.

There has been much excitement about the possibility of using nano-sized structures for real-world applications.[648] There are now a number of techniques for fabricating ultra-small structures, which can have industrial and biological applications. AFM and STM techniques are now being used for atomic- and molecular-scale manipulations. The development of characterization techniques in the regime below 1000 Å coupled with enhanced understanding of the role of more recent advances in biology have given insights into the way in which nature uses self-assembly to construct microstructures (nanotechnology).

Nanotechnology science is expected to provide support for industrial development that will encompass integration of economics, technology, health, and environment. These nanotechnological devices will have an important impact on various activities of mankind. In fact, this clearly indicates that the most significant role scientists play in everyday life is of today and the future. In the following, some important inventions will be described, with literature references and practical examples. This area of science is expanding rapidly, as found from a large number of references.

For example, SAMs on gold can be used to create a variety of chemically well-defined, functionalized surfaces, interesting for the study of many different interface processes.[20] They are particularly suitable for the study of antibody–antigen interactions,[649] cell growth, and protein adsorption.

Many recent advances in chemical analysis have involved the incorporation of biomolecules onto functional surfaces of new devices. Avidin–biotin binding was investigated by molecular recognition system for patterning surfaces.[650]

7.1 STM AND AFM IN ORGANIC CHEMISTRY

In a recent study, a three-dimensional array of gold nanoparticles was assembled on a nanostructured TiO_2 template by subjecting the colloidal gold suspension to a dc electric field (50 to 500 V).[651] The procedure used was based on controlling the concentration of gold colloids in toluene and the applied voltage — it is possible to control the thickness of nanostructured gold film without inducing aggregation effects. As found from AFM, these gold films are highly porous and consist of nanoparticle assembly of fairly uniform size. The surface plasmon characteristics observed in these nanostructured gold films suggest that these particles retain the identity of individual nanoparticles. In recent years, research activities have been seen in casting nanostructured films of semiconductors and metals that display size-dependent optical, electronic, and chemical properties.[652–655]

Few efforts have been made to cast nanostructured films of metal nanoparticles by direct electrodeposition.[656–661] or by using thiol-derivatized gold surfaces.[661,662] Assembling nanoclusters as thin films provides a two- or three-dimensional network of nanoparticles that can suitably be functionalized with photo or electroactive groups. Of particular interest are the nanoscale metals, which have important applications in biological nanosensors and optoelectronic nanodevices.[663–665]

Ultrathin polarizing films cast from gold islands are considered to be useful micropolarizers for fiber-embedded inline optical devices, microoptics, and hybrid integrated optics.[666] Previous efforts to deposit metal nanoclusters as thin films using Langmuir–Blodgett (LB) film or the self-assembled monolayer approach have provided low coverage of these particles.[667–672]

Although these techniques are effective for assembling gold nanoparticles as two-dimensional arrays on an electrode surface, their low surface area limits their use for further chemical modification. More recently, the ordered assembly of gold nanoparticles has been achieved on copper grids[673,674] and conducting glass plates.[675] It was shown that interparticle distance could be controlled by the size of the alkane chains on the stabilizers used in the preparation of the sols. Monodisperse fractions of thiol-stabilized gold nanoparticles have also been crystallized into two- and three-dimensional superlattices.[676]

The deposition of thicker films of metal nanoparticles is often difficult, because close packing of nanoclusters leads to interparticle interactions. Such aggregated nanoclusters exhibit optical properties that are different from those of dispersed nanoparticles. A simple method of casting a three-dimensional assembly of gold nanoparticles that exhibits strong surface plasmon resonance characteristics was reported. The feasibility of achieving relatively thick nanoporous gold films with minimal aggregation effects was shown for the first time. Highly concentrated (12 mM) gold nanoparticles in toluene were prepared according to the previously reported procedure.[677–679]

The solution in water was extracted into the organic phase using a phase-transfer reagent, tetraoctylammonium bromide (TOAB), and was reduced with $NaBH_4$. The colloids were then washed with dilute HCl and later with water. Na_2SO_4 was added to remove traces of water from the colloidal solution. The gold suspension in toluene was diluted to achieve the desired concentrations. TEMs of gold solution in THF showed nearly spherical particles with a diameter of 5 to 8 nm. Nanostructured TiO_2 films were cast on optically transparent electrodes (OTE, conducting glass with 20 ohms) by spreading a colloidal TiO_2 suspension over an area of 2 cm^2. The electrode was further annealed at 673 K. The colloidal TiO_2 particles were found to have a diameter of 10 to 15 nm, and the film thickness was about 1 μm. The TiO_2 film-deposited OTE electrode (connected to positive terminal) and another plain OTE electrode were immersed in a small glass cell containing the colloidal gold in toluene. The distance between the two electrodes was maintained at 3 mm. A dc voltage (50 to 500 V) was applied to initiate the electrodeposition process. The electrophoretic deposition process was stopped at 4 min intervals, and the absorption spectrum of the electrode was recorded after being washed with toluene. This can be attributed to the deposition process to an electrophoretic phenomenon in which the charged gold nanoparticles migrate and become deposited on the OTE/TiO_2 electrode. The charging of particles during the application of the dc electric field was also apparent from the settling of gold particles when the application of dc field for an extended period was continued. The particles can be resuspended by shaking the suspension. The presence of porous TiO_2 film on the OTE surface was found to be essential for achieving good deposition of gold nanoparticles. No significant deposition could be seen when excluding TiO_2 films from the OTE surface. The gold particulate film cast on the nanostructured TiO_2 film was transparent enough to perform spectroscopic measurements. It was found that one of the major problems encountered during the direct deposition of gold nanoparticles (e.g., by solvent evaporation technique) was the interparticle interaction leading to aggregation effects. Upon aggregation, the metal nanoparticles lose the surface plasmon absorption characteristic of individual gold nanoparticles. These films usually exhibit blue coloration caused by the red-shift in the absorption band.[680]

Similar data were reported by means of surface charge neutralization and pressure-dependent studies in LB films.[681–682] These interparticle interactions can be avoided with the use of procedures to help functionalize the gold nanoparticles with alkyl-thiols. By employing the electrophoretic deposition method,[674] success was achieved in assembling alkyl-thiol-capped gold nanoparticles as a two-dimensional array on the copper grid. In this investigation, the system used was based on casting a three-dimensional network of gold nanoparticles on a nanostructured TiO_2 film using a similar electrophoretic approach. The absorption spectra of gold films cast on nanostructured TiO_2 films were recorded at different time intervals following the deposition of gold nanoparticles using a spectrophotometer. At early stages of electrophoretic deposition, the absorption was dominated mainly by the TiO_2 film. Also, there were scattering effects that contributed to absorption in the visible domain. With increasing deposition time, an increase in the absorption in the visible domain was observed. The absorption maximum was observed at 516 nm arising from the surface plasmon band gold nanoparticles. The increase in the absorption of the surface plasmon band with time represents the deposition of gold nanoparticles on TiO_2 surface as a three-dimensional array. By continuing the electrophoretic deposition for an extended period of time, one can achieve relatively thick gold films with an absorbance of the surface plasmon band as high as 2. Such a high value of absorption is an indication that a large number of gold nanoparticles (~1.13 mmol Au/cm^2) are assembled as nanostructured film.

The rate of absorption spectra of gold films cast on nanostructured TiO_2 films is explained as follows. The spectra were recorded at different time intervals (0, 2, 10, 18, 26, 34, 46, 50, 58, and 70 min). A dc electric field of 100 V was applied to OTE/TiO_2 electrode in contact with 2.5 mM gold colloidal suspension in toluene containing ~2.4 mM of tetraoctylammonium bromide (TOAB) (absorption spectrum of 0.3 mM of Au colloids in toluene).

The main absorption band of the gold particulate film showed a spectral feature similar to the absorption spectrum (maximum 530 nm) of the gold colloids in toluene solution. However, the tail absorption in the red region (600 to 700 nm), as well as the broader feature of the surface plasmon band, indicated that the interparticle interaction and altered medium of the immediate surroundings might influence the optical properties of the gold nanoparticles in the film. The aggregation could dominate these films, and one would have observed a broad band in the IR region (700 to 900 nm). The surface plasmon band is very sensitive to immediate surroundings.[682–684]

The binding of chemical species to the gold surface has a pronounced effect of dampening of the plasmon band. On the surface, adsorbed TOAB in the nanostructured gold film is likely to have some influence on the dampening of the absorption band. The electrophoretic deposition mechanisms as described were found to be quite effective in assembling gold nanoparticles as a three-dimensional array. It was also found that these films were quite stable and did not undergo any visible changes under ambient conditions. The surface capping of TOAB was likely to act as a spacer between the adjacent particles in the film. A similar role of surfactants in suppressing the aggregation effects of dye molecules on TiO_2 surfaces was discussed earlier.[685] The alkane chains of the stabilizers that were used in the preparation of the gold sols were also shown to control the interparticle distance in the ordered assembly of gold nanoparticles.[681] The nanostructured TiO_2 film acted as a template for the assembly of these gold nanoparticles and, thus, influenced the overall deposition process. The fact that the gold nanoparticles suspended in TOAB/toluene became deposited on positively charged TiO_2 film indicated that the gold nanoparticles carried a net negative charge. Although the gold nanoparticles were capped with the positively charged TOAB, this organic cap was rather labile and could be readily replaced with other substituents (for example, complexation of 1-pyrenemethylamine with gold nanoparticles.[679]

Due to the negative charge, increased adsorption could take place during an encounter between particles and the dc electric field. The TOAB provided the necessary screening during the transit of nanoparticles to the TiO_2 film. As the gold nanoparticles approached the positively charged TiO_2 surface, a few of the TOAB molecules were exchanged on the TiO_2 surface, and the rest protected gold nanoparticles from interparticle interactions. The rate of gold nanoparticle deposition was dependent on the applied voltage and the initial concentration of gold solution. The rate of gold nanoparticle deposition on TiO_2 films gave rise to the increase in the plasmon absorption at 516 nm with time. At a lower dc voltage, the deposition occured with a rate of 0.022 s^{-1}. The initial deposition rate increased with increasing concentration of gold colloids and deposition voltage. At higher concentration, 6 mM and dc voltage at 400 V, an increase in the deposition rate constant of 0.067 s^{-1} was observed. In all cases, the identity of the nanoparticles in the film was retained. At gold concentrations of 12 mM, the films showed aggregation effects with a broad absorption extending into the red-infrared region.

The growth of the plasmon absorption band of the nanostructured gold film during the electrodeposition process was measured. The experimental conditions for the deposition of gold nanoparticles on OTE/TiO_2 electrode were (a) 2.5 mM gold and 100 V; (b) 2.5 mM gold and 400 V; and (c) 6 mM gold and 400 V.

In these studies, the AFM images of nanostructured gold films were obtained following the deposition of gold nanoparticles on OTE/TiO_2 electrode. The AFM images exhibited that a particle growth process was dependent on the initial gold concentration, as expected. On the other hand, the AFM image of the nanostructured gold film exhibited an assembly of gold nanoparticles of fairly uniform size (with an average diameter of 20 to 25 nm). The AFM image showed larger aggregated clusters of 50 to 60 nm diameter. Significant dampening of the surface plasmon band for the films containing these large-size nanoclusters was also observed.

These films were cast using 0.25 mM and 12 mM colloidal gold solution, respectively. Obviously, a higher concentration of gold colloids leads to rapid growth of gold nanoparticles during the electrophoretic deposition. Moreover, the particle diameters of the two examples were significantly larger than the initial diameter (5 to 8 nm) of suspended gold nanoparticles. This further supports the argument that several gold nanoparticles from solution must be coalescing to form a larger particle during the electrophoretic deposition process. Choosing a lower gold concentration during electrophoresis can minimize the growth process and yield films that exhibit strong surface plasmon absorption. Although these particles are deposited as closely packed as arrays, they retain the properties of individual particles and remain separated without undergoing aggregation or inducing bulk film effects.

AFM images were studied of gold nanoparticles assembled as thin films on OTE/TiO_2 electrode using 0.25 mM and 12 mM gold colloids in toluene. Imaging was done using AFM employing etched silicon tips as probes. The image was recorded in the tapping mode after electrodepositing gold nanoparticles on an OTE/TiO_2 using a dc field of 100 V.

The nanostructured film obtained by electrodeposition method was highly porous, thus providing a large surface area for anchoring electroactive or photoactive molecules. The ability to assemble gold nanoparticles as a three-dimensional array of clusters gives a new possibility for designing sensors and optoelectronic nanodevices. Nanostructured gold films of high surface area also have potential applications in catalysis and photoelectrochemistry. For example, threefold enhancement in the photocurrent generation efficiency can be achieved by coupling gold nanoparticles to the nanostructured semiconductor electrode surface.[686]

7.1.1 Imaging Liquid Crystals by SPMs

Liquid crystals make up a series of compounds that exhibits diffuse transition states between liquid and solid phases. Liquid crystals are an important class of substances found in biological systems and in synthetic molecules. In biological systems, the liquid crystal systems might indicate the necessity for life-important systems to not be dependent on sharp liquid–solid phase transitions, as delineated in the following:

Phase	Transition	Comment
Liquid	Solid	Sharp
Liquid	Liquid-Crystal	
	Solid	Diffuse

Many ferroelectric polymers were found to be of importance in various industrial applications. The liquid crystalline ferroelectric polymers were investigated by AFM.[378] The polymer films were transferred onto gold-coated glass substrates. AFM images were obtained by the tapping mode. Furthermore, the image was modified only by zero-order flatten and first-order plane fit software procedures. Additionally, to reduce the influence of scanner nonlinearity, the mica substrate was imaged, and this was subtracted from the polymer image.

Self-organization of a dinuclear metal complex in lyotropic liquid crystal gave rise to *ribbon-like* supramolecular assemblies.[687] Organization of metal complexes has been attracting considerable interest in relation to the development of novel materials with unique physicochemical properties such as magnetization. The self-organization occurs in water as well as in crystals. The self-organization behavior of a dinuclear metal complex, $Na(Cr_2(L\text{-}tart_2H)(phen)_2)$ (NaCrP), in water was investigated by cryotransmission EM and AFM. Aqueous solutions of NaCrP were dried on freshly cleaved mica substrates and observed by AFM. Ribbon-like structures were found with width of the assemblies as 40 to 100 nm, while the thickness was 1/10th of the width. Ribbon-like assemblies were observed in 6 mM and 0.06 mM solutions.

The structure of molecular assemblies in the nematic liquid crystal of NaCrP was recognized as ribbon-like by TEM and AFM. The molecular length of NaCrP was about 16.5 Å, and the thickness was about 9.5 Å. The ribbon-like structure can be expected to be stabilized by π–π stacking, resulting in the formation of a sequential helical array.

7.1.2 Tunneling Mechanism through Organic Materials

The characterization of metal-molecule-metal tunnel junctions made by contacting Au-supported SAMs of alkanethiols were studied with a conducting AFM tip.[688] The electrical properties of individual molecules and molecular assemblies are currently of heightened interest because of potential applications in molecular electronics and new opportunities for understanding charge transport in organic systems.[689] Molecular-level electrical transport studies require innovative approaches for making electrical contacts to oriented molecules. STM[690–700] and electrochemical methods[701–706] have been used for a number of years to examine transport in surface-confined molecules. In STM and electrochemistry, the molecules are in direct contact with one metal electrode, and charge is delivered by vacuum tunneling or redox molecules, respectively. More recently, metal-molecule-metal junctions were fabricated by assembling molecules inside metal-capped nanopores[707–709] and mechanical "break junctions,"[710] or between mercury drops,[711–713] nanofabricated electrodes,[714] and crossed wires.[715–719] The latter investigations showed that rotaxane-based molecular switches could be strung together to form logic gates, although the state of the switches could only be changed one time. Later reports, however, showed that catenane-based molecular switches could be reconfigured many times on and off.[717] The difference between the on and off states is much too small (i.e., resistance) to be useful for logic circuits at this stage. Metallic nanoparticles have also been used as electrical contacts to molecular monolayers supported on metal surfaces.[720] SAMs and LB film techniques are commonly employed in these studies, because they are convenient approaches to immobilizing molecules at metal surfaces.[20]

7.1.2.1 Conducting Probe Atomic Force Microscopy (CP AFM)

CP AFM provides an attractive approach to electrically contacting monolayer films and the formation of metal-molecule-metal junctions.[721–728] In CP AFM, a metal-coated AFM tip is placed in direct contact, under controlled load, with the material to be probed. Si_3N_4 cantilevers (nominal force constant 0.06 N/m) were coated with 40 Å Cr followed by 400 Å of Au.

Data of I to V traces were obtained sequentially for SAMs of $CH_3(CH_2)5SH$, $CH_3(CH_2)6SH$, $CH_3(CH_2)7SH$, $CH_3(CH_2)8SH$, and $CH_3(CH_2)9SH$, respectively. After completing the series of measurements, rerecording the I to V on SAMs of $CH_3(CH_2)6SH$ showed that no substantial changes to the tip occurred. The standard deviation was calculated for each chain length using 20 resistance values taken from five different points on the sample.

The technique differs from STM in that the probe is positioned using normal force feedback, which decouples probe positioning from the sample conductivity and facilitates interpretation of the tip location with respect to the sample (i.e., in contact or out of contact). In earlier studies, CP AFM was used to measure the resistances of individual carbon nanotubes,[728] semiconductor nanoparticles,[727] LB films,[724] adsorbed molecules on graphite,[725] and organic semiconductor microcrystals.[721] By using metallic levers with high force constants, the current and cantilever deflection was measured simultaneously as the lever was brought into contact with a SAM of $CH_3(CH_2)11SH$.[721–729]

It was shown that CP AFM may be used to make mechanically stable electrical contact to SAMs of alkanethiols on Au. The current–voltage (I to V) characteristics were investigated as a function of the number of methylenes in the alkane chains and the load applied to the tip–sample contact. The resistance of these junctions increased exponentially with monolayer thickness, as expected for tunneling through a dielectric film. For these studies, SAMs were prepared of alkanethiols, $CH_3(CH_2)nSH$, 5 n 9, on polycrystalline Au and were contacted with an Au-coated Si_3N_4 AFM cantilever. Si_3N_4 cantilevers (nominal force constant 0.06 N/m) were coated with 40 Å Cr followed by 400 Å of Au. Measurements of I through V were obtained in ambient conditions under a constant, controlled load at a fixed point on the SAM using external electronics, described previously.[720]

The I–V curves for an Au probe in contact with a decanethiol SAM for applied loads ranging from 2 to 96 nN were measured. The I–V traces were measured with the same tip at the same location and were reproducible at different locations on the sample surface. These I–V traces were found to be linear in the –0.3 to +0.3 V range. When the tip was in contact with the SAM at a load of 2 nN, the resistance was found to be 1.3×10^{10}. It was found that application of higher loads resulted in the probe penetrating the SAM, thereby reducing the resistance. One could apply enough force (typically ~100 nN for decanethiol) to punch through the film to make Au–Au contact. In these systems, the magnitude of resistance was ~20. The significance of these data was that while the junction resistance was clearly load dependent, changes in load on the order of a few nN resulted in small changes in resistance. It was estimated that the load precision in these experiments was approximately 0.2 nN, which corresponded to resistance variations of a few percent. Current–voltage (I–V) characteristics as a function of applied load for an Au tip in contact with a SAM of $CH_3(CH_2)9SH$ were measured (between –0.3 and +0.3 V). Further, I–V traces were recorded sequentially for SAMs of $CH_3(CH_2)5SH$, $CH_3(CH_2)6SH$, $CH_3(CH_2)7SH$, $CH_3(CH_2)8SH$, and $CH_3(CH_2)9SH$, respectively. The series of measurements was rerecorded on SAMs of $CH_3(CH_2)6SH$, data of which showed that no substantial changes to the tip had occurred (load was 2 nN). The data showed that within the sweep range, the I–V traces were linear. The data were plotted as a semilog of the average resistance for each type of junction as a function of the number of carbon atoms in the chain. It was found that the resistance increased exponentially with chain length. An exponential increase was not surprising, because the SAM thickness increased linearly with the number of methylenes[729] and the transport mechanism was likely to be tunneling, based on previous STM[725] and electrochemical studies.[722] The decay constant calculated from the data gave a value of 1.45 per methylene unit. This value was found to be in agreement with the data of 1.5, 1.21, and 1.14 per methylene as found in the literature.[711] These studies showed that it is possible to construct a conducting probe in reproducible mechanical contact with SAMs and to obtain I–V characteristics that show predictable dependence on the SAM thickness. An advantage of CP AFM measurements is experimental simplicity, providing a convenient alternative to labor-intensive microfabrication methods for making metal-molecule-metal junctions. In comparison to STM studies, interpretation of the I–V characteristics is simplified by the absence of an additional tunneling gap between the CP AFM probe and the SAM.[698–725] Thus, CP AFM is a promising approach for studying transport through molecular junctions and its dependence on conjugation, functional group distributions, orientations, and molecular dimensions. A potential limitation of this apparatus is the irradiation of the junction with scattered light (~670 nm) from the optical beam used to detect the lever deflection. The optical gap of alkanethiols (~6 eV) is much greater than the photon energy (1.8 eV), so direct excitation of the molecules cannot take place.

It was concluded from these analyses, that in future studies, optical absorption may be of much importance. Furthermore, it may be desirable to eliminate the optical beam or to use the beam to intentionally stimulate electronic or vibrational transitions of molecules in the junction. It was concluded that it is possible to make stable electrical contact to SAMs of alkanethiols by AFM. The resulting metal-molecule-metal junctions behave as tunneling junctions, in which the resistance increases exponentially with SAM thickness. Contacting SAMs with metallized AFM tips should be a general approach to probing the conductance of molecular junctions as a function of the bonding and functional group architecture of the constituent molecules.

7.2 SEMICONDUCTOR STUDY BY SPM

Controlled and reproducible modifications of solid surfaces on a molecular scale or even atomic scale have large implications in technical applications. STM and AFM are sensitive instruments that have been successfully developed to study semiconductors, where molecular modifications have been attempted.

The generation and manipulation of atomic-scale structures with STM were reported.[730] For example, a compact disc is constructed of polycarbonate plastic and has μm-sized slots impressed in one side by a metal tamper. After aluminizing, these pits allow light to be reflected off the surfaces by a playback laser in a CD player. STM images of a single pit were obtained. Technological applications may be like novel high-density data storage devices or the field of molecular electronics.[731] With the STM, a device has become available that is capable of positioning a local probe with an accuracy on the length scale of atoms. In computer technology, over the past decade, storage capacity has become a limiting hardware bottleneck. SFM offers great potential to increase storage data density by the possibility of using a scanning capacitance microscope (SCM) for charge storage.[731] A metallic cantilever was positioned over a nitride oxide silicon (NOS) structure. An electrical pulse of –40 V, 10 μs, allows the storage of charges in the NOS system. The SCM measures capacitance as a function of the bias voltage and can detect the storage charge by the displacement of the CV curve. The magnitude of capacitance was found to be dependent on the tip radius and became smaller for sharper tips. This method allowed a data density of more than 180 bits/μm^2. Besides the high data density storage possibility, a high data rate is also an important mass storage device, as high as Mbits/s. Unlike bacteria, fish, or even humans, which rely on exhaustible food supplies, the bits and bytes of digital information can proliferate without limit. In fact, they do not die until their creators kill them. On the other hand, computer memory capacity has turned out to be of unlimited capacity as demanded by the needs of humans, therefore, future applications in this area are useful for nanotechnology.

In a recent study, the *in situ* STM data of Cu (111) surface structure and corrosion in pure and benzotriazole containing sulfuric acid solution were reported.[732]

A general procedure was reported about manipulating atoms on solid surfaces by the help of a tip in a STM.[733]

The electrochemical structures of well-defined semiconductor thin films are being investigated by different methods. The thin-film structures of CdS monolayers selectrodeposited on Au (111) substrates were studied by STM and electrochemical methods.[733]

7.2.1 Composite Materials Investigations by SPMs

Surface modification chemistry, often referred to in the literature as molecular self-assembly (MSA), was developed and used extensively in electrical materials, metals, semiconductors, and insulators. Carbon-black-filled polyethylene composites were the focus of much research attention in recent years because of the strong dependence of mechanical and electrical properties on composite morphology.[734–738] This system exhibits a positive temperature coefficient (PTC) effect, where composite resistivity can be a strong function of temperature and is used as self-regulating heaters and resettable fuses.[739–741] Because of this system's commercial importance, its microstructure was extensively examined by various scattering and microscopy techniques.[742–743]

Although a variety of techniques have been applied, questions remain concerning the mechanisms of morphology development during the processing of these composites. In spite of their important role in composite function, the formation and structure of voids have not been thoroughly studied. A new combination of techniques is needed to investigate the mechanisms of void formation during the processing of carbon-black-filled polyethylene composites. Voids in these materials were first detected through comparison of small-angle neutron and x-ray scattering (SANS and SAXS) experiments and quantitatively confirmed by pycnometry.[734] Subsequently, a formalism was developed that enabled the quantification of void size and volume fraction in bulk composites.[742] This method used SANS and relied upon the comparison of SANS profiles from multiple composites separately formulated with protonated and denaturated polyethylene (hPE and dPE). The sensitivity of SANS to isotopic substitution made it an ideal technique for discriminating and measuring the bulk void phase in these conductive composite materials. Subsequent investigations have used the same method of SANS contrast variation to investigate the void phase in composites of varying polymer and filler morphologies.[743] This study demonstrated an empirical relationship between the volume fraction of the void phase and filler surface area, suggesting that voids form as a result of interfacial composite morphology, an interpretation reinforced by the observation that voids are not present at temperatures above the melting point of polyethylene. Presumably, these voids result from stress generated by the crystallizing polymer matrix during composite processing.

The neutron scattering/contrast matching approach, however, is sensitive only to bulk morphology and is incapable of identifying the specific location of void development. Whether voids form at the carbon black–polyethylene interface or within the polyethylene matrix may have a strong effect upon electrical performance, because conductivity through the material occurs as a result of tunneling between adjacent particles in the matrix.[734] Voids at these junctures could lead to decreased performance, reliability, or catastrophic failure. At this stage, another analytical method is needed for determining specific information regarding local morphology. It was shown that a variation of TM AFM, known as phase imaging, is capable of imaging the nanostructure of blended or multiphase materials with high resolution at submicron length scales. SANS analyses showed *voids* in a size range between 50 and 250 nm, therefore, AFM should be an useful technique for directly observing the void phase in carbon-black–polyethylene composites.[743,744]

AFM was utilized to study several aspects of filled polymer composites, such as particle microdispersion, fiber–polymer interfacial morphology, polymer blend crystal morphology, and composite deformation under shear.[745–748]

AFM studies were used to obtain topography and phase images of conductive composites. Complementary information was acquired by combining the modes that allow for comparison of AFM data to SANS results as well as the direct investigation of carbon-black–polyethylene interfaces. With this procedure, AFM study can be compared with previous SANS investigations to identify the mechanism of void morphology development in these materials. The morphology of voids within carbon-black-filled polyethylene conductive composites was studied by AFM and SANS.[743] Void size and polyethylene crystalline lamellae thickness were measured with both techniques, and the results were compared. AFM indicated that voids form at the carbon-black–polyethylene interface as previously hypothesized from the interpretation of neutron scattering experiments. This result supports the conclusion that voids provide a stress release mechanism during polyethylene crystallization and that variables, such as filler morphology, volume fraction, and polymer crystallinity may be used to quantitatively predict the extent of void incorporation as composite composition is varied.

Experiments were conducted with composites of carbon black volume fractions ranging from 0.085 to 0.429 using a lamp black of high structure and furnace blacks of medium and low structures.[748] Each carbon black at a particular volume fraction was separately compounded into hPE and dPE for void measurements via SANS. Composites were prepared in a DACA small-scale twin-screw minicompounder, a mixer/extruder designed for formulations of 1 to 5 cm^3. Carbon black was added to each polyethylene to achieve the appropriate volumetric composition at a constant total mass of 3 g and was mixed at 190°C and 100 rpm for 5 min in the minicompounder.

From the extruded composite material, 76.2 mm × 76.2 mm slabs were pressed at 205°C under 44 N to a thickness of 0.356 mm using a hydraulic press. These slabs were then inserted into another press at the same pressure and cooled to 25°C. The slabs were subsequently punched into 16 mm diameter disks and annealed by heating from room temperature to 170°C at a rate of 2°C/min under atmospheric pressure and composition. The magnitude of 170°C was selected as an annealing temperature, because it is high enough to remove the thermal history of the polymer, yet low enough to prevent significant oxidation.[749] Additionally, it has been demonstrated that carbon black serves to significantly suppress thermal oxidation in polyethylenes even at low weight fractions.[750–752] After being held for 90 min at 170°C, the samples were first heated to 100°C at 0.5°C/min and then cooled to room temperature at 2°C/min. Samples were prepared from these slabs for AFM characterization by cutting an approximately 2 mm × 2 mm section from the slab with a razor blade. This small sample was then imbedded in an epoxy [bisphenol A (epichlorhydrine)] mold to facilitate microtoming of the composite face. Microtoming produced thinly sliced composite slices (<25 m) that were analyzed by AFM. Inspection of initial AFM images revealed the effects of microtoming on the sample surface. To remove such artifacts and achieve morphology similar to that imparted by the original annealing cycle, the samples were once again annealed using a high-temperature AFM stage. Due to the limitations of the stage (T_{max} = 120°C), the original heating cycle was approximated as closely as possible. The sample was heated from 25 to 120°C at a rate of 2°C/min and held for 2 h, at which point they were brought to 85°C at 1°C/min and then returned to 25°C at 2°C/min. AFM images of the postannealed composite samples showed no remnants of the microtoming process. The samples were analyzed by using SANS. The incident neutron beam of wavelength 5 Å was collimated by source and sample slits separated by a distance of 7.5 m. The sample-detector distance was 17.5 m, providing an effective q range of 0.002 $nm^{-1} < q < 0.4864\ nm^{-1}$. The net intensities were converted to an absolute differential scattering cross section, [$d/d(q)$], per unit sample volume (in units of cm^{-1}) via precalibrated secondary standards.[753] Incoherent scattering, due primarily to protons in the hPE, was calculated and subtracted according to a procedure described previously.[754] The analytical procedure to determine void size and volume fraction from reduced SANS data was described.[743]

AFM experiments were performed on a multimode instrument equipped with an extender electronics module. Topography and phase images were collected using tapping mode in air at room temperature. Measurements of lamellar thickness and void size were performed using section analysis on AFM analysis software package. As the feature size being measured was larger than the probe dimensions in the x and y directions and significantly smaller along the z axis, no deconvolution was required to correct for tip effects in the AFM before measuring lamellar widths.[755–756]

AFM images of a freshly microtomed composite and the same composite immediately after heater stage annealing were obtained. It was found that postmicrotome annealing was effective at removing shear-induced orientation at the sample surface. From these images, it was concluded that the benefits of phase imaging were useful. Specific composite features such as polyethylene lamellae and carbon black aggregates can be distinguished in height mode, however, the phase images are much clearer and remove artifacts due to surface roughness that could skew size or position analyses. One uses the phase imaging method to determine whether the composite has regained morphology typical of the bulk, specifically by comparing lamellae thickness, a feature measurable by SANS and AFM. Though phase imaging highlights the location of interfaces within the composite, it cannot be used to locate or identify voids, which do not appear in this mode. Height mode was therefore used in combination with phase imaging to detect sudden discontinuities in topography along the correct length scales, which one may interpret as voids.

SANS profiles for carbon-black-filled hPE and dPE were investigated. The hPE-filled composite profile was divided by the scattering invariant ratio 12 in order to normalize for the contrast difference between hPE and dPE. Deviations of this curve from the dPE-filled composite represent morphological features that are not apparent in hPE-filled composites, namely, voids and the crystalline lamellae phase.

From these data, the crystalline lamellae long-spacing Bragg peak was analyzed by first dividing by a correction factor to account for the unoriented state of lamellae.[757,758] The correction for the Lorentz and polarization factors was applied to these data and was found to be appropriate for perfectly random crystals in a disk or rod geometry.[759] After the long-spacing Bragg peak position was established, the mean lamellar thickness (L_m) was calculated from, Lc, the long spacing obtained from Bragg's law. Xc, the weight percent of crystalline polymer in the matrix phase, was determined by wide-angle x-ray scattering (WAXS).

Values for L_m in annealed polyethylene were determined by SAXS to be between 10 and 30 nm depending upon processing conditions.[757–760] These values obtained for L_m by SANS and AFM compared well to those values. Similarities in the results determined for lamellar thickness suggest that, despite the fact that AFM is inherently a surface technique and SANS is a bulk approach, one may study similar microstructures with both. Additionally, this implies that morphology development mechanisms at the surface are similar to those of the bulk, and that results from AFM measurements can be extended to interpretation of bulk experiments.

Typical SANS data from carbon-black-filled hPE and dPE composites showed two kinds of curves: the first peak is attributable to the void phase, and the second peak is the crystalline lamellae long-spacing Bragg peak.

From these analyses to the voids, height and phase images of 36% medium-structure carbon black in hPE at increasing magnification, allows for close inspection of a composite microstructure, including the void phase.

The data of phase imaging clearly showed the interface between polyethylene lamellae and carbon black aggregates, though voids were not clearly evident. However, the height image revealed distinct discontinuities at these locations. This free space at the carbon-black–polyethylene interface, as revealed by the comparison of height and phase images, was the void phase. Although the determination of void sizes by AFM and SANS is made in two and three dimensions by fundamentally different methods, this comparison demonstrates that surface and bulk morphology are of similar length scales.

In the image data, the stacked lamellar geometry of polyethylene and the aggregate or "beads on a string" morphology of carbon black are evident. In this case, it appears that the crystalline polymer excluded carbon black during the crystallization process and dewetted the surface at a contact point between primary carbon black particles, generating voids.

In these studies, variations of the mechanism through which crystallization appears to induce void formation were observed. These images showed lamellar structures running perpendicular to the carbon black surface. From this, it was concluded that polyethylene molecules are adsorbed to particle surfaces, known as bound polymer or carbon gel,[761,762] do not fully desorb upon crystallization, and therefore, influence composite morphology development. It was also found that polyethylene was attached to the carbon black surface at particular locations of higher density polymer lamellae.

The AFM images of local conductive composite morphology showed the connection of lamellae to carbon black particle surfaces and to the formation of a void between adjacent carbon black aggregates. In those systems where many carbon black aggregates are present in the same area, and lamellar propagation occurs in multiple directions from particle surfaces, their growth may eventually become hindered at intersection points.

Thus, voids appear to form as polyethylene dewets from the carbon black surface. Similarly, the images showed a large void between adjacent excluded carbon black aggregates perpendicular to the local direction of lamellar growth. These hypotheses for void formation were interpreted in the context of a phenomenological scaling argument developed from SANS data.[743] It was shown that void volume fraction can be predicted with knowledge of carbon black structure (DBP), carbon black volume fraction, polymer weight percent crystallinity, and volume fraction. Mathematically, this argument was presented as being only valid at room temperature. The predicted data from this theoretical model fit with the experimentally determined void volume fraction as estimated by AFM. It was concluded that the prediction of the origins of void formation were identified correctly. It was also concluded that void content is proportional to the carbon-black–polyethylene interfacial surface area present. AFM images presented here, however, allow for a more thorough understanding of the void formation mechanisms suggested by these SANS results. The AFM images showed that voids formed at the carbon-black–polyethylene interface. Furthermore, voids appear primarily at points where polyethylene lamellae nucleate at or tangentially contact a carbon black aggregate. These observations, coupled with SANS results that correlate the appearance of voids with the onset of matrix crystallization, lead to the conclusion that void formation as a method of stress relief is correct. The disappearance of voids in the melt is also accounted for by diffusion and readsorption of polymer chains to the carbon black surface, resulting in uniform particle coverage. Finally, the linear increase in void volume fraction with carbon black content is readily explained as the result of increased matrix-filler interfacial surface area where voids form. The data of void content were analyzed for the system DBP12, as a function of annealed and quenched composites. The increase in void volume fraction with carbon black structure observed in these data was also found for void formation at the interface. More highly structured carbon blacks, characterized by a higher DBP number, were found to exhibit a greater amount of surface area, thus providing an opportunity for the formation of more voids. Void size, however, has also been shown to increase with structure, suggesting that it may be individual void volume and not number contributing to the increase in void volume fraction with carbon black structure. Because the carbon blacks used in these experiments increase in primary particle and aggregate size along with structure, the corresponding increase in void volume fraction is likely a combination of possibilities.

Although previous small-angle-scattering-based experiments have identified and quantified voids within bulk carbon-black-filled polyethylene composites, the local morphology of voids, and conductive composites in general, has remained unknown. It was demonstrated that AFM can be used to study conductive composite bulk morphology with excellent resolution. This technique allows identification of voids through a comparison of complementary height and phase images. With this technique, it was determined, as previously hypothesized, that voids form at the carbon-black–polyethylene interface. It was also suggested that voids form as a mechanism of relieving stresses generated during polyethylene crystallization and can be attributed primarily to polyethylene dewetting at the carbon black particle surface.

7.3 STM AND AFM IN INORGANIC CHEMISTRY

7.3.2 Corrosion Phenomena Studies by SPMs

Corrosion phenomena in everyday life are an important problem for mankind, with extensive economical consequences. In most advanced countries, corrosion costs some percent of the gross domestic product. Most of these corrosion problems occur at solid interfaces.[732] Application of STM and AFM in these systems can thus be expected to add new insights and help in prevention and control of corrosion phenomena. The most common corrosion reaction takes place by the coupled electrochemical dissolution of metal (for example, iron) and reduction of substances in the surroundings [oxygen (in air or as dissolved in water), water, etc.]. This leads to the flow of current in the metal from reduction sites (cathodes) to oxidation sites (anodes). The electronic flow is completed by the solution. Due to a thin layer of oxide in stainless steel, the degree of corrosion is reduced to less than 0.1 μm per year. The studies of such oxide surface films are of much recent interest. Atmospheric controlled erosion is one of the most common corrosion processes dependent on wet/dry recycling. Furthermore, many corrosion problems are mitigated by the use of alloys. This arises from the fact that during alloy corrosion, the surface becomes enriched by the more noble alloy components. In a recent study, STM was used to such alloy corrosion processes. Almost atomic resolution was achieved under these investigations. Alloy of composition 60% gold to 40% silver was studied by STM after corrosion with time. As silver atoms were removed from the surface, the gold atoms were seen clearly to be able to reorganize.

In a recent study, the surface morphology of thin (60 nm) Fe film irradiated with 243 MeV Au ions was investigated by using AFM.[763] The degree of surface roughness was found to increase from 2.5 nm to 4.5 nm by the irradiation. This allows one to conclude that increased surface roughness arises from the rearrangement of the atoms on the surface.

The *underpotential deposition* (UPD) af Ag on an iodine-coated Au (111) surface was examined in perchloric acid solution by electrochemical STM.[764] Two different structures of (3 × 3) were found for the rust UPD layer of Ag. It was expected by

STM measurements that the iodine exchanges and exists above the layer of bulk deposited Ag. The structure of iodine and bulk deposited Ag is simple, with no regularly arrayed phase boundaries.

The *adsorption of pyridine* on metal surfaces, especially on coinage metal (Ag, Au, and Cu) surfaces, was investigated from surface science interest. The main interest focused on the orientation of pyridine molecules. In a recent study, the orientation and structure of pyridine adsorbed on Au surface from pyridine aqueous solution was investigated.[764] STM tips of tungsten were sealed with clear nail polish to minimize the occurrence of Faradic currents of the tips. STM images of evaporated Au thin film showed particles of average size 80 nm. A high-resolution STM image analysis showed pyridine as blobs of size 0.6 nm. These images showed flat-lying, tilted, and vertically standing pyridine molecules.

PYRIDINE..............
SILVER (Ag)..........

Noble materials, freshly dispersed and supported on oxides, are useful in a broad class of applications, including catalysis, photocatalysis, and gas sensing. In these systems, the primary = area-to-volume ratio that is economically necessary when using noble metals. Titania and other reducible transition-metal oxides have become interesting systems, because they chemically interact with the dispersed metal. For example, annealing noble metals supported on titania greatly reduce their capacity for CO and H_2 adsorption, an effect not seen on supports such as Al_2O_3 or SiO_2.[764] This behavior was termed a strong metal-support interaction and was explained in terms of an electronic effect. A wide variety of techniques has been applied to this interesting system, including EM, electrical conductivity, thermal desorption, and UHV ion and electron spectroscopy.

In this study, rhodium films were vapor deposited on TiO_2 rutile and examined *in situ* at various coverages and following various annealing procedures by using STM and auger electron spectroscopy (AES). Rh deposition at ambient temperature leads to STM images that indicate nucleation and growth of three-dimensional particles. Size varies around 30 Å. The AES data corroborate the STM thickness data. These studies indicated that useful information can be obtained by STM for such gas sensor devices.

7.3.2 Diverse Systems

Mercury (Hg) has attracted much attention throughout history due to its unique characteristics, e.g., a fluid with metallic properties (conduction, etc.). However, until recently, there was no method that allowed one to study the molecular surface structures of Hg. STM studies of Hg sessile drops were reported in a recent study.[765]

Magnetic nanoparticles, i.e., quantum dots, organized in two dimensions were found to permit the investigation of particle-to-particle transport and nanoscale magnetooptic phenomena.[766,767] The most important application of such assemblies has been suggested to be memory storage possibility. The combination of SiO_2-coated and -uncoated nanoparticles offers a unique opportunity to investigate the effects of charge transport and magnetooptic phenomena in two-dimensional assemblies. SAM of magnetic particles was initiated by adsorption of PDDA [poly(dimethyldialylammonium chloride)] on silicon wafers. This wafer was immersed into magnetic nanoparticles. AFM images were performed after drying. The force versus distance curves were taken in aqueous media using contact mode. TEM images showed that magnetic nanoparticles were oval to spherical, and the average diameter was found to be 12 nm. AFM image of naked magnetic film revealed mutiple aggregates composed of 3 to 10 nanoparticles.

7.3.3 Silica Particle Size and Shape Analyses

The properties of all colloidal particles are determined basically by the size and shape of the particles.[3] Size effects on the properties of nano-sized colloidal particles have been receiving great attention, due to their interesting properties of quantum confinement and superplasticity.[768,769] On the other hand, the access to monodisperse, well-defined particles is facilitated by many achievements during the past decades concerning the preparation of uniform (micro-sized) colloid dispersions.[770,771] It is currently possible to obtain many particle samples, prepared following a uniform procedure and covering a broad range of particle sizes. These particles are expected to have a uniform chemical composition, but this is seldom demonstrated experimentally. There is circumstantial evidence for nonuniform, size-dependent particle behavior, which is not easily understood. For instance, a recent attempt to prepare amorphous and crystalline monolayers of silica particles with diameters of 100, 200, 300, 400, 500, and 1000 nm met with a greater success when the 300 nm particles were used, as compared to the smaller or larger particles. This may be due to chemical differences among the particles of the various sizes or to some other factor, which has not yet been elucidated.[772] Furthermore, the unique physical properties of Au nanoparticles are of much interest for applications in technologies.[773] Au nanoparticles of size in the range 30 to 100 nm were prepared using a new method. The procedure used was by seeding solutions of NH_2OH/Au^{3+} with 12 nm particles. A new method of quasi-monodisperse spherical silica particle preparation was reported[773] many years ago, based on the hydrolysis of tetraethoxysilane (TEOS) in alcoholic media under catalysis by ammonia. Silanol groups were first formed by TEOS hydrolysis, and siloxane bridges were then formed by a silanol condensation reaction. Spherical particles were obtained when enough ammonia was present in the initial reaction mixture. The final particle size depends mainly on the initial water and ammonia concentrations, and the particles thus obtained cover the whole colloidal range. Stöber silica particles have been used as model colloids in a large number of experimental investigations.[774–776]

Significant chemical heterogeneity was observed within another important group of model colloids, the monodisperse polymer latex particles. This chemical heterogeneity affects particle aggregation, macrocrystallization, and film formation.[777,778] AFM images of latex films were studied, and under high magnification, it was found that particles were roughly spherical (Figure 7.1). The surface of particles was covered with small protrusions, like raspberry, with 40 to 60 nm diameter.

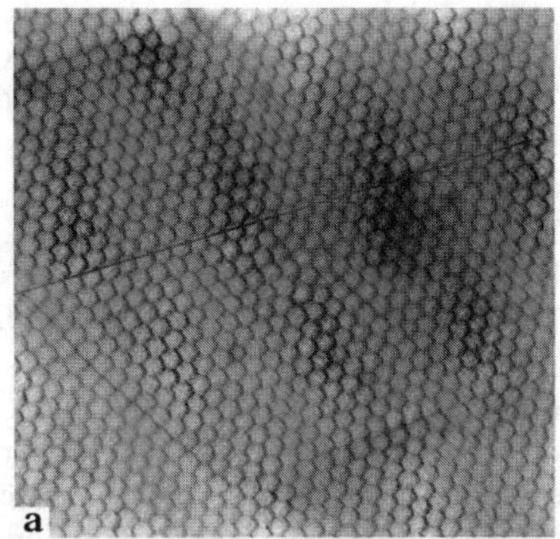

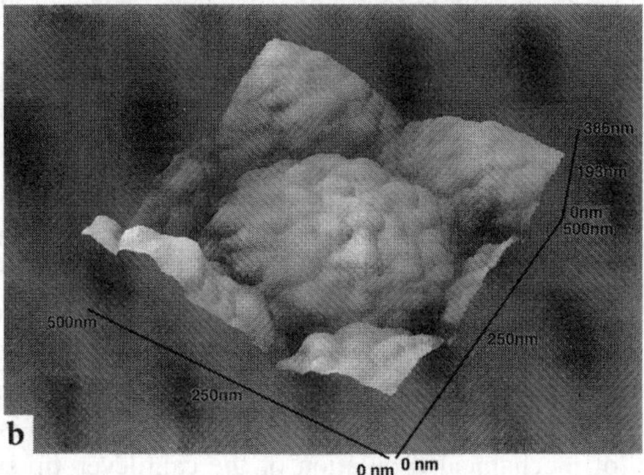

FIGURE 7.1 AFM images (noncontact mode) of PS-HEMA copolymer latex: (a) a self-assembled array of latex particles (10 mm × 10 mm); (b) a raspberry surface morphology of the particle. (From Cardoso et al., *Langmuir*, 14, 3187, 1998. With permission.)

This information is not yet available for other model colloidal particles, and therefore, a detailed examination of Stöber silica particles using different microscopy techniques was carried out.

In a recent study, two different samples of monodisperse silver silica particles were examined using three different microscopes: energy-filtered analytical transmission electron microscopy (EFTEM), high-resolution field-emission scanning electron microscopy (FESEM), and SPM, in the noncontact AFM and scanning electric potential microscopy (SEPM) modes.[778]

After drying the silica dispersions, the larger (ca. 141 nm) particles were only partially deformed by capillary adhesion, whereas the smaller particles (ca. 36 nm) were strongly deformed and closely packed into dense films of a low porosity, which is evidence of their larger plasticity, or superplasticity. Electric potential distribution maps obtained by SEPM showed a significant interparticle as well as intraparticle contrast, especially in the case of the smaller particles. Examination by electron backscattering also revealed a larger contrast among the smaller particles, thus evidencing a nonuniformity of chemical composition.

In these reports, silica particles were prepared as follows.[778] The volumes of the different components were as follows: (a) 4 mL TEOS, 4 mL ammonium hydroxide (sat.), and 50 mL ethanol; and (b) 4 mL TEOS, 2 mL ammonium hydroxide (saturated), and 50 mL ethanol. The extent of particle formation was related to the increasing opalescence of the mixture, starting between 1 and 5 min after the addition of TEOS, and reaction completion was evidenced by the stability of sample turbidity. The effective diameter was measured using photon correlation spectroscopy (PCS). The samples were also analyzed by using an EM equipped with a energy filter spectrometer EFTEM. For individual silica particle examination, one drop of the silica dispersion (1% solids content) was deposited on carbon-coated parlodion films supported in 400-mesh copper grids. High-resolution FESEM and SEM were also used for particle analyses. Silica dispersion (1% w/w, two drops) samples were placed on top of mica sheets, previously mounted on a brass sample holder with double-face adhesive tape. The dry silica films were carbon coated.

AFM microscopy was used to obtain topographic information on the films formed by drying silica dispersions. The sample was prepared by placing 0.05 mL of 1% w/w silica dispersions on top of a mica sheet and allowing it to dry at room temperature. Topography changes were measured by monitoring detector signal amplitude, at a 300×300 pixel resolution and 1.81 m/s scan rate. The tips were made of silicon coated with platinum and had a pyramidal form with a spherical 20 nm nominal radius tip, which was verified by imaging the tip in the SEM.

SEPM in this instrument uses the standard noncontact AFM setup, but with the following modifications: the Pt-coated conducting tip was fed with an AC signal, 10 kHz below the frequency of the normal AFM oscillator, which is the same as the natural frequency of mechanical oscillation of the cantilever–tip system (40 to 70 kHz). The mechanical oscillation of the tip was tracked by the four-quadrant photodetector and analyzed by two feedback loops. The first loop was used in the conventional way to control the distance between tip and sample surface, while scanning the sample at constant oscillation amplitude. The second loop was used to minimize the electric field between tip and sample: a second lock-in amplifier measured the AC frequency oscillation while scanning and added a DC bias to the tip to recover the undisturbed AC oscillation. This technique differed from those procedures where the phase displacement of the AC voltage was employed, using the phase displacement by DC biasing.[779] The image was built using the DC voltage fed to the tip, at every pixel, thus detecting electric potential gradients throughout the scanned area. This technique is similar to the oscillating electrode technique for monolayer study.[780] The major difference between the techniques is the detection technique used, because SEPM uses phase detection of the perturbations on the applied AC voltage. Particle effective diameters determined by PCS were 141.5 and 36.5 1.0 nm, and the larger particles were obtained in the presence of a higher ammonia concentration, as expected.[778]

The particle size analysis was made from EFTEM images. The larger particles were found to be spherical, and some interstitial material was observed as dark-gray areas (or “necks”) amid the particles in the bright-field picture. The silicon and oxygen maps for the larger and smaller particles showed a major difference. The borders between neighboring particles were clearly seen in the Si map, but these were almost undistinguishable in the O map.

In these scanned analyses of the Si and O maps, the observed elevations corresponded to the brighter domains that relate to the Si- or O-rich regions, and the valleys were related to the Si- and O-depleted interparticle regions. It was concluded that (a) the sharp ups and downs in these curves correspond to large composition fluctuations, which are probably due to pores and to local changes in the Si/O elemental ratio; (b) the particle to the left appears broader in the O map than in the Si map, because the corresponding half-height widths in the line-scans extend for, respectively, 33 and 26 pixels; and (c) the smaller particle shapes are less regular than in the case of the larger particles. These data showed that the larger particles have a well-defined core-and-shell nature, and the outer shells appear brighter in the O map than in the Si map and have a higher O/Si content than the particle bulk. It was concluded that the shells consisted of silica with a higher degree of hydration than the bulk.

The data from FESEM images were analyzed. In FESEM, the backscattered electron image contrast depends largely on chemical composition fluctuations across the sample, as well as on local specimen surface inclination, crystallography, and internal magnetic fields. Elastic scattering efficiency increased with the atomic number of the target atom, and sample areas with a larger average atomic number scattered more strongly than areas with a lower average atomic number.[781] The beam electrons were found to penetrate to a significant sample depth before reversing their course and returning to the surface to escape as a backscattered electron. Electrons that travel along such trajectories may clearly be influenced by subsurface features of the specimen structure (e.g., inclusions of different composition, voids, etc.), and they carry information on that structure upon escaping. Therefore, the BEI signal is known to be sensitive to bulk and surface composition at conventional SEM beam energies (>10 keV). It was thus found that the brighter areas in AFM images indicated where the heavier elements were expected to be present in a noncrystalline sample. Compared to the backscattered electrons, the secondary-electron coefficient is relatively insensitive to the local average atomic number.[781] Field-emission SEMs of the silica particles obtained in the SEI and BEI modes were investigated. As opposed to the TEM pictures, these samples were prepared with higher particle loadings, which is allowed in SEM but not in TEM due to the experimental peculiarities of both techniques (TEM detects unscattered electrons, or electrons inelastically scattered at low angles; SEM detects electrons scattered at large angles). Comparing the two SEI data, it was observed that the larger particles are closely packed, but the individual particles are still distinguished in the aggregates. On the other hand, the smaller particles are coalesced, forming a dense film with few pores that allows the identification of only a few individual particles. The BEI images provided three important results:

1. In the larger particle sample, the particle borders could be clearly seen, and the particles appeared smaller and more clearly separated than in the SEI picture. It was concluded that the particle cores are richer in Si (the heavier element) than the particle shells. This was found to be in

agreement with the results from TEM elemental mapping, indicating that the shells are made out of a more hydrous silica than the cores.
2. In these images, bright areas were observed in the background of the BEI micrograph of the larger particles in which particulate material was not detected. This is independent evidence for the existence of nonparticulate, nondializable solute, thus confirming the TEM results.
3. The BEI image of the smaller particles showed small brighter dots, from which it was concluded that the silica film is made out of domains with varying degrees of hydration. AFM and SEPM images were obtained by noncontact AFM and SEPM. These samples were also prepared at high particle loading. It was found that the larger particles were strongly deformed in the AFM picture. Particles were deformed in every direction; consequently, their deformation does not follow a preferential direction that could be assigned to a scanning artifact. However, it was noted that this image was taken from a border region, from the area covered by the dispersion droplet used for sample preparation.

The film obtained by drying the smaller particles was smooth and nonporous, and its low roughness was found to be related to that of a mirror-like reflective surface (less than 1/20 visible light wavelength). From image analysis, it was concluded that the domains containing less hydrous silica were more negative than the domains made with more hydrous silica. This was found to be in accordance with the observation of darker cores in the SEPM image for the larger particles. A comparison of the results of particle examination by these three techniques is presented as follows:

1. EFTEM, bright-field
   a. Coarsely spherical, smooth surfaces, thin "necks" joining clusters of particles; uniform sizes, irregular shapes, marked "neck" formation
2. EFTEM, Si and O maps
   a. Core-and-shell particles: the outer shells have a higher O/Si content than the particle bulk; particle borders are seen in the Si map, but they are almost indistinguishable in the O map
3. FESEM, secondary electrons
   a. Contacting but discrete particles, few coalescing particles; highly coalesced particles form a film with some pores; few individual particles are discerned
4. FESEM, backscattered electrons
   a. Contrast between bulk and borders (particles appear to be smaller than in the SEI mode), broader interparticle voids; significant contrast with bright domains smaller than the particles and within domains with large numbers of coalesced particles; also, nonparticulate material
5. AFM, noncontact
   a. Particles appear strongly distorted, without a preferential direction for distortion.

Formation of a continuous, densely packed film with low roughness and without morphological features identifiable as the original particles. Particle cores are electrically negative, relative to particle shells. There is significant contrast within particles. Black (negative) and gray dots in a bright (positive) matrix cam be seen in variable dot sizes, but most extend for a few tens of nanometers, in the same range as the particle diameters.

The measurements of two Stöber silicas in different size ranges, by different microscopy techniques, showed that the particles are quite different in their chemical nature and plasticity. In most sols, that consist of discrete spherical particles of amorphous silica, the interior of the particles is made out of anhydrous SiO_2, and the surface holds silanol groups that are not lost when the silica is dried to remove free water.[782] These data showed that silica particles have a core-and-shell nature. However, comparison of the images from SEI and BEI showed that the thickness of the shell layer extends over many nanometers. This showed that silanol groups were found not only in a silica particle surface monolayer but also over a thick particle shell. The smaller particles were found to be quite heterogeneous, as detected by SEPM. The shapes of these particles departed strongly from spherical, even when they were well apart. This means that a low particle–solvent interfacial tension exists. As expected, a larger interfacial tension would drive particles more strongly toward aggregation, thus hindering their existence as nanoparticles. The small particles exhibit plastic characteristics, because they easily form dense films. The larger particles could also undergo distortion, but they were able to retain their identities without merging into a coalesced film. The core-and-shell nature of the larger particles and the chemical heterogeneity of the smaller particles may be understood considering the particle synthesis procedure and the intervening reactions.[783,784]

It was proposed that the first hydrolysis of an alkoxide group be the rate-limiting step in the formation of small nuclei, and particle growth after the formation of the stable seed particles occur mainly by the addition of a monomer to the surface, not by aggregation of the small nuclei particles.[785,786]

It was postulated that the hydrolyzed monomer reacts to form microgel polymers that collapse upon reaching a certain size and cross-linking density.[787] The denser seed particles grow by adding hydrolyzed monomer or polymer to their surface. It is proposed that the polymeric particles are better described as mass fractals,[788] in which the nucleating backbones or seeds are used to build compact and stable particles observed later in the growth.[789]

These studies showed that small particles are highly heterogeneous and plastic, whereas the larger particles are more uniform and less easily deformable. Furthermore, both types of particles have internal domains, characterized by different degrees of hydration: in the larger particles, the more hydrated domains concentrate in a thick surface layer, but these are randomly scattered throughout the smaller particles. Nonparticulate material was detected in the finished dispersions using two independent techniques, which adds a new element to the understanding of these silica dispersions. These results provided information on the formation of different kinds of siloxane chains in solution followed by their heteroaggregation, forming small gel particles of nonuniform chemical composition. However, many chains persist as polymeric solutes. The particles are strongly deformable due to their hydrated domains, and the more hydrophilic (more branched) chains accumulate at the outer particle shells, driven by the minimization of interfacial tension. Deformation has also been shown by previous studies on other polymer particles. These conclusions are in agreement with the mechanism described elsewhere for core-and-shell latex formation.[790]

7.4 NANOLITHOGRAPHY AND NANOMACHINING

As one can visualize from the above studies, there is an increasing interest in the science and nanotechnology of nanometer scale designed useful structures. In other words, instead of imaging a surface, one creates a surface morphology as required for some purposeful application, electronics or biological. The complexity of such an experiment can be realized by considering the molecular perfection needed to create a working nanostructure. Lithographic technologies will be superior for the fabrication of model nanostructures. Micro-sized lithography is an ancient technique known to mankind. For instance, in some parts of the world, a holy word or even sentence, has been written on a grain of corn or rice. There has been an increasing interest in the surface science of nanometer scale supported structures. This has resulted from the extensive research of uniform and flat (at molecular scale) solid surfaces. Lithographic technologies are superior for the fabrication of model nanostructures, for instance, in semiconductors or biosensors. A conventional lithographic process is carried out as follows:

SUBSTRATE......................
RESIST COATING.............
....
ELECTRON/PHOTON EXPOSURE
. . . .
. . . .
. . . .
. . . .
DEVELOPMENT/LIFTOFF

However, one may use the AFM and STM technology after some necessary modifications for nanolithography.[791,792] It is easily realized that by nanolithography, one can print over one million pages of text on one regular page. This is remarkable in various ways, especially when considering if such lithography will be reproduced on platinum surfaces. This would leave an imprint of lasting property.

In a recent study, lithographic methods were demonstrated as an application to fabricate model systems for surface science.[793] There were different model systems reported:

- Lithographic fabrication of model supported catalysis
- Lithographic fabrication of model nanostructures using polymers

It was mentioned that resolution will be the major challenge in lithographic fabrication. This may even turn out to be the limiting step in nanoscale surface science applications. In the case of industrial supported metal catalysts for hydrocarbon conversion reactions, the size distribution is 1 to 100 nm. It is not a trivial task to attain such a range of resolution. Photolithography is known to have an ultimate resolution of 0.1 μm (100 nm). Therefore, it is seen that the latter cannot be used for fabrication of model catalysts. Lithographic technology is expected to be superior for fabrication and design of model nanostructures. This is basic for the semiconductor industry. However, it is accepted that in due time, sub-10 nm structure size will be achieved. Nanotechnology is becoming a science that will be common in most laboratories in the upcoming decades.

STM has been used extensively for imaging and for surface modification of such substances as $MoSe_2$.[793] In these studies, a variety of etching patterns were constructed of $MoSe_2$ (Figure 7.2).

STM in air was used to investigate the topographic and diffusion properties of the surface formed by melting gold and platinum wires into spheres.[794]

(a)

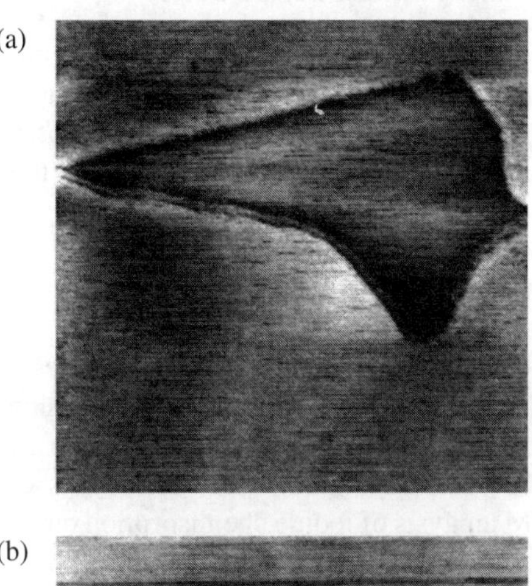

(b)

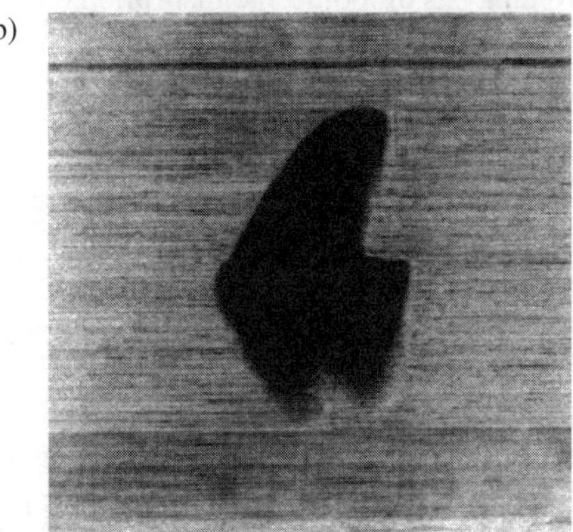

FIGURE 7.2 Etching patterns observed in the modification of $MoSe_2$: (a) 600 nm × 600 nm; (b) 500 nm × 500 nm. (From Ohmori et al., *Langmuir*, 14, 6287, 1998. With permission.)

Nanolithographic parameters, including thresholds for surface modification of the gold and platinum surfaces, were studied, and coefficients for self-diffusion were determined.

This is carried out by increasing the force applied to the cantilever; the surface can then be scraped, leaving behind a rectangular depression. In the literature, more results were reported using this procedure. In a recent report,[795] a procedure for writing and reading by STM at a nanoscale was described. The substrate used was gold and an electrochemical cell.

It was also shown that atomic-resolution tunneling microscopy can be done under nonpolar liquid.[796] Two possible areas of application were explored: topography and imaging air-sensitive materials. The lithography was demonstrated by writing a hole 20 Å × 40 Å (wide) × 20 Å (deep) onto a gold surface. The imaging of air-sensitive materials was demonstrated by imaging GaAs.

In another report, a general procedure of manipulating atoms on solid surfaces with the help of a tip in a STM was described.[733]

7.4.1 Atomic Switch (Nanoscale) by STM

In general, the ultimate size of any component in an instrument is almost a wish to make it as small as possible (nanoscale). Reactions taking place at or near a single molecule could be classified as the smallest nanoreactor possible. The latter is almost within reach by using STM and AFM. In order to reach this goal, STM and AFM have contributed much in the recent decade. By combining various other techniques with STM and AFM, one may expect a large variety of industrial applications in nanoscale instruments. An atomic switch was realized by using the principles of STM, according to a recent report.[797]

A multiple photochemical switching device was described by using an LB method consisting of molecules with switching, transmission, and working units.[798] It was shown that by using different switching units to assemble the device, its conductivity could be controlled by four different types of light irradiation. The potentiodynamic STM analysis of iodide chemisorption on Au (100) was reported.[799]

STM studies of Au etching of the electrode surface in the presence of adsorbed tetramethylthiourea (TMTU) were reported.[799] In Figure 7.3, the STM images of etching of Au electrode are given as a function of time.

7.4.2 Solid Surface Manipulation at Molecular Scale

The STM may be used as a solid surface modification tool at the molecular level. The phenomenon of metal surface reconstruction was investigated, whereby the top atomic layer assumed ordered structures that differed markedly from the bulk-phase crystal lattice.[800]

STM has been useful in obtaining real-space/time insight into the local structural changes associated with surface reconstruction.[801] The capacitance of a substance determines the ability to add electrons or to extract them. STM exhibits stepwise charging (Columbic staircase) which is expected for such materials as metal films. Single molecules (liquid crystals) have also been found to exhibit this incremental charging behavior.[802]

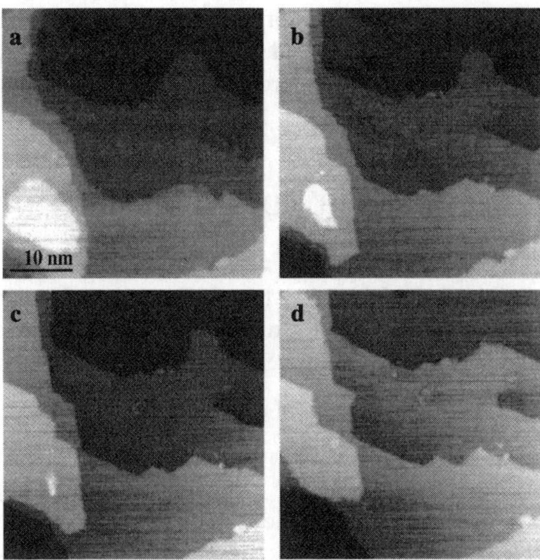

FIGURE 7.3 STM images of Au in 0.1 M H_2SO_4 + 0.00001 M TMTU. Images B through D were taken after 1.25, 2.5, and 3.75 min after image A. (From Bunge et al., *Langmuir*, 13, 85, 1997. With permission.)

A method was described that allowed one to study the electrochemical behavior of a single molecule by using the principles of STM and the scanning electrochemical microscope.[78]

The kinetics of graphite oxidation that showed monolayer and multilayer etch pits in HOPG were studied by STM.[803]

Recently, a STM study of Ag on Pt(lll) surfaces was reported, and it was concluded that upon annealing, Ag submonolayer films, at temperatures above 620 K, Ag atoms at Pt top edges, diffuse into the adjacent Pt terraces forming small Ag clusters, diameter ~10 Å, within the top substrate Pt terrace.[804] Furthermore, Pt clusters of smaller dimensions were found to build up within the Ag adsorbate layer at $0.5 < Ø < 1$. This phenomenon was found to be restricted to the first layer, i.e., the second and following Ag(Pt) layers do not exhibit this mixing effect. These findings suggest that at half monolayer Ag coverage, Pt clusters develop in Ag islands.

7.5 QUALITATIVE AND QUANTITATIVE ANALYSES AT NANOSCALE

As already mentioned above, one can easily observe from the images that in STM and AFM methods, one needs virtually only a few molecules. In other words, extremely small amounts of material are needed for analyses, which can be less than µg. Nanoscale analyses (qualitative and quantitative) are now widely performed using STM and AFM. These analyses are qualitative and quantitative. The surface morphology has already been seen to provide information on the purity of substances. The quantitative analysis is seen in the case of cholesterol collapsed monolayer analyses (see Figure 3.3). The analyses of water (distilled tap water) as mentioned elsewhere could be used qualitatively and quantitatively. The images could be subtracted such that any water purification process, for example, could be analyzed by these AFM image analyses.

7.6 APPLICATION OF SPMS UNDER DYNAMIC CONDITIONS

It has not been possible in the laboratory to see dynamic molecular reactions, until the advent of SPMs. SPMs have been extensively used to create movies of "live" reactions between molecules. For example, adsorption–desorption of gas molecules on solid surfaces have been studied.

7.6.1 STM Studies of Adsorption of Gas on Solid Surfaces

The deformation of a stressed crystal implies essentially elastic, inelastic, and plastic-type deformations leading to simultaneous movements of molecules.[805] These kinds of elastic deformations are instantaneous and reversible. Plastic deformations produce localized skip lines and bands from dislocation motions. In hydrogen charging and in the use of hydrogen storage in different metals, mechanical deformations have been observed. In a recent study, STM measurements of Pd single-crystal domains during hydrogen charging/discharging cycles were studied in order to follow the formations produced by the Pd.[806] These STM surface roughness studies proved to be related to hydrogen adsorption.

7.7 APPLICATION OF AFM TO IMMUNODIAGNOSTIC SYSTEMS

Diagnostic applications are becoming more widespread as the need increases for common use of such devices, for example, diagnostic applications in third-world areas is a much needed market. Furthermore, techniques such as STM and AFM have added even more possibilities. At this stage in the literature, one finds that the areas of such applications are almost unlimited. The ability to precisely position and orient biological molecules on microengineered substrates is an enabling technology critical to the long-term goal of integrating biomolecular motors with nanoelectromechanical systems.[807] Electron beam lithography and nanoimprinting techniques were used to produce arrays of nickel dots (50 to 200 nm diameter). Analysis using fluorescence (of microspheres attached to the subunit) and AFM of these arrays demonstrated precise positioning, spacing, and orientation of individual F1-ATPase molecules.

Emergent nanofabrication technologies enabled the production of nanoscale structures and devices with features as small as 7 nm.[808,809] However, the ability to power such devices has been hindered by the inability to produce silicon-based nanoscale motors with similar size and compatible force generation. Biophysical analysis of a variety of enzymes revealed that many act as nanoscale biological motors. Enzymes such as kinesin,[78] dynein, myosin,[810–812] DNA helicase,[812,813] and RNA polymerase[814] act as linear, stepped motors moving along or shuttling stationary tracks. In contrast, the bacterial flagella protein complex[815,816] and adenosine triphosphate synthase (ATPase) act as true rotary motors.[817–821] The force generation and size scale of these biomolecular motors are compatible with nanofabricated structures; thus creating the potential exploitation of these enzymes to power hybrid nanoscale devices.[819,821] The rotary nature of ATPase was conclusively demonstrated by attaching an actin filament to the subunit of a thermostable F1-ATPase.[817,818] The subunit of F1-ATPase transitions among three equidistant catalytic sites during hydrolysis.[817,818] Analysis of these experiments suggests that the work performed by this enzyme exceeds 80 pN/nm and is approximately 100% efficient.[818–823] Similar findings were also reported for an F1-ATPase from *Escherichia coli* and spinach (*Spinacea oleracea*) chloroplasts.[824–826] On the basis of its force generation, no-load velocity of ~17 rps,[827] and an overall diameter of 10 nm, the F1-ATPase is a tailor-made molecular motor. Thus, current research being conducted in our laboratory is focused on the integration of F1-ATPase with nanoelectromechanical systems (NEMS) to produce a hybrid nanoscale device.[819–821] Several prerequisite technologies, however, must be established prior to realizing this goal. One fundamental technology centers on the precision attachment of biomolecular motors to engineered, nanofabricated substrates. The future evolution of the underlying mechanisms for interfacing biological molecules with NEMS will provide the impetus for further development of advanced hybrid devices. Two methods, electron beam and nanoimprint lithography, were used to create nanofabricated substrates capable of precisely and accurately positioning individual biological motors. In the first method, a standard silicon wafer was coated with two layers of electron beam-sensitive polymer poly(methylmethacrylate) (PMMA) resist, with the lower layer slightly more sensitive to the electron beam exposure. Arrays were patterned into the bilayer using electron beam lithography, followed by development. A 5 nm tantalum (Ta) adhesion layer and a 10 nm metallic nickel (Ni^{2+}) layer were deposited using electron-beam physical vapor deposition. Liftoff was performed using a 1:1 solution of methylene chloride:acetone for 15 min followed by ultrasonic agitation for 10 s. Regular arrays of 60 to 600 nm size dots on 1 m pitch with circular profiles were confirmed by SEM. Arrays were stored under appropriate conditions (e.g., vacuum) to prevent oxidation of the Ni surface prior to experimental use.

The precision attachment of individual F1-ATPase biomolecular motors on the Ni dot arrays was demonstrated using two methods. In the first method, a recombinant F1-ATPase from the thermophilic bacterium, Bacillus PS3, was expressed in pKK223-3 and purified using ion exchange and size exclusion chromatography.[821] The recombinant F1-ATPase was specifically engineered for integration with nanomechanical systems and contained a 10× histidine (His) tag on each of the subunits, as well as a unique cysteine residue at the tip of the subunit (Cys). The His tags were engineered onto the subunits due to the high affinity of poly-His for cationic Ni conjugated to a variety of media, as well as metallic Ni.[828–830]

F1-ATPase molecules were specifically biotinylated through disulfide linkage between cystine and biotin malemide in the presence of N',N'-dimethylformamide. Biotinylated F1-ATPase molecules (100 g/mL in 10 mM phosphate buffer, pH 7.0) were allowed to attach to Ni arrays through diffusion and adhesion to dots for 15 min at room temperature, followed by washing with phosphate buffer. Subsequently, 1 μm streptavidin-coated, fluorescent microspheres were incubated on the arrays, followed by washing with phosphate buffer to remove all unbound microspheres. Data were also reported on arrays incubated with streptavidin-coated microsphere in the absence of F1-ATPase.

Attachment of F1-ATPase-microsphere complexes was examined using epifluorescence microscopy, demonstrating attachment of the complexes in specific array patterns. F1-ATPase molecules did not attach to all dots in the array; however, at least 50% attachment was observed over large areas. Streptavidin-coated microspheres were not observed on arrays that had not been coated with F1-ATPase. Oxidation of Ni dots potentially may have limited the number of chemically active dots capable of binding F1-ATPase motors. Other factors such as diffusion rates and F1-ATPase concentration may be important in regulating the efficacy of attachment. In order for adhesion to occur, the F1-ATPase must attach to the Ni surface through the His-tags engineered on the subunit of the molecule, thus leaving the biotinylated subunit perpendicular to the substrate and able to bind a streptavidin-coated microsphere. Because the subunit of F1-ATPase rotates in response to ATP hydrolysis, the molecule is correctly oriented to utilize the rotary nature of the subunit. Overall, this experiment demonstrates the ability to precisely position F1-ATPase motors in two dimensions (i.e., x and y), with the proper orientation for utilizing the work performed by the subunit during ATP hydrolysis. Because the previous experiment could not eliminate the potential adhesion of more than one F1-ATPase molecule per dot, AFM was used to verify the precise attachment of individual biological molecules on nanofabricated arrays. Biotinylated F1-ATPase molecules (10 to 1000 g/mL in 10 mM phosphate buffer, pH 7.0) were allowed to attach to Ni arrays, for 15 min at room temperature. Following extensive washing with phosphate buffer, arrays were imaged in TM AFM to reduce the force between the tip and the sample. Arrays prior to addition of F1-ATPase, as well as arrays coated with phosphate buffer, were used as negative controls. AFM imaging of nickel dot (60 to 600 nm diameter, 4 to 20 nm high) arrays produced by electron beam lithography demonstrated the presence of individual F1-ATPase molecules (14 nm high and 8 nm diameter) on the nickel dots. The apparent diameter of the F1-ATPase molecules was 30 to 50 nm due to AFM tip–F1-ATPase convolution. Deflection in the AFM tips corresponding to the presence of F1-ATPase motors was easily observed due to the relatively smooth surface of the dots. Nickel arrays incubated with 10 mM phosphate buffer (pH 7.0) were used as controls; no protrusions were observed, substantiating the identity of F1-ATPase with the AFM. Depending upon F1-ATPase concentration and Ni dot size, the density of F1-ATPase varied from one molecule per 30 dots to 2 to 3 molecules per dot. In general, smaller dot sizes preclude the adhesion of more than one F1-ATPase molecule per dot.

It was shown that optical lithography, on the other hand, cannot produce such small dimensions due to the wavelengths employed. An alternative strategy, nanoimprint lithography, has been used to obtain large numbers of nanoscale structures including nanometer-size dots.[831–833] Consequently, this process is well suited for producing large nickel dot arrays for precision attachment of biological molecules.

The enhanced production capabilities associated with nanoimprint are appropriate for subsequent application of these arrays for constructing integrated nanomechanical devices powered by biomolecular motors.

One may even propose the possibility of studying antibody–antigen interaction mechanisms by attaching antibody to the tip, and studying force curves over substrates with antigen:

TIP
BONDED
ANTIBODY
:::
:::::::::::FLUID MEDIUM::::::::::::
:::
ANTIGEN
SUBSTRATE

Furthermore, *nanoimprint lithography* involves transferring engineered patterns from a mold by physically pressing it against a surface previously coated with a polymer (e.g., PMMA). In this case, the mold consisted of atomically sharp tips and was fabricated on a monocrystalline silicon substrate. In this context, dots, 500 nm in diameter, were defined by photolithography using an i-line stepper. The pattern was transferred onto the silicon dioxide layer through fluorine- and chlorine-based reactive ion etching (RIE), followed by thermal oxidization of the silicon for initial definition of the tip structure. Because of a stress effect around the neck of the post, a nanometer-scale, sharp tip was obtained.[834] This tip was used again to remove the oxide from the floor of the structure, followed by chlorine-based RIE to further etch the silicon and create a shaft for the tips. The oxide was removed using a 1:6 buffered hydrofluoric acid solution. Optical lithography was used to create the nanoimprinting mold (A), which subsequently was used to create large arrays of nickel dots (B). Dots were 50 to 250 nm in diameter and 5 to 15 nm high. To form the patterned arrays, the mold was pressed against a quartz substrate or silicon wafer coated with PMMA and heated to 120°C (above the glass transition temperature of PMMA). A mold release agent (e.g., a 10% solution of lanolin in mineral oil) was used to facilitate the withdrawal of the mold. The tips only penetrated partially into the PMMA to prevent damage to the structure. An oxygen-based RIE was used to extend the printed pattern down to the surface of the substrate to form the base of the dots. Later, a metal film (i.e., Ni) was deposited and then lifted off to expose the nickel arrays. Arrays produced using nanoimprint lithography contained uniform nickel dots of 50 to 250 nm diameter. Arrays were missing a small percentage of dots depending upon the variables in imprinting. Larger arrays (10 to 20 times larger) could be produced with nanoimprint, as compared to electron beam lithography. Attachment of F1-ATPase biomolecular motors was confirmed using fluorescence microscopy as described above; similar patterns and efficiency of F1-ATPase attachment to Ni arrays were observed. Hence, nanoimprint lithography offers a viable alternative for production of biocompatible arrays capable of precisely attaching biomolecular motors. Both nanofabrication methodologies, nanoimprint and electron beam lithography, were capable of producing nanoscale, biocompatible arrays for specifically attaching biomolecular motors. The ability to precisely position and orient individual biological molecules represents an enabling technology fundamental to creating hybrid NEMS devices powered by biomolecular motors.[819,820] The processes and chemistries outlined above have helped in establishing a system for integrating organic and inorganic materials in an engineered system. Further refinement of the processes will permit the construction of Ni arrays with objective and design-specific patterns. In addition, NEMS structures, such as valves and pumps, can be integrated into this system and permit the construction of novel, useful nanomechanical devices powered by biomolecular motors.

In a recent report, the conventional nanoparticles based on acrylic compounds that are lipophilic and possess a negative surface charge were studied.[605] This is due to their manufacturing process and to the chemical structure of the polymer. Hence, these particles are not suitable for the adsorption of hydrophilic anionic drugs. In the present investigation, positively charged copolymer nanoparticles prepared from aminoalkyl- and methylmethacrylates were evaluated with regard to their physical properties. This report provides a detailed description of the synthesis of the non-commercially available monomers and their polymerization procedure. Various parameters were investigated, such as comonomer content, total amount of monomer, concentration of the radical initiator, and the composition of the polymerization medium. The resulting particle diameter and the surface charge were found to be strongly dependent on the polymerization conditions and on the pH. Optimization of the polymerization procedure yielded nanoparticles of about 200 nm exhibiting a positive surface charge. The charges of the different copolymer particles were then compared at different pH values. *N*-trimethylaminoethylmethacrylate (TMAEMC) nanoparticles with quaternary ammonium groups located at their surfaces, possessed a nearly constant positive zeta potential at various pH values and, consequently, pH-independent particle diameters. The physical characteristics of the other aminoalkyl copolymers correlated with the basicity of the monomers employed and were found to be strongly dependent on the pH of the dispersion medium. Aminoethylmethacrylate (AEMC), methylaminoethylmethacrylate (MMAEMC), aminohexylmethacrylate (AHMC), and aminoethylmethacrylamide (AHMAC) copolymer nanoparticles exhibited strong positively charged surfaces even at physiological pH and, thus, can be useful for the adsorption of anionic drugs.

The application of the LB method to assemble biomolecules on solid surfaces provides a useful procedure for biological sensors.[20] Such LB films are advantageous, because the biological activity can be optimized by tailoring the layer compositions. LB films are highly structured and reproducible. Factors like charge density, density of the linking group, and composition of the monolayer matrix have particularly large effects on the activity. In a recent study, oxidase enzymes were immobilized on SAMs, and conductive layers of thiophene oligomers were placed on gold clusters.[836]

The enzymes were bound by electrostatic interaction of the negatively charged enzyme with the positively charged pyridinium groups. The enzymatic activity of glucose oxidase was dependent on the surface density of the conduction and on the ionic strength during adsorption. Further studies were carried out on Fab-fragments bound to LB films of various lipids, and the immobilization efficiency (relative amount of bonding sites) was optimized by variation of the constitution of the lipid LB film. The films were deposited onto various substrates by vertical contact transfer of a preformed mixed Langmuir film of the linker lipid and a matrix lipid. The immobilization efficiency was determined with radioassay, quartz crystal microbalance (QCM), and surface plasmon resonance (SPR) measurements, while AFM was used to image the structure of the films at different stages of binding. After Fab-binding, AFM images showed globular structures. The immobilization efficiency of the LB film appeared to be much higher, as that of more conventional methods was less than 10%. Films with di-myristoyl-phosphatidyl-lecithin (DMPL) as the matrix lipid and a maleimide derivative of di-myristoyl-phosphatidyl-ethanolamine (DMPE) showed the highest sensitivity (and lowest detection limit) in QCM measurements.[836]

A chemical device is a sensor that converts a quantity of a solute that has to be detected into an electrical or an optical or any other suitable signal. Among these sensors, biosensors use biological device molecules, mainly enzymes, as recognition detecting molecules. It is essential that the enzymes used be immobilized under a controlled way in order to reproduce the detection signals. Immobilization can be performed by covalent adsorption on a transducer or on a suitable polymer matrix. The LB film method can also be used in these sensors. For alcohol sensing, an alcohol dehydrogenase (ADH) which necessitates a coenzyme, a nictonamide adenine dinucleotide (bet-NAD) in its oxidized form or NADH in its reduced form, can basically be applied.[837] The detection principle is known to proceed by ADH-catalyzed conversion of the nondetectable ethanol (CH_3CH_2OH); the detectable element is a reduced form of coenzyme NADH:

$$CH_3CH_2OH^{+}b\text{–}NAD^{+} - ADH - CH_3CHO\ (\text{aldehyde}) + NADH + H^{+} \quad (7.1)$$

The species NADH is then reoxidized into b-NAD^+ at the electrode. The activity of enzyme in the LB film was investigated.

Acetylcholinesterase (AChE) is an intrinsic membrane-bound enzyme that is essential for nerve tissue.[838] This enzyme is responsible for the rapid hydrolysis of the cationic neurotransmitter acetylcholine after its release at cholinergic synapses. AChE is a target site for a variety of commonly used insecticides in widespread use and nerve gases. The hydrolysis reaction of acetylcholine catalyzed by the enzyme, AChE, was studied at the air–liquid interface by spreading the enzyme as a monolayer and dissolving the substrate in the subphase. The AFM images of AChE on HOPG as LB films revealed that the enzymes are well-resolved, and their shape tends to be ellipsoidal. This analysis agrees with the x-ray data.

In a recent study, the site-specific immobilization of human serum albumin (HSA) to an ordered array of nanometer-sized pits drilled in a gallium arsenide wafer was studied by AFM (SFM).[839] Before the adsorption of has, the wafers were cleaned and studied by AFM. HSA solution was placed on a wafer and after drying was studied using AFM. The pits were imaged before and after HSA adsorption. After adsorption of HSA, the rim of each pit in wafer was found to have features corresponding to clusters of several HSA molecules. This behavior is in agreement with hemoglobin images as mentioned earlier (Figure 4.4).

7.8 APPLICATIONS OF STM AND AFM IN INDUSTRY

Throughout this text, application areas of STM and AFM were mentioned, mostly indirectly for different industries. In this section, more pertinent and direct applications to industrial inventions will be described. For example, STM has become a valuable tool in investigating systems where conduction is involved at interfaces. For instance, in a recent study, the benzotriazole adsorption and inhibition of Cu (100) *corrosion* in HCl was investigated.[710] In this study, a combination of *in situ* STM and *in situ* FTIR spectroscopy was used.

The surface coordination chemistry of the cuprate superconductor, $YBa_2Cu_3O_7$, was investigated using cyclic voltammetry in conjunction with a series of redox-active ferrocenyl adsorbates.[840] AFM images of these superconductor films indicated roughness on the 100 nm scale. Images of a typical film showed grains and particles with diameters in the range 10 to 100 nm. Patterned SAMs of alkanethiols were used to create patterned surfaces that exhibit gradients in chemical properties.

The studies of inorganic and organic hybrid molecular materials have recently become of interest. These structures provide an opportunity to combine the functionalities associated with each type at close proximity. The LB film method is the most stable technique for constructing such structures.[20] Inorganic and organic hybrid ultrathin films of MoS_2 and a cationic amphiphile, dihexyadecyldimethylammonium bromide (DHA^+Br^-), were prepared using the LB film technique.[841] The surface pressure versus area isotherms of DHA^+Br^- changed upon addition of MoS_2 in the subphase. Quartz plates were used for LB films. AFM was used in noncontact mode with cantilever of 1.5 N/m. AFM images showed that the LB films were flat and plate-like. The typical step height was 2 nm, corresponding roughly to the interlayer spacing determined by x-ray data. From these data, it was concluded that assemblies consisting of hybrid DHA^+/MoS_2 LB films can be constructed in which inorganic and organic components are stacked alternately.

Nanoparticles of metals and semiconductors have attracted much interest for potential applications as photonic materials based on their nonlinear optical properties.[842–846] Furthermore, electron tunneling and charging phenomena of metal quantum dots have been opening a new basis for future single-electron devices.[847–849] To fabricate these nanometer-scale devices, at this stage, a simple method for nanoparticles to be embedded and arranged in a suitable thin film medium is needed, instead of other complicated processes such as lithography procedures. Several techniques, such as ion implantation,[850] rf-sputtering,[851] plasma deposition,[852] etc., were reported for preparing metal nanoparticles in dielectric matrixes; however, they are not applicable for fabricating and patterning particles in a desired arrangement. Au nanoparticles in silica/titania (SiO_2/TiO_2) glassy films were prepared using a sol-gel process.[853–856] Photochemical reduction of Au(III) ions doped in thin films dip-coated from SiO_2/TiO_2 precursor solutions yielded Au particles in the film.[857–859] The particle formation occurred by diffusion and reduction of doped Au(III) ions, both of which were promoted at higher TiO_2 contents. Therefore, the size and morphology of the particles were changed by the SiO_2:TiO_2 ratio. This controllable photogeneration process enabled us to fabricate micropatterns of Au nanoparticles in glass films using photomasks. However, the patterning resolution is limited within the wavelength of irradiated ultraviolet light, and it does not allow us to fabricate a single particle in a nanometer-scale arrangement.

Studies have been made to achieve lower dimensional structure — quantum wires and quantum dots (zero-dimensional).[710] In combination with lithography, one can carve out or electrostatically squeeze thin lines or dots or patterns. The potential of this application of STM is in the initial stages of research. In another application, STM was used to measure the wave patterns in a metal.[710]

The progress of SPM has extended its application as a nanofabrication tool. STM has allowed manipulation of individual atoms on a clean metal surface in an UHV.[860] AFM has permitted measurements of local properties in air that are imaged under an independent atomic force feedback control using a cantilever. Nanoindentation is carried out by mechanical contact of a cantilever to the sample surface.[861] A metal-coated cantilever enables us not only to image the electric conductivity at a constant gap between probe and surface but also to induce local chemical reactions by applying bias voltages.[862–864] Nanostructures on silicon surfaces have been fabricated by AFM-field-enhanced oxidation.[862] Furthermore, *NSOM* has broken through the wavelength limit, and one can obtain optical imaging at a resolution below several tens of nanometers.[865,866]

Using NSOM, the surface plasmon band of the Au particles photogenerated in the SiO_2/TiO_2 film was detected.[853] Introduction of UV light into a fiber probe of NSOM would be an alternative tool to locally generate Au particles and simultaneously characterize their optical properties by the near-field probe light, although the resolution of fabrication and characterization of the particles by NSOM would be limited to tenths of the wavelength, depending on the fiber probe aperture.

In view of the higher resolution and easy feedback control with the laser-controlled cantilever, AFM has been used to perform nanofabrication of Au particles in the SiO_2/TiO_2 films prepared by a sol-gel method.[853] Gold (Au) nanoparticles were fabricated in thin glass films by a sol-gel process and AFM.

Using a conductive AFM cantilever, local reduction of Au(III) ions doped in a dip-coated silica/titania gel film generated Au particles embedded at a certain position of the film. The sizes of the particles were controlled by the voltages applied between the cantilever and an indium-tin-oxide-coated glass substrate. Scanning the biased cantilever produced Au particles dispersing in the scan area, which exhibited a visible absorption spectrum corresponding to the typical surface plasmon band of nanometer-scale Au particles. Further scanning resulted in Au clusters heaped at the scan area, which gave a red-shifted absorption band due to aggregation.

Single-particle formation was carried out by AFM-assisted local reduction using a metal-coated conductive cantilever and a conductive glass substrate. The size of the generated Au particles was controlled by the bias voltage and distance between the cantilever and specimen film. Scanning reduction was also performed to pattern the Au particles in a specific area, and their densities and morphologies were changed by the number of scan rates and of cycles.

Thin films of SiO_2/TiO_2 containing Au(III) complex ions [abbreviated as Au(III)-SiO_2/TiO_2] were prepared by the sol-gel process with an ethanol solution of chloroauric acid ($HAuCl_4Ã_4H_2O$) and a tetraethyl orthosilicate [$Si(OC_2H_5)^4$]/tetraethyl orthotitanate ($Ti(OC_2H_5)^4$)] mixture under acid catalysis.[853] A 6.7 mmol sample of $Si(OC_2H_5)^4$ and 3.3 mmol of $Ti(OC_2H_5)^4$ was added to 6 mL of a 0.05 M $HAuCl_4$ ethanol solution, and then 0.2 mL of 2 N HCl was added to the mixture. This *sol* solution was hydrolyzed overnight at room temperature. An indium-tin-oxide (ITO)-coated glass substrate used for film deposition by the dip-coating procedure was cut to 25 mm × 10 mm and cleaned with detergent, distilled water, and ethanol in an ultrasonic bath. In a desiccator box, the substrate was immersed into the sol solution and then pulled from the solution at 3 mm/s. For gelation by dehydration condensation, the coated Au(III)-SiO_2/TiO_2 was dried at 150°C for 1 min using a tube heater inside the desiccator. The film thickness measured by a stylus profiler was around 200 nm. The formation of Au particles in the Au(III)-SiO_2/TiO_2 films was performed using an AFM with a commercially available gold-coated Si_3N_4 cantilever. The film-containing ITO substrate was mounted on a piezo scanner, and electric contact was made on one edge of the ITO surface with silver paste. Prior to particle formation, topographic images of the film surface were recorded without application of bias voltages. Then, the cantilever was approached to a certain point of the surface. The reduction of Au(III) ions in the film was carried out by applying a negative voltage, *E*, to the ITO substrate using the cantilever as ground. Reduction of Au(III) ions was monitored from cathodic currents in the current–voltage (I–E) and current–distance (I–Z) curves, where *Z* denotes the distance between the cantilever and the film surface. Formation of Au particles was also carried out by scanning the cantilever in a certain square area while the ITO substrate was biased at a negative potential. The scanning rates and cycles were used to control the population and morphology of Au particles. To confirm the AFM-assisted formation of Au particles, visible absorption spectra were taken from the films in which multiple Au particles were formed in a square area (A) using the scanning procedure. Transmission measurements were carried out using an optical microscope equipped with a 40× objective lens, a 100 W halogen lamp, and a CCD multichannel spectrometer.

The present sol-gel process produced optically transparent Au(III)-SiO_2/TiO_2 films in which Au(III) ions were homogeneously dispersed in the matrix film. Dry films had a faint yellow color due to the UV absorption tail of the ligand-to-metal charge transfer (LMCT) band of the doped $AuCl_4$ ions.[867]

The surface morphology of the Au(III)-SiO_2/TiO_2 films was first examined by conventional AFM topography. Random scanning with the maximum scan range (35 × 35 m^2) showed that the surface roughness was less than a few nanometers. The AFM-assisted reduction of the Au(III)-SiO_2/TiO_2 films was carried out by scanning the cantilever in a smaller range after moving the cantilever to the center of the last topographic image, while the ITO substrate was biased at a negative potential. In these experiments, the center square area of the film was reduced by scanning. The images thus obtained showed particle-like protrusions having a height of 50 to 100 nm. On the other hand, the highly reduced area appeared as a square plateau with a height of ca. 100 nm. To characterize these reacted films, the scanned regions were examined by transmission spectroscopy using optical microscopy. The square region exhibited a faint purple color. The transmitted light through these images was transferred to the CCD spectrometer, and visible absorption spectra were obtained by calibrating the instrument with a reference spectrum corresponding to an unreacted region of the film. The film with the bumpy particles exhibited an absorption band at around 550 nm. This absorption spectrum corresponds well to the surface plasmon band of Au nanoparticles.[868–870] Therefore, it confirms that Au nanoparticles are produced in the scanned region by the AFM-assisted reduction of the Au(III) ions doped in the SiO_2/TiO_2 film, and a considerable number of the particles grow to stick out on the film surface. A possibility of particle formation by Au diffusion from an Au-coated cantilever can be excluded, because no particle formation was observed in experiments using a SiO_2/TiO_2 film in which Au ions were not doped. When further cycles of scanning reduction are employed in order to produce the square plateau, the absorption spectrum exhibits a broad, red-shifted band at 550 to 750 nm. These red-shifted absorption spectra are found to be typical of networks of Au particles.

In earlier studies, a similar red-shifted absorption band for Au particles photogenerated in the SiO_2 film was found, and TEM revealed joined large triangular plate-like and needle-like particles forming networks. This suggested that the protruded plateau structure may be formed by large networked Au clusters, in which an excessive amount of Au(III) ions is supplied from the surrounding region during successive reduction cycles.

It was also reported that when positive bias voltages were applied to the ITO substrate, there was no current response. It was concluded that the reduction of the Au(III) ions starts to occur at the ITO/film interface, and then an Au nucleus is formed on the ITO surface beneath the cantilever position. Thus, the nucleus grows to a particle, while Au(III) ions are supplied by diffusion from the surrounding region of the film. As reported for photogeneration of Au particles, the diffusion of Au(III) ions is known to be affected by the gel structure of the SiO_2/TiO_2 matrix film, which is dependent on the SiO_2:TiO_2 ratio.[870] At a higher content of SiO_2, formation of Au particles was not observed, probably because the matrix film consists of a tightly cross-linked structure, and ion diffusion is slow. On the other hand, when the content of TiO_2 exceeds 50%, the film surface became too rough to perform the AFM-assisted local reduction, due to the higher hydrolytic reactivity of the TiO_2 precursor. Consequently, a controllable fabrication of a single Au particle was achieved on films with SiO_2:TiO_2 ratios of 2:1, which most probably yield a loosely cross-linked structure. As the local reduction proceeds further, the particle grows bigger and then protrudes out of the film surface. The I–E data were obtained for AFM-assisted local reduction of the Au(III)-SiO_2/TiO_2 film in which E was scanned from 0 to –3 V with a sweep time of 50 ms. This concern of particle growth is also supported by results from a second and different procedure used to fabricate a single Au particle.

After the surface topography was imaged, the cantilever was moved to a certain position and kept in contact with the film surface. Then, while a constant bias was applied to the ITO substrate, the cantilever was pulled up in the Z direction from the surface at a constant rate of 16 nm/ms.

AFM images of the surface were obtained, where such local reduction was carried out at four corner points, at E = –1, –2, –3, and –4 V, respectively. At E = –1 V, no morphological change was seen on the surface. When E was increased to –2 V, a small particle with a diameter of 30 nm appeared. The particle grew 120 nm in diameter and stuck out of the surface with a height of 15 nm at E = –3 V. When E was raised to –4 V, the diameter further grew to 250 nm, while the circumference of the particle caved in below the surface. The current responses corresponding to the particle growth at the above four positions were recorded in the I–Z modes. These reduction processes exhibited cathodic current peaks when the cantilever was moved out at a Z range of 300 to 500 nm from the surface. The appearance of the low current peak (at $E = -1$ V) was suggestive that the reduction occurs to form an Au particle embedded in the film. This behavior was assumed to support the idea that particle growth starts from the substrate.

These Z values seem considerably higher when compared to the typical distance at which the cantilever is attracted to the surface by adsorbed water under air. A previous infrared spectroscopic study[12] revealed that the Au(III)-SiO_2/TiO_2 films still contained a considerable amount of unreacted alkoxides, alcohol, and water due to the low dehydration condensation temperature in the present sol-gel process. These species probably mediate the reduction process at long Z ranges.

An AFM image of the Au(III)-SiO_2/TiO_2 film surface was studied, where the AFM-assisted local reduction was performed at four corner points with $E = -1$ (A), –2 (B), –3 (C), and –4 V (D), respectively. After the surface topography was imaged, the cantilever was moved to the respective position and kept in contact with the film surface. The cantilever was moved up in the Z direction from the surface at a constant rate of 16 nm/ms, while the respective bias was applied to the ITO substrate.

From these studies, it was concluded that the AFM-assisted local reduction of Au(III)-SiO_2/TiO_2 films prepared by a sol-gel method provides a controllable fabrication of Au nanoparticles in a thin glass film medium. Embedding individual metal particles and arranging them on the nanometer scale is a useful method for nanofabrication of electrodes, nonlinear optical devices, and single-electron devices. Further developments of combined SPM are expected to enable us to fabricate an integrated architecture of nanoparticle materials in thin-film devices.

7.8.1 Domain Images by Scanning Force Microscopy (SFM)

In the current literature, the studies of the domain structure in ferroelectric thin films on a nonmetal substrate have shown direct assessment of the electrostatic mechanism contribution to the domain contrast in the SFM contact mode. It has been shown that the polarization charges of the ferroelectric film are effectively compensated. These investigations suggest secondary effects of the electrostatic tip–sample interaction on the domain imaging mechanism in ferroelectric thin films in contact SFM compared to the major contribution of the piezoelectric effect. SFM has become a powerful tool for the characterization of ferroelectric materials, providing crucial information on their dielectric properties at the nanoscale level. In particular, SFM has been reported to allow nondestructive nanoscale visualization of domain structures in ferroelectric thin films.[871–873]

Direct imaging of ferroelectric domains by means of SFM has proved to be particularly effective in elucidating the microscopic mechanisms of polarization reversal and degradation effects such as fatigue and retention loss in ferroelectric films.[874–880]

Most of the SFM studies of ferroelectric domains in thin films are based on the application of a modulation voltage to a probing tip in contact with the film surface.

It is assumed that the applied voltage generates a local film surface vibration due to the converse piezoelectric effect (that is why this method has often been called piezoresponse SFM).[871–880]

The phase and the amplitude of the vibration signal detected by the lock-in technique provide information about the polarization direction and the value of the piezoelectric coefficient, respectively. However, there is an alternative view on the mechanism of domain imaging in contact SFM, which could be related to the electrostatic tip–sample interaction due to the presence of surface polarization charges in the ferroelectric sample.[881–884] From the physical point of view, both effects could take place during domain imaging. It is also expected that their contribution to the measured signal is quantitatively different depending on the sample properties and experimental conditions. It has been shown that the piezoresponse and electrostatic methods produce almost identical images of the poled domain patterns in thin films.[885,886]

There have been studies reported on domain imaging in PbZr(0.20)/Ti(0.80) (PZT) films without the bottom electrode, which is known to exclude the contribution of the piezoelectric effect to the domain image. These data showed that the polarization charges were effectively compensated, and from this, it can be concluded that in contact SFM, the piezoelectric effect plays a dominant role in the ferroelectric domain imaging mechanism in film/electrode heterostructures.

Experiments were performed by means of a commercial force microscope using 210 nm thick PZT films deposited by laser at 600°C and 100 mTorr oxygen pressure on a $LaAlO_3$ substrate (LAO). The thickness of the LAO substrate was 500 μm.

Domain images were obtained by applying an ac modulation voltage with an amplitude of 10 V and a frequency of 10 kHz. The modulation voltage was applied to a standard Au-coated silicon nitride cantilever. The microscope was operated in a contact mode (repulsive force regime). Topographic images of the PZT film were obtained. A domain structure consisting of two different domains, with in-plane (INP) and out-of-plane (OOP) polarization, respectively, was observed in epitaxial ferroelectric thin films as a result of strain energy relaxation.[887]

The rectangular INP-OOP domain pattern was observed in the topographic image due to the tilting of the surfaces of domains.[871,888]

Because the applied voltage drops across the LAO substrate and no piezoelectric effect is induced, this domain image could be attributed only to the electrostatic interaction between the biased probing tip and surface charges.

The procedure for quantitative estimation of the bounded surface charge density is a known method.[883–890] The surface charge density can be estimated by measuring a dc tip bias as the electrostatic force vanishes. When a dc bias is applied to the tip during scanning along with the ac voltage, an additional component of the electrostatic force is needed.

A charge, induced in the gold layer of the tip by the electrostatic field of the sample and the applied voltage, interacts with the bound charges of these domains. By adjusting the dc bias, this induced charge can be compensated for, and the electrostatic interaction can be suppressed. The electrostatic domain image was observed at a zero dc bias. Two dark circular spots were found in this image after an ac voltage of 10 V was applied to these sites for about 10 min by the fixed probing tip. The data were analyzed by assuming that for contact SFM, a parallel plate capacitor model could be applied in PZT films. Even from this oversimplified model, it can be understood that the polarization charges of the PZT film are effectively compensated by accumulation of free charge carriers from the environment on the film surface. Application of the external dc or ac voltage for a sufficiently long time could change the surface charge distribution and, as a result, alter the electrostatic image of the domain pattern. Application of the ac voltage of 10 V to two sites on the film surface for about 10 min through the fixed SFM tip resulted in the appearance of two dark circular spots, which was ascribed to the change in distribution of surface charges. The layer of the film could be expected to reduce the electrostatic interaction between the tip and the sample. Hence, it was concluded from these analyses that, in the SFM contact mode, under typical experimental conditions used for domain observation, the ferroelectric domain imaging mechanism in film/electrode heterostructures is not due to the electrostatic interaction but is mainly due to the piezoelectric effect. However, the situation might be different in ferroelectric crystals with the low value of the piezoelectric coefficient.[884]

In this case, given that the field generated by the SFM tip is highly inhomogeneous, i.e., it is strong at the film surface but quickly decreases with distance in the film polar direction, which leads to the vanishing of the piezoresponse, the contribution to domain contrast from piezoelectric and electrostatic effects could be comparable. From these studies, it was concluded that the SFM imaging of the ac domain pattern in PZT films on the nonmetal substrate allowed direct assessment of the electrostatic signal contribution to the domain image in the contact mode. The results showed that in ambient conditions, the polarization charges are effectively compensated by free charge carriers from the environment and cannot significantly affect the domain contrast. Thus, in the SFM contact mode, the main imaging mechanism of ferroelectric domains in thin-film/electrode heterostructures is the piezoelectric effect.

7.8.2 Glassy Carbon (GC) Electrodes

Glassy carbon (GC) is used extensively in electrochemical analysis (pH glass electrodes, etc.) because of its chemical inertness. The activation of GC by one of several established pretreatment protocols is a common requirement prior to its application. The physical structure, electrochemical reactivity, and response stability of hydrogenated GC (HGC) electrodes were investigated by using AFM, Raman spectroscopy, XPS, mass spectrometry (MS), voltammetry, and chronocoulometry.[891] Si_3N_4 tips mounted on gold cantilevers (200 μm legs, 0.12 N/m spring constant) were used for obtaining AFM images. The GC and HGC were examined by AFM. It was found that GC surfaces are grainy in texture due to the fine carbon microparticle polishing layer that remains after ultrasonication. The mean roughness was in the range of 38 to 50 nm. This apparent roughness was almost the same as 50 nm diameter alumina grit used in the final polishing step. HGC was found to have a completely different surface morphology. The grainy character was absent due to the gasification (e.g., CH_4 or C_2H_4). The plasma-treated surface develops ridge-shaped features. The mean roughness of HGC was in the range 85 to 115 nm.

Graphite is an excellent choice for the negative electrode of "rocking-chair" batteries[892] due to its high charge density and superior cycling behavior. However, despite its rapid diffusion and commercialization in modern batteries, several problems remain unresolved. One major problem concerns the dimensional changes occurring during electrochemical processes. Knowledge of these changes is fundamental for an understanding of the electrochemical intercalation/insertion mechanism and for the evaluation of potential practical applications. The macroscopic changes occurring in the mechanical properties of graphite during insertion have been investigated for many years.[893]

While the ion exchange between an electrode and the bathing electrolyte may be monitored by techniques such as probe beam deflection (PBD) or electrochemical quartz crystal microbalance (EQCM), an *in situ* observation of dimensional changes of the electrodes is difficult.[894] Neutron,[895] x-ray,[896] diffraction, and dilatometry techniques can provide average information about the overall results of the intercalation processes but are completely useless on a local microscopic scale.[897]

It has been shown that SPM techniques can be used to investigate the surface of graphite electrodes under potential control and to probe thickness changes or local electronic properties, even on a nanometer scale.[898] The Li+ intercalation has been observed *in situ* by STM[899] and AFM,[900] which proved to be powerful techniques in the context of battery research. Several authors studied the electrochemical behavior of graphite in aqueous electrolytes.[901]

It has been shown that the perchlorate ion is one of the best intercalating species at relatively low acid concentrations (2 to 4 M $HClO_4$), while hydrogen sulfate ions can intercalate efficiently only at higher molarities (10 M H_2SO_4).[902,903]

The relationship between intercalation processes and side reactions such as oxidation has been investigated.[904] The Raman evidence for graphite lattice damage occurring as a consequence of anion intercalation suggests that the graphite intercalation compounds (GIC) initially formed will subsequently oxidize carbon to form graphite oxides. The observation of surface blister formation on graphite in concentrated acid electrolytes at high potentials by AFM[905] and STM[906] was taken as further evidence for a correlation between initial intercalation and subsequent oxidation. At low electrode potentials, no damage of the graphite surface was observed.[907] The latter observation suggests that the contribution of side reactions can be minimized by selecting the right combination of acid concentration and electrochemical potential.

In the context of ion transfer batteries, HOPG was studied as a model in aqueous electrolytes to elucidate the mechanism of electrochemical intercalation into graphite.[908] The local and time-dependent dimensional changes of the host material occurring during the electrochemical intercalation processes were investigated on the nanometer scale. AFM, combined with cyclic voltammetry, was used as an *in situ* analytical tool during the intercalation of perchlorate and hydrogen sulfate ions into HOPG electrodes, and their expulsion from the HOPG electrodes. This is the first time that a reproducible, quantitative estimate of the interlayer spacing in HOPG with intercalated perchlorate and hydrogen sulfate ions has been obtained by *in situ* AFM measurements. The experimental values are in agreement with theoretical expectations, only for relatively low stacks of graphene layers. After formation of stage IV, HOPG expansion upon intercalation typically amounts to 32% when tens of layers are involved but to only 14% when thousands of layers are involved. Blister formation and more dramatic changes in morphology were observed, depending on the kind of electrolyte used, at higher levels of anion intercalation. The data of cyclic voltammograms (second cycle) were obtained at HOPG in 2 M $HClO_4$ and in 1 M H_2SO_4 (scan rate: 25 mV/s).

The system of HOPG in perchloric acid was examined as a model in order to elucidate the mechanism of electrochemical anion intercalation in graphite. Perchlorate ions were used, because these are known to be excellent intercalating properties at relatively low acid concentrations. Furthermore, investigations were made with sulfuric acid in order to see whether the type of electrolyte has any influence on the intercalation mechanism. Studies were performed on local and time-dependent investigation of dimensional changes of the host material on the nanometer scale during the electrochemical intercalation processes. AFM, combined with cyclic voltammetry, was used as an *in situ* analytical tool during intercalation and deintercalation of perchlorate and hydrogen sulfate ions in HOPG. AFM investigations were performed with an atomic force microscope calibrated according to the procedures as described in the literature.[908]

This instrument was equipped with an electrochemical cell and a potentiostat/galvanostat for *in situ* measurements. A platinum quasireference electrode (Pt-QRef)[909] showing a drift of less than 10 mV/h was used in the AFM electrochemical cell.

All potentials are quoted against the Pt-QRef, which in turn was +700 ± 20 mV versus the saturated calomel electrode.

The increase in interlayer distance between graphene layers (percentage value) after formation of stage IV in 2 M $HClO_4$ as a function of the thickness of the step was investigated. Potential was scanned from the open circuit value to 920 mV (scan rate 25 mV/s) and held for 4 min. As a working electrode, a HOPG crystal was used with a mosaic spread of 3.5, according to x-ray diffraction measurements. All the HOPG electrodes were prepared in rectangular shape, with a length and width of 3 to 4 mm and a thickness of 1 mm. They were freshly cleaved with adhesive tape prior to each experiment. The total side faces of the sample were painted with an insulating varnish in order to prevent exposure of the HOPG edge planes to the electrolyte.

The electrolytes used were 2 M $HClO_4$ and 1 M H_2SO_4 solutions prepared by mixing water with 60% concentrated $HClO_4$ or 98% concentrated H_2SO_4. Cyclic voltammograms of the HOPG electrodes were measured with the AFM electrochemical cell at a scan rate of 25 mV/s. The AFM images were obtained with a 100 m scanner in the contact mode at constant force (scan rate 1 Hz/silicon cantilevers).

The height of an HOPG step was obtained by calculating the difference between the mean values of height measured in two marked areas on the top and bottom layers of the step. This procedure was essential in order to average the noise contributions and the statistical fluctuations in height due to the local aspect of the process.

All theoretical values for the dimensions of stacks sequences of HOPG layers were calculated by considering an interlayer spacing of 3.35 Å for unfilled gaps (nominal spacing for the ABAB structure of HOPG), 7.95 Å for a gap filled with perchlorate ions, and 7.98 Å for a gap filled with hydrogen sulfate ions.[910,911] In the line measurement mode, the AFM tip was always scanning across the same line (scan speed 0.5 Hz, pixel resolution 256 × 256), while the potential of the sample was cycled between 550 mV and 950 mV for seven cycles, with a scan rate of 25 m.[912] An equilibration time of 8 s before scanning the potential and a waiting period of 20 s at both scan inversion points were also introduced. The resulting total acquisition time was 8 min 32 s. The changes in thickness and bias of the sample were recorded simultaneously. The perchlorate ion intercalation into the HOPG basal planes was induced under well-controlled conditions in 2 M $HClO_4$.

These studies showed that the different stages of graphite intercalation compounds (GIC) could be easily seen in the cyclic voltammograms. Over a potential range of only about 200 mV, up to three different stages of intercalation of anions and associated solvent or acid molecules into the graphite lattice were visible in the form of anodic peaks in the positive scan.

AFM images of an HOPG were obtained in the step region in 2 M $HClO_4$ with the probe scanning a line while cycling the potential of the sample between +550 mV and +950 mV. The histogram data were compared with the experimental and theoretical values.

The acquisition time for AFM images was about 4 min. The main reaction involved in these chemical species was described extensively (dependent on graphite oxide formation, carbon dioxide formation, water electrolysis).[913]

Dimensional changes of the HOPG steps were studied over a range of heights between 6.7 Å (corresponding to a bilayer of HOPG planes) and a micrometer, corresponding to thousands of HOPG planes, and a bilayer step on an HOPG crystal was imaged under potential control. Before intercalation (+200 mV), the height of the step was 6.8 Å. After formation of stage IV (+900 mV), the height of the step had increased to 11.3 Å. Because the distance between carbon layers upon incorporation of perchlorate ion is 7.95 Å, this increase in step height is exactly what one might expect of such a system.[914] Thereafter, the potential was returned to the original value and maintained for 10 min. As a result, the height of the bilayer decreased to 6.7 Å, which is close to the initial value. When applying this method to other steps, sometimes the time dependence of the two processes (intercalation and deintercalation) appears to be different. This aspect has been reported elsewhere.[915] The same experimental approach was repeated at steps with different heights, that is, different numbers of carbon layers. According to the dimensions obtained for the perchlorate ion, one might have expected an increase in height of about 34% for every sequence of four or multiple of four graphene layers (stage IV). However, this expectation was only confirmed for steps smaller than 50 Å.

In agreement with literature data, these results were not surprising,[916] when studying graphite as a negative electrode for lithium-ion cells, the intercalation of species into graphite generally requires considerable energy, because the gaps between the graphene layers held together by van der Waals forces must be expanded. The expansion energy depends on the mechanical flexibility of graphene layers deformed by the intercalation process, and the energy increases with the number of adjacent graphene layers present on both sides of a particular gap.[917] Therefore, intercalation generally begins close to the surface, in the gap adjacent to the top basal planes. The intercalation process was found to progress toward internal layer gaps. The effect of oxidation on the morphology of a fresh HOPG crystal in 1 M H_2SO_4 after two potential cycles between 0 and 1.25 V was studied. A second additional explanation for this result resides in the fact that the probability of defects and vacancies increases with the number of graphene layers, that is, with the thickness of the step. The presence of these defects can prevent the perchlorate ions from intercalating, because the adjacent layers are pinned together and difficult to move with respect to one another. As a consequence, the usable capacity for insertion is reduced, which was actually shown in recent studies of the properties of synthetic and natural graphite.[918] The movement of graphene layers, for instance from AB- to AA-stacking, can also influence the interlayer distance. However, this is mainly important for stage I compounds, as suggested in the literature, it should be less effective for stage IV compounds, which are investigated here.[918]

AFM data of relative expansion were found to be in discrepancy with previous dilatometry measurements of crystals in highly concentrated sulfuric acid, which showed a relative thickness variation of 34% during formation for 0.01-mm-thick HOPG.[919] This may be ascribed to a different sample of the HOPG. In fact, the importance of a highly graphitized structure for the effective formation of GIC32 is well known. In an effort to study the reproducibility of the intercalation process, images of a step on an HOPG crystal with the AFM working in the *xzt* mode were obtained. The AFM tip was forced to cross a step of 146 nm, corresponding to 438 HOPG layers, at the same fixed line position, while the electrode potential was scanned between 550 mV and 950 mV.

The data showed that the thickness variation clearly reflected in the alternating peaks and valleys, which correlate with the intercalation and deintercalation of perchlorate ions in HOPG, respectively. The thickness change observed was related to the intercalation of perchlorate ions into a second step outside the field of view.

The correlation between thickness variation and HOPG potential was more clearly visible when examining the height profile plotted across a line parallel to the HOPG step together with the potential transients. The contribution of the second step was subtracted in order not to overestimate the change in thickness. The magnitude of the change was 26 nm. It was thus found that the number of graphene layers involved in the process was high, and the expansion estimated from AFM (18.7%) data was lower than that theoretically expected (34%). Further, a step of four graphene layers on an HOPG crystal was imaged at different potentials. Initially, the sample was imaged at 0 V. After the formation of each stage, the potential was returned to 0 V in order to dissolve the sample, and then the step was imaged again to test the reversibility of the process, and the histograms were analyzed. Unfortunately, it was not possible to study the formation of all the steps, because the potential required was so high that side reactions took place, and as a result, one observed blister formation or etching of the step.

To study the influence of the type of electrolyte on the intercalation process, a second system was investigated by *in situ* AFM. A freshly cleaved HOPG electrode was exposed to a 1 M H_2SO_4 solution. In these images, a step of 21 layers was found before and after intercalation of sulfate ions. The corresponding histograms show the increase in step height of 33.1%, which is close to the expected expansion of the interlayer spacing after formation of stage IV. However, after a few (two to three) cycles, the process cannot be reproduced. For instance, the surface of an HOPG crystal was found to be damaged after two cycles in potential between 0 and 1.25 V. It was reported that the surface of HOPG was destroyed much more easily in sulfuric acid than in perchloric acid.

7.8.3 Blister Formation

The phenomenon of blister formation was occasionally noticed on basal planes after the intercalation of perchlorate ions. Blisters were observed becoming larger or smaller, depending on the applied potential. The AFM images showed the dynamics of formation and partial disappearance of a group of blisters as a result of several cycles of potential between 0 and 0.95 V. In the first image, two groups of blisters were clearly visible at 0 V. When the potential was moved to +0.95 V, the smaller group of blisters was no longer present, and part of the larger blister disappeared. When the potential was moved back to 0 V, the blister group reappeared and some blisters merged.

This blister formation is similar to that observed on an HOPG surface in less concentrated electrolytes and is probably associated with the same mechanism of formation.[920,921] A model was proposed in which blister formation reflects the intercalation of electrolyte and water into HOPG followed by subsurface gas evolution (electrolysis).[921] For the first time, a reproducible, quantitative estimate of the interlayer spacing of graphene layers could be obtained by *in situ* AFM measurements for the $HClO_4$_GIC and for $H2SO_4$_GIC formation in HOPG. The data in perchloric acid were found to be in agreement with the theoretical expectation only for small stacks of graphene layers. Furthermore, blister formation in perchloric acid was observed as a function of the applied potential. The corresponding experimental values for the change of interlayer spacing were also found to be in agreement with theoretical reports.

7.9 SPM STUDIES OF NANOSCALE REACTORS

The physical size in regard to volume of any reactor apparatus is variable as found in everyday life. For example, this is easily visualized when comparing the volumes involved in the biological reactions that take place in the stomach or in the tiny blood capillaries inside the eye. It is mainly in the biological systems that the size varies many orders of magnitude, in fact, to as small as a single molecule (retina and light signals). The main difference will be related to the container wall effects in the microreactors, which will be negligible in macrosystems. Nanotechnology will demand whether reduction of reaction vessels will imply different reaction mechanisms or would be without any effect. STM and AFM allow one to design apparatus that can monitor reaction vessels, where dimensions can be as small as a few microliters (or even nanoliters) in volume. Furthermore, one can easily visualize that in many applications, such as biological, these ultrasmall volumes are already of necessity, such as eye fluids. In recent literature, an increasing number of studies are being reported on such nanoreactor systems. These experiments are based upon monitoring surface reactions at an AFM tip, a novel approach to study reaction kinetics in SAMs.

The kinetics of alkaline hydrolysis of ester groups in SAMs were monitored by a combination of AFM on the nanometer scale and FTIR spectroscopy in the continuum limit.[922] The main objective was to study surface reactions *in situ* with chemical specificity, from the nanometer perspective, using an AFM. This could not be achieved by conventional AFM friction or force measurements due to insufficient resolution, and instrumental or thermal drift, respectively. These problems were circumvented by a novel approach, which was termed "inverted" chemical force microscopy (ICFM). In ICFM, chemical reactions, which take place at the surface of the tip coated with reactants, are probed *in situ* by force–distance measurements on a scale of less than 100 molecules. The pull-off forces of different reactive SAMs were shown to vary with the extent of the reaction. Reactivity differences for these monolayers observed in this manner by AFM on the nanometer scale agree well with macroscopic behavior observed by FTIR and can be related to differences in the SAM structure. These results, together with additional force microscopy data, support the conclusion that for closely packed ester groups, the reaction spreads from defect sites, causing separation of the homogeneous surfaces into domains of reacted and unreacted molecules.

SAMs of organic molecules on solid substrates are becoming increasingly important for various technologies.[20] Areas of possible application range from surface modifications for wettability control,[923] tribology, or lubrication,[923–935] devices, or surface patterning (soft lithography).[928,936–941] Beyond their practical importance, SAMs offer unique opportunities to enhance our fundamental understanding of interfacial phenomena. They can serve as well-defined model systems to study the behavior of surfaces at the ultimate limit of atomic detail. Although surface science techniques are generally developing rapidly and allow one to characterize surfaces with improving lateral spatial resolution, *in situ* molecular-level studies of chemical reactions have largely eluded surface scientists until recently.[942]

Therefore, to obtain a better understanding of these processes at the level of single molecules, novel approaches must be developed to examine chemical reactions occurring at surfaces and interfaces. When functional groups are confined in closely packed molecular arrays, their reactivities often change. It has been found that significantly enhanced catalysis of acetone hydrogenation probably occurs because of enforced (favorable) orientation of a rhodium complex incorporated in corresponding LB films.[943] Penetration of external reagents to functional groups buried in a well-packed monolayer is usually restricted, and reactivity is thus reduced. As shown earlier, monolayers of aliphatic esters with the carboxyl group buried 10 methylene groups below the surface show remarkable stability toward trans-esterification.[944,20] Well-packed monolayers of isonicotinate esters were shown to hydrolyze only very slowly.[945] Many biological reactions occur at similarly well-ordered interfaces. The course of such important reactions, however, remains largely unknown, and they could be susceptible to the procedures described here, where AFM was used. Recent developments in the field of scanning probe techniques promise to give novel insights into processes that occur at surfaces or interfaces.

Surface studies with nominal nanometer resolution can be carried out in different media or environments. It is, therefore, not surprising that surface processes have been studied in the past with high-resolution scanning probe microscopes. STM is, in general, limited to (semi)conducting samples[946] and references therein. In addition, the STM tip can induce changes in the surface.[947] In this study, the hydrogen sulfide, H_2S, was observed by STM to easily corrode an Au surface. The interaction of sulfur-containing compounds with metal surfaces continues to be a subject of much interest. In many cases, atomically well-ordered sulfide and hydrosulfide monolayers have been discerned on metal surfaces. Hydrogen sulfide sensors based upon the interaction of H_2S with gold are much studied. The formation of a monolayer of a substance delivered from the gas phase has the advantage that this procedure can be performed under controlled and well-defined conditions. STM was performed under a helium atmosphere. STM images of Au showed a rough and pitted surface, after exposure to 100% H_2S. It was concluded that this was due to the formation of gold sulfide salts.

In the case of a thick layer of adsorbate, the feedback loop, which keeps a constant tunneling current during scanning, will cause the tip to penetrate the material, resulting in high shear forces. *Ex situ* AFM using chemically modified tips has been shown to provide information about surface energy changes related to interfacial reactions.[948–950] Previously, *in situ* AFM was primarily applied to monitor topographical changes, e.g., in crystal growth or dissolution, in crystallization of polymers from the melt, and in solid-state isomerization reactions.[951–955] All the aforementioned observations are, in general, limited by the finite contact area between the AFM tip and the sample surface. Depending on the radius of curvature of the AFM tips, on the materials properties, and on the imaging force, the radius of the contact area can be estimated to be between 2 and 5 nm. The estimation is based on Hertz theory as described in the literature.[956]

In the Hertz model, the applied force, F_{ap}, the elastic modulus of the substance, E_{elas}, the Poisson number, ν, and the radius of curvature of the tip, R_{tip}, are related by:

$$z^3 = 9/16(1 - \nu^3)/(E_{elas})^2 F_{ap}(2/R_{tip}) \tag{7.2}$$

In a typical case, $z = 6$ Å, $F_{ap} = 2$ nN, $R_{tip} = 5$ nm, and $\nu = 0.3$ (a reasonable value for most materials), and one gets the elastic modulus of $E_{elas} = 1.3$ GPa. For a normal force of 2 nN and a tip radius of 5 nm, a contact radius of 1.8 nm can be estimated. However, typical radii of modified AFM tips were of the order of 50 to 75 nm. This leads to a radius of the contact area of 4.0 to 4.5 nm. As a consequence, the true resolution is often much lower than the nominal accuracy of positioning the specimen and the sensitivity of height or lateral force data. True atomic resolution has been reported for experiments with custom-made force microscopes that operate under UHV conditions.[957–960] Defect motion was observed on InP (110) surfaces with noncontact AFM.[961,962] Intermittent contact modes are often used to obtain true atomic resolution; see, e.g., a report on true atomic resolution obtained by AFM on isolator surface.[963] The "high" resolution obtained with conventional contact mode AFM (e.g., for periodic lattice structures found for SAMs) can be considered a "lattice resolution." SAMs were subject to numerous studies by AFM.[31,964–972] As shown recently, lattice resolution can be considered an average over the periodic electron density distribution of the actual contact area.[956] The expected resolution of AFM in imaging of surfaces in liquid can be assumed to be no better than several nanometers. In addition, thermal and instrumental drift make high-resolution imaging of chemical reactions difficult or impossible. Pronounced scatter was observed in the data obtained in a study of kinetics of inhomogeneous reactions in SAMs by conventional force–distance measurements. The scatter was attributed to the aforementioned drift which results in interactions between different areas of the SAM with the AFM tip. Because the area imaged is, in general, changing constantly, and furthermore, "reference points" such as topographical features might change, as well, as a consequence of the reaction, the lateral resolution may be poorer than the true physical resolution discussed above.

A novel approach to studying surface reactions of organic thin films with nanometer-level adhesion measurements using AFM was described. In these studies, the reactants were immobilized on the AFM tip rather than on the sample surface. The use of chemically functionalized probe tips in AFM is known in the literature as chemical force microscopy (CFM). It has been shown that the terminal functional groups of SAMs of functionalized thiols on gold-coated AFM tips can dominate the adhesive or frictional interactions with the surface.[973] The variation of pull-off forces between the tip coated with the reactant and an inert surface is consecutively monitored as a function of time. Compared to pull-off force measurements on Au (111) substrates modified with reactants, this procedure eliminates the problem of thermal and instrumental drift which results in lateral displacement of the sampling area (contact area at pull-off). Thus, adhesion is measured between the same molecules (functional groups) and the homogeneous inert substrate.

The contact area of the tip at pull-off in such experiments using nonreactive SAMs (as inert samples) deposited on Au (111) varies between approximately 10 and 100 effectively interacting molecular pairs.[974–977] In recent reports on functionalized carbon nanotubes as tip apexes, it was demonstrated that smaller tip radii and contact areas are, in principle, accessible. The tip chemistry, degree of functionalization, and lateral confinement of reactive groups are, however, better controlled using SAMs. In general, as predicted by contact mechanics theories, the contact area at pull-off is smaller than the contact area during scanning.[978] This suggests that the pull-off force measurements should be expected to have higher resolution.

In ICFM, the pull-off forces of a reactant-covered AFM tip were studied *in situ* during the conversion of the reactive groups. The interaction between tip and inert surface varies with the extent of the reaction. Depending on the changes in the pull-off forces with the extent of the reaction, the effective force per interacting molecular pair, as well as the number of interacting pairs, can vary.[296,950]

Thus, one can, in principle, monitor chemical reactions with a resolution of the number of contacting molecular pairs in AFM force measurements, provided that the conversion of the reactants to products is accompanied by changes in the pull-off forces. The force contrast can be controlled, because the magnitude of the pull-off force depends on the choice of the inert substrate. Since the original report on CFM, numerous articles have been published that have made investigations using the same principles.[566,974–982]

As a consequence, a large number of functional group interactions in different media are known, and tip functionality can be chosen accordingly to yield high contrast. Thus, a large variety of different systems can, in principle, be studied. It was shown in these studies that the kinetics of surface reactions can be followed *in situ* on a noncontinuum level (10 to 100 molecules), which was termed ICFM. ICFM and FTIR spectroscopy were applied to study the kinetics of alkaline hydrolysis of ester groups at the surface of SAMs. From these data, the kinetic measurements were averaged over a large area (by FTIR), nanometer-scale kinetic information (by inverted CFM), and additional high-resolution force microscopy from which the reaction mechanism was estimated.[983,984] For closely packed ester groups, the reaction was concluded to spread from defect sites, causing separation of the homogeneous surfaces into domains of reacted and unreacted molecules.

11-Mercaptoundecyl acetate 1 was synthesized as described in the literature.[985] Acetyl chloride (0.65 mmol) in ethyl acetate was added to the solution of bis(11-hydroxyundecyl) disulfide (0.43 mmol) in dry ethyl acetate (10 mL) at 30°C. The mixture was stirred at 50°C for 4 h, and solvent was evaporated. Crude product was purified by flash chromatography from ether:CH_2Cl_2 1:6.

Histograms of pull-off forces were measured with a single octadecanethiol-functionalized tip on neat SAMs of thiol 1- and 11-mercaptoundecanol in water. SAMs for FTIR analysis were self-assembled onto evaporated gold as described elsewhere.[986] In these procedures, the gold substrates were prepared by thermal evaporation of 5 nm of Cr followed by 100 nm of Au onto silicon wafers. Before the deposition of monolayers, the gold substrates were cleaned with concentrated HNO_3, and subsequently washed with water and ethanol. The procedure for monolayers preparation was by dipping the substrates into 1 mM solutions of these compounds in dichloromethane for 16 h at room temperature. The substrates were later rinsed with dichloromethane and ethanol and dried. SAMs for lattice imaging and *in situ* imaging of the reaction on Au (111) substrates were prepared similarly on annealed Au (111) substrates as described earlier.[987] These substrates consisted of triangular Au (111) terraces as was found from images.

TM AFM height image was obtained of Au (111) surface covered with inert octadecanethiol SAM (z-scale, 5.0 nm). In these studies, the IR spectra of monolayers were measured using an instrument with a grazing angle accessory. SPR data were also measured. The magnitudes of thicknesses were estimated from the shift of the resonance angle using the standard Fresnel equation, assuming the dielectric constant of the film to be equal to 2.1.[987] Contact angles of water drops (1 μL) were measured with a CCD camera and digitized for analyses.

The pull-off forces were measured during hydrolysis as a function of reaction time t of SAMs of 1, 2, 3, and octadecanethiol (blank) in 0.1 M NaOH followed by inverted CFM. Results of representative individual experiments for hydrolysis [1 − x] were analyzed as a function of reaction time t, with extent of reaction (x) of SAMs of 1 in 0.1 M NaOH followed by inverted CFM. From these data, it was found that following the induction periods, the reactions proceed at similar rates.

Pseudo-first-order rate constants for the hydrolysis reaction were observed for 3 by ICFM, and the mean value of second-order rate constant k_{AFM} was 2.4×10^{-2} L mol^{-1} s^{-1}. In these studies, triangular-shaped silicon nitride cantilevers and silicon nitride tips were covered with 50 to 70 nm of gold in a sputtering machine or, alternatively, with ca. 2 nm of Ti and ca. 75 nm of Au under high vacuum. The gold-covered tips were later functionalized with SAMs of 1, 2, or 3 following the procedures as described earlier.[948,950,972,986] The AFM measurements were carried out with a multimode AFM (DI) using a liquid cell. Force measurements were performed with modified tips, and imaging was done with unmodified silicon nitride tips, respectively. Tip functionalization is concomitant with an increase in tip radius. For imaging of the lattice as well as of the hydrolysis reaction of the SAMs on Au (111) substrates, the sharper unmodified tips were used.

Cantilever spring constants were calibrated as described in literature. Calibration was performed using micromachined cantilevers, with a known spring constant. The procedure as well as the specifications are given, following.[988] TM AFM images of the Au (111) substrate were acquired as described in the literature.[973,987]

The functionalized AFM tip was placed in the liquid cell. After a brief equilibration period in ultrapure water, the tip was engaged on an octadecanethiol SAM on Au, and a set of force–distance curves was recorded. After withdrawal of the tip, the cell was flushed with more than 20 times the cell volume of aqueous NaOH of known concentration. When a stable photodiode reading was obtained, the tip was engaged again. Immediately after engaging, force–distance curves were recorded at 30 (60)-s intervals. The mean of 10 individual pull-off events, measured after each 30 s (60 s for slow reactions), was calculated. AFM data were based on averaging over many individual experiments.

In situ AFM (friction mode with unmodified tip) of hydrolysis of a SAM of 1 on Au (111) in 1 M NaOH was performed. The images were obtained after 4, 10, 13, and 24 min, respectively. The conversion of ester groups to hydroxyl groups was calculated from the measured average pull-off forces at $t = 0$, $t = t$, and $t = \infty$, respectively.[948] The calculations were based on the assumption that the forces change linearly with the work of adhesion. The surface free energy of the tip and the interfacial free energy were assumed to be influenced only by changes in the end-group, while the surface free energy of the inert substrate was assumed to be constant.[948–950]

The structure of the SAMs of compounds 1 through 3 was first investigated in detail using contact angle, FTIR spectroscopy, and SPR spectroscopy. The alkaline hydrolysis of the SAMs was studied by *ex situ* FTIR spectroscopy. The hydrolysis reaction was further investigated *in situ* by ICFM. Finally, additional results were obtained by *in situ* AFM.

7.9.1 Self-Assembled Monolayer Structure

In this study, SAMs of the ester-terminated thiol ET, the corresponding symmetrical disulfide SD, and the half-ester disulfide HED were investigated. Close packing in the monolayers of esters ET1 and SD2 was established by thickness (SPR spectroscopy) and wettability measurements.

The FTIR data were consistent with literature data on similar films.[989,990] The asymmetric stretching vibrations a(CH_2) at 2919 to 2920 cm^{-1} are typical of well-packed alkane chains in an all-trans conformation.[991] The (C–O) vibration at 1259 to 1262 cm^{-1} shifted to higher frequencies as compared with bulk spectra (1240 to 1242 cm^{-1}), showing strong lateral interactions between adjacent carbonyl groups in ordered quasi-crystalline environments.[992] SAMs of HED3 showed a higher frequency a(CH_2) vibration (2922 cm^{-1}), and the water contact angle of SAMs of HED3 was lower than those of ET1 and SD2, indicating partial exposure of hydroxyl groups at the monolayer surface.[993] Molecular (lattice) resolution AFM measurements performed in water revealed a hexagonal tail group lattice structure with a lattice constant of 5.2 Å.[956] Monolayers of ET1, SD2, and HED3 possess a well-packed structure. AFM friction force measurements (vide infra) were consistent with the interpretation that the hydroxyl and ester groups in the monolayer of HED3 are evenly distributed to the level of the lateral size range of the contact area between AFM tip and sample surface. There is ample evidence that mixed disulfides do not phase separate at room temperature.[993,994]

The ester hydrolysis was also studied *in situ* by FTIR (in 1.0 M aqueous sodium hydroxide at room temperature). The kinetics were analyzed from FTIR spectroscopy data by following the decrease of the integrated intensity of the (C=O), (C–O), and s(CH_3) vibrations *ex situ*. SAMs of thiol reacted much more slowly than SAMs of the mixed disulfide. Kinetics of the mixed disulfide HED3 was exponential, while ET1 and SD2 showed sigmoid behavior. The half-reaction times were estimated. For the mixed disulfide HED3, the pseudo-first-order rate constant kFT-IR was 1.00×10^{-2} s^{-1}.

In monolayers in a close-packed surface state,[20] access of hydroxide ions to the carboxyl groups appears to be hindered, even though the reactive carboxyl groups are buried only a few angstroms below the surface. This is what one would also expect in solid state. Similar sigmoid kinetics are found, for example, in surface reactions of perfect crystals which often exhibit low initial reactivity. There is ample evidence that mixed disulfides do not phase separate at room temperature.[994,995]

In these cases, defects at the crystal surface may allow a slow initial reaction, and acceleration is observed as more reactive sites become exposed. A recent study showed that imaging the monolayer of a succinimide ester with AFM disrupts the order strongly enough to appreciably accelerate hydrolysis.[996,997]

SAMs of esters 1 and 2, at half-hydrolysis (i.e., with 50% of the ester groups transformed into hydroxyl groups), would at first appear structurally equivalent to the mixed disulfide HED3. Reaction of esters ET1 and SD2 at 50% conversion is, however, ca. 10 to 20 times slower than that of ester HED3. Intramolecular general base catalysis [which could be responsible for the autocatalytic (sigmoid) behavior] can be excluded on the basis of control experiments with mixed disulfides, in which one chain carries a terminal ester group and the other a terminal methyl group. Further, the origin of the main reactivity difference was ascribed to the structure of the surface of the SAMs. Thus, it is of interest to obtain structural and compositional information with high resolution, preferably on the molecular scale. *In situ* information on the composition of the reacting monolayers was obtained by the ICFM approach described above. In these experiments, the force required to pull the AFM tip coated with ET1, SD2, or HED3 away from contact with an inert octadecanethiol SAM on flat Au (111) was followed in real time *in situ* during the hydrolysis.

The change in surface composition could be measured accurately, because the hydrophobic force between ester (the reactant)- and alkyl-terminated surfaces is large. Average pull-off forces measured between neat ester-terminated SAMs and methyl-terminated tips were found to be 9 nN, whereas neat hydroxyl-terminated SAMs and methyl-terminated tips showed an average pull-off force of 0.4 to 0.3 nN. These forces follow the trends previously observed and explained by the Johnson–Kendall–Roberts (JKR) theory of contact mechanics.[997] As the contact sampling area of the functionalized AFM tip is between 10 and 100 molecules, depending on the actual tip radius, the reaction can be studied in a highly localized fashion. The use of chemically functionalized probe tips in AFM is known in the literature as CFM. It has been shown that the terminal functional groups of SAMs of thiols on gold-coated AFM tips can dominate the adhesive or frictional interactions with the surface.[973] The Au (111) substrates were found to be atomically smooth over distances of several hundred nanometers, with only occasional steps and depressions present. With highly ordered octadecanethiol SAMs on these substrates, interaction between the same functional groups at the tip apex and an homogeneous inert substrate is ensured. The individual reaction profiles for thiol ET1 and disulfide SD2 showed significant induction periods that depend on the hydroxide concentration. It may indicate the presence of an activated process of nucleation and growth. For SAMs of the mixed disulfide, no induction periods were observed. Test experiments with (inert) octadecanethiol-coated tips showed that SAM damage can be excluded, and that the SAM/gold assembly stays intact over the typical experiment times. This has also been observed in the author's laboratory on SAMs on Au substrates. The results of four different experiments for thiol ET1 performed under identical conditions were analyzed. The individual reaction profiles showed a wide distribution of induction periods.

The contact area between the tip and the surface at pull-off was found to be of smaller size than the typical domain size in SAMs. Typical domain sizes of SAMs on gold which were not heat-treated after assembly were found to be of magnitude 5 nm.[998] It was concluded that if a defect was present in the contact area, the reaction would be observed immediately. On the other hand, if there were no defects in the contact area, the reaction will not be observed until a defect is formed. This interpretation was found to be in accord with an estimate of the domain size based on the observed induction times and reaction kinetics. From the analysis of these data, a maximum domain size of 3.2 to 0.2 nm for thiol ET1, and 2.6 nm for the symmetric disulfide SD2, were estimated. These values are in agreement with literature data.[983] It can be concluded at this point that the observed distribution of induction periods is a strong indication of separation of reacted (–OH) and unreacted ($-O-CO-CH_3$) terminal groups during the hydrolysis. Thus, the profiles of the different experiments showed differences, because ICFM follows the reaction with high lateral resolution. The kinetic profiles in the continuum limit were estimated by averaging many AFM experiments. These profiles were found to be similar to the FTIR profiles. The reaction of SAMs of half-ester disulfide HED3 gave an average second-order rate constant $k_{AFM} = 2.4 \times 10^{-2}$ L mol^{-1} s^{-1}, which was found to be in agreement with the value obtained by FTIR. Quantitative differences between AFM and FTIR results were explained by the fact that the sampling area for AFM is more than ca. 1000 times smaller than for FTIR and that reactions, e.g., at grain boundaries, would not affect the AFM results. By contrast, the SAMs of thiol 1 and symmetrical disulfide SD2 showed sigmoid profiles for the average kinetics. Macroscale kinetics determined by FTIR can be reproduced by averaging snapshots of the reaction of 10 to 100 molecules from the AFM experiments.

7.9.2 *In Situ* AFM Imaging of SAMs during Hydrolysis

Further indication for the formation of domains of reacted and unreacted molecules during the hydrolysis reaction was obtained from *in situ* contact mode AFM experiments with uncoated tips on SAMs of ET1 and HED3 on flat Au (111) substrates. It was found that the magnitude of the friction force observed on Au (111) terraces covered with a SAM of ET1 changed significantly during hydrolysis. After ca. 4 min, the smooth image that showed only the step edges of the Au surface began to change to an image exhibiting a "rippled" appearance. The inhomogeneous friction producing the ripples increased and then disappeared during the course of the reaction. These data could be explained, as the observed "ripples" were related to inhomogeneous adhesion, and hence friction, because of the formation of domains of reacted and unreacted molecules during the course of the reaction.[974,976] The use of chemically functionalized probe tips in AFM is known in the literature as CFM.

These kinetics data of *in situ* hydrolysis of SAMs on gold-covered AFM tips were as follows:

Concentration NaOH, (mol/L)	1
Average half-reaction time 1/2, (s)	365
Range of induction periods (s)	0–560

It was shown that the terminal functional groups of SAMs of functionalized thiols on gold-coated AFM tips can dominate the adhesive or frictional interactions with the surface.[973]

The inhomogeneous friction producing the ripples can be evaluated by analyzing the deviation of the friction for each pixel from the mean value similar to a roughness analysis. The mean roughness RA of a surface is defined as the standard deviation of the height with respect to the center plane within the scan area selected. The RA values were obtained according to the following relation:

$$RA = M(zxy - xy)^2/[(N - 1)(M - 1)])^{1/2}$$

where N and M are the numbers of pixels in the x and y directions, respectively, and zx, y is the image pixel height with respect to the center plane height xy for the pixel (x,y).

The data showed that the friction contrast peaks after ca. 15 min for thiol ET1, while there is no contrast change for mixed disulfide HED3. These observations were found to be in agreement with FTIR and ICFM results.

The lattice of the SAMs could be imaged with molecular (lattice) resolution prior to and after the hydrolysis on the Au (111) terraces. The corresponding high-resolution lattice images were obtained. The fact that lattice imaging was possible prior to and after the hydrolysis demonstrates that SAM damage can be excluded as a basis for the disappearing inhomogeneity of the friction images of ET1. The true resolution of the friction measurements can be assumed to be in the order of 2 to 5 nm, which is a typical size for domains in SAMs. For this rough estimate, a circular domain shape was assumed. The longest induction period was assigned to the idealized situation in which the circular contact area (with radius r_0) is centered on this domain. As there are no defects in the contact area, the reaction is not observed until the reaction front, which starts at the domain boundary, reaches the contact area. On the basis of JKR theory, the original contact area was calculated [$K = 4.7 \times 1010$ Pa; W12(t = 0) = 50 mN/m]. A reaction rate was defined as the ratio of *ro* and reaction time *tR* (*tR* is the time elapsed between the first decrease in adhesion forces and the end of changes). Using this rate (in nm/s), the original domain size can be calculated on the basis of the observed induction time. It is reasonable to assume that the domain sizes on sputtered or evaporated gold are smaller than those on flat Au (111). Experimental evidence for small crystalline patches further supports this assumption.[983]

Quantitative differences between the AFM and FTIR results can be explained by the fact that the sampling area for AFM is more than ca. 1000 times smaller than for FTIR, and that reactions, e.g., at grain boundaries, would not affect the AFM results. The true resolution of the AFM is not sufficiently high to unequivocally enable one to image subdomain details. In addition, it is not *a priori* clear how the friction measured on a few molecules can be related to exposed functional groups. Typical domain sizes of SAMs on gold that has not been heat-treated after assembly are in the order of 5 nm.[998] As the friction contrast did not change significantly in SAMs of the mixed disulfide HED3, the inhomogeneous friction observed for thiol ET1 strongly suggests that the reaction proceeds inhomogeneously for SAMs of ET1.

From these investigations, it was found that the reaction of ET1 starts at defect sites and creates domains of reacted and unreacted molecules. The reaction then proceeds at these domain boundaries until all molecules have reacted, while for HED3, the reaction occurs homogeneously. The mechanism proposed was consistent with the AFM and FTIR data and accounts for the reactivity differences between half-reacted monolayers of ET1 and SD2 and the monolayers of HED3. While the layer of HED3 remains topographically homogeneous, ET1 and SD2 separate into domains during the hydrolysis.

This demonstrated for the first time that average macroscopic kinetics of reactions in SAMs can be correlated and explained by nanometer-level force (adhesion) measurements in the confined environment of the monolayer studied. Data were obtained using a modified CFM method (ICFM). In this procedure, one could measure reactions of as few as 10 to 100 molecules *in situ*. Structure–reactivity differences, arising from differences in monolayer structure, as observed on the nanometer scale, agree with macroscopic behavior observed by FTIR. These data showed that reagents penetrate functionalized monolayers at specific defect sites or at domain boundaries.

7.10 NANOSCALE EVALUATION OF SURFACE ROUGHNESS BY SPMS

Understanding the mechanisms, energetics, structures, and dynamics underlying the interactions and physical processes that occur when two material bodies are brought together, separated, or rub against each other is of fundamental importance in many basic and applied processes in daily life. The word tribology stems from the Greek tribos, meaning to rub. Typical examples include adhesion, capillarity, contact formation, surface deformation, elastic and plastic response, characteristics, hardness, micro- and nanoindentation, friction, lubrication, wear, fracture, atomic scale probing, and modifications and manipulations of material surfaces. Surface roughness reflects physical structures on solid surfaces that were placed during crystal growth. Fractal surface roughness was reported.[788,835] The term fractal was extensively described in the literature. In simple terms, one may describe the fractal of a straight line as follows. A straight line has fractal equal to one, because the length of the line between two points A and B is the same no matter how small the fractions taken to measure its length. However, if the line between the same two points A and B is jagged, like the borderline between two countries, then it is easily seen that the length will be dependent on the steps used to measure this distance. The fractal dimension can be measured by plotting log of number of step versus log of step length. It will be found to be greater than 1. In other words, the more jagged the borderline, the larger the fractal value. The same is defined for surfaces (magnitude of fractal is two for smooth surfaces) and volumes (fractal is three for spheres and ellipsoids). In the case of irregular lines, areas, or volumes, the fractal would deviate from these exact values, and hence, the designation fractal.

Solid surfaces were found to exhibit fractal morphology.[999] A study of the quality and accuracy of the methods based on frequency analysis for the fractal characterization of solid surfaces was carried out by STM. The study was based upon computer simulation of images of fractal surfaces. Measurement of the fractal character of a surface in the microscope range has specific problems. The most important is that the images of a given solid surface arise from the projections of the surface topography in which the vertical dimension is unknown, whereas from STM and AFM, one can obtain the vertical dimensions.

Recently, STM images were used for the fractal surface characterization of gold films prepared by vapor evaporization.[835] The gold deposits were grown in an evaporator on smooth glass substrates.[536] The average thickness was 200 nm. These data were analyzed based on the dynamic scaling method. The root mean square roughness, S_{stm}, measured by STM on scale lengths of L_s is as follows:

$$S_{stm}(L_s, h_x) \propto L_s^{\alpha} f(x) \tag{7.3}$$

where $f(x) + (h)/L_s \exp(\gamma)$, fx + constant for $x \rightarrow \infty$, and $fx + x \exp(\alpha/\gamma)$ for $x \rightarrow 0$, and S is defined as:

$$S_{stm}(L_s) = ((1/\mathrm{N})\Sigma(h(xi) - (h)^2)^{1/2} \tag{7.4}$$

where $h(xi)$ is the deposit height measured along the x direction at position x, and $\langle h \rangle$ is the average height of the sample formed by N points. The adsorption can be defined as follows after a certain period:

$$S_{stm}(L_s) \propto L_s^{\alpha} \tag{7.5}$$

This relation allows one to evaluate α from a $\log(S_{stm})$ versus $\log(L_s)$ plot. This is a general procedure used for fractal analyses.[894] The magnitude of α_l was estimated at 0.77 to 0.07, depending on L_s. The smoothness was found to be related to α_l (in the case of a perfect smooth surface, $\alpha_l = 0$).

In a recent study, LFM was used for measuring friction, shear, and adhesion properties of monolayers.[1000] LFM was developed in order to provide information on semiquantitative character of organic SAMs by correlation of surface friction and image contrast to the composition. The LFM is found to be relatively cheap as compared to other methods, such as XPS, SEM, and secondary ion mass spectroscopy (SIMS). The principle of operation of LFM is similar to that of ordinary AFM. In LFM, the interaction between the probe tip and the features on the surface of the substrate are measured. A silicon nitride tip is placed in contact with a sample. A piezoelectric scanner moves the sample beneath the tip in the x, y, and z directions. The reflected laser beam from the tip is focused on a laser diode. The reflected beam is detected on a four-quadrant photodetector. The tip movement is registered as a function of the lateral movement of the tip on the surface, which is dependent on the tip–surface roughness.

Quadrant Photodiode.. LASER...
...........CANTILEVER
...........SAMPLE
PIEZOELECTRIC SCANNER..............

The friction forces were measured by sliding the tip across the sample surface in the x direction, while the y direction can was disabled.

7.11 APPLICATION OF STM AND AFM IN POLLUTION CONTROL

As described above, in addition to producing images of molecules at surfaces, STM and AFM can provide information about composition of molecules. The analysis of pollutants globally is an ever-demanding problem, especially in such systems as drinking water or sea water. The nanoscale investigations of materials involved in any pollution area is a possibility today. STM and AFM methods can be used as a nanoscale qualitative and quantitative analysis tool for materials on surfaces (e.g., pollution control). This was mentioned earlier, when 60×10^6 cholesterol molecules were estimated in the image (Figure 3.4). The lower limit in detection by AFM or STM is theoretically one atom or molecule (i.e., 10^{-23} mole). For example, the morphology of amino acid crystals clearly shows how AFM can be used for such nanoanalyses

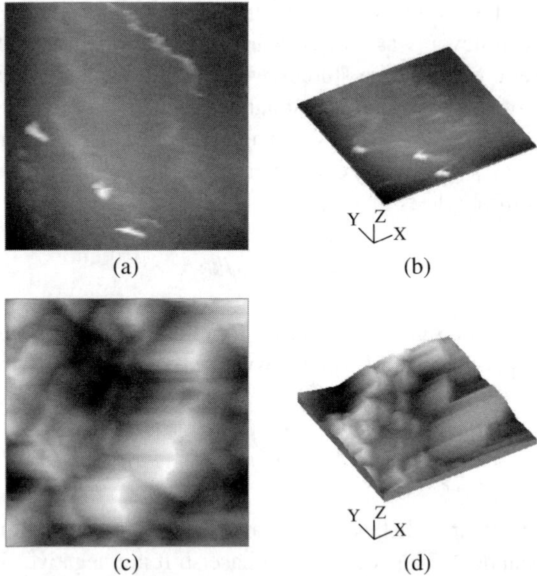

FIGURE 7.4 AFM images of distilled water (a; b) and tap water (c; d) (60,000 Å × 60,000 Å). A drop of water (10 μL) was placed on HOPG, and AFM images were obtained after evaporation.

(Figure 5.1). This means that quantitative analysis has an extreme lower limit for detection. This has importance in pollution control, etc., where very small amounts must be monitored. As shown above, from morphology of images, one can estimate qualitatively the degree of impurity (as for pectin, shown in Figure 4.3).

The control of CO (an important pollutant) in atmosphere or of similar gas molecules was described elsewhere.[764]

Another important application could be the control of drinking water. AFM images of tap water and distilled water were carried out after evaporation of 10 μL drops on HOPG. AFM images of tap water and distilled water are given in Figure 7.4. It is clearly seen (preliminary data) that salt crystals in tap water are much more pronounced than in distilled water.

The investigations on the hydrous aluminium oxide particles precipitated from wastewater were reported by AFM.[1001] Alum [$Al_2(SO_4)_3.nH_2O$] is the most widely used coagulant in the wastewater treatment process. The use of alum requires that sufficient alkalinity be present in the wastewater to produce solid hydrous aluminum oxide species. These species remove orthophosphates by forming insoluble aluminium hydroxyphosphate or other related complexes. Alum is added in any activated sludge plant under different stages of treatment: before biological treatment or after aeration and prior to the clarification process. In the former case, it is possible that phosphate ions are coprecipitated with alum. In this study, the changes in alum products during water treatment were investigated by AFM. The synthetic wastewater was made up of $NaHCO_3$ (0.2 mol/L), $KHPO_4$ (3.7 mol/L), and tannic acid $C_{76}H_{52}O_{46}$ (0.1 mol/L). Alum was added to precipitate complexes. AFM was made on a 30 μL sample of wastewater on mica after air drying. AFM operated as tapping mode, except that the cantilever was magnetically coated and driven by an externally oscillating magnetic field. These studies were caried out under cantilever that had a force constant of 0.5 N/m and a resonance frequency of 100 kHz. The images were analyzed as follows. The phase shift angle, delphi, can be shown to be a function of different factors. Assuming the cantilever is at its resonance frequency in the free state, the phase shift angle is given by:

$$dp = Q_q \sigma_i / k_{fc} \tag{7.6}$$

where Q_q and k_{fc} are the quality factor and force constant of the cantilever, respectively. The term σ_i is the overall force derivative experienced by the cantilever:

$$\sigma_i = \sum_i (dFi/dz_{ts}) \tag{7.7}$$

where z_{ts} is the tip–sample distance of separation. When the net force acting on the tip is attractive, van der Waals, σ, and the phase shift are negative. The first problem is to determine which raised objects in the image correspond to the particles and which to the substrate. The images were found to be dependent on the precipitation procedures, thus showing the application of AFM to such analyses.

In order to determine the qualitative AFM image of a natural product, such as ginseng (extract of a root), the following procedure can be followed. Ginseng sample was extracted in hot water, and after cooling, a drop of 10 mL was applied to the surface of HOPG. After evaporation of water overnight, the AFM image was obtained (Figure 7.5). It is clearly seen that besides lipid-like flat SAM structures, there are also larger structures. Ginseng is known to consist of a variety of substances (such as lipids, fatty acids, and larger plant molecules). It can be concluded that by such AFM qualitative analysis, one may compare ginseng of different origins.

7.12 FRICTION FORCE MICROSCOPE (FFM)

In many industrial applications, boundary lubrication is important in modern technologies. For example, in the case of oil exploration, tunnel construction, and similar constructions, enormous amounts of energy are expended in mechanical input. Furthermore, in a wide variety of chemical, biological, and physical processes, interfacial molecular architectures at nanometer length scales are involved. This energy input is reduced if the friction is reduced by additives. The molecular structures of additives used determine the degree of friction reduction. The AFM has become a key research tool in the surface characterization of materials, such as polymers and organic thin films, where low conductivity precludes the use of the STM.[645] Surfaces with relief deeper than ~100 nm, such as on compact disks or integrated circuits, the AFM accurately yields the surface topography.

AFM differs from other forms of microscopy in that a controlled force is applied to the specimen while the image is being taken. This imaging force can be a

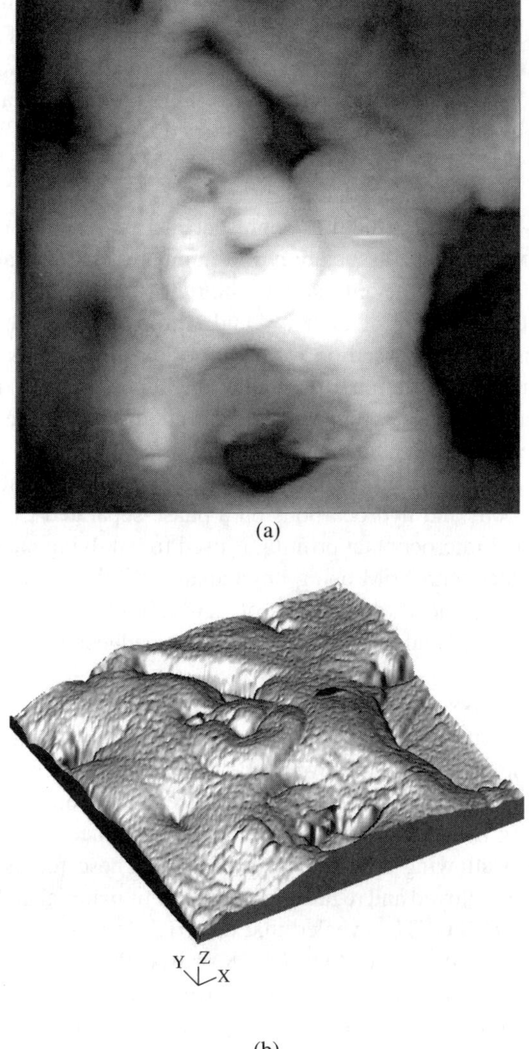

FIGURE 7.5 AFM image of a water extract of a ginseng root. 10 μL water extract was placed on HOPG, and an AFM image was obtained after water had evaporated: (a) two-dimensional; (b) three-dimensional (90,000 Å × 90,000 Å).

limitation, under certain circumstances, if one is seeking to image weakly bound adsorbates or soft materials, as described earlier. On the other hand, if the magnitude of the force is measured as a function of tip–surface distance, the chemical and physical properties of the surface can be measured. A preferred method of lubrication is by the deposition of organized (as SAMs) and dense molecular-scale layer(s) of preferably long-chain organic molecules. Obviously, the common method with which to produce such organized SAM-like monolayers or thin films is LB method, or grafting procedures.

In a recent study, humidity and temperature effects on frictional properties of mica and of octadecytrimetheoxysilane (OTE) were studied by FFM.[1002] Commercially available Si_3N_4 cantilevers with force constant of 0.5 N/m were used. For the friction measurement, the feedback loop was disabled to allow changes of load via the applied voltages on the piezotubes. Mica showed lower friction with high humidity. OTE SAMs reduced the surface friction of mica. However, the effect of humidity on OTE SAMs was complex, as one might expect.

Because self-assembled structures represent thermodynamic minima, because they are formed by reversible association of a number of individual molecules, and because the enthalpies of the interactions holding these molecules together are relatively weak, the interplay of enthalpy and entropy in their formation is more important than in syntheses based on formation of covalent bonds.

In surface chemistry, friction forces play an important role. For example, in construction of large tunnels, energy costs can be substantially reduced if friction forces could be reduced by suitable chemicals at surfaces involved. A modified type of AFM, an FFM, was described.[1003] This method was shown to be able to distinguish between fluorocarbons and hydrocarbons on a phase-separated LB film.

The technique of microcontact printing is used for studying patterned lubricants on flat iron substrates with FFM using Fe-coated tips.[1004] This technique gives the possibility of microscopic investigations of new lubricants under realistic conditions. The samples are analyzed by friction force, adhesion, and dynamic force measurements.

An STM based upon an Inchworm motor for coarse expansion coefficients to minimize thermal drifts has also been developed.[1005] The inchworm pushes a ceramic sled that carries the tunneling unit toward the sample until it is in range of the tube piezoscanner. When its motion is reversed (to disengage from the sled) interfacial forces at the sliding interface move backwards over distances between 100 nm and 20 gm) rather than allowing it to remain stationary. These forces vary depending upon the materials employed and regarding the choice of material and the sled design.

FFM data of OTS/FOETS revealed that the OTS domain had a higher fictional coefficient compared with that of the FOETS matrix. Phase-separated morphology of an immobilized organosilane monolayer gave much useful information. These phase structures were suggested to be useful for microbioreactor or biosensor or memory devices.

The force curves of grafted C18 were investigated by FFM.[1005]

The cantilever force curves (Figure 7.6) showed that, when treated, silicon surfaces exhibited changes in surface forces.

The application of AFM force versus distance curves were reported on polymer lubricants [Demnum: $(OH–(CF_2CF_2CF_2O)n–CF_2CF_3)$] on hard disks.[1005]

On bare carbon surface, the hysteresis loop between the inwards and outwards paths in the curve was much smaller than in the case of when lubricant was added. The effect of lubricant molecules on the tip–carbon surface was ascribed to long-range attractive force. This application of AFM to the lubrication industry is thus an imortant area.

In a recent study, FFM was used to resolve spatially the chemical composition at two types of test interfaces.[980] The basis of this kind of characterization is the

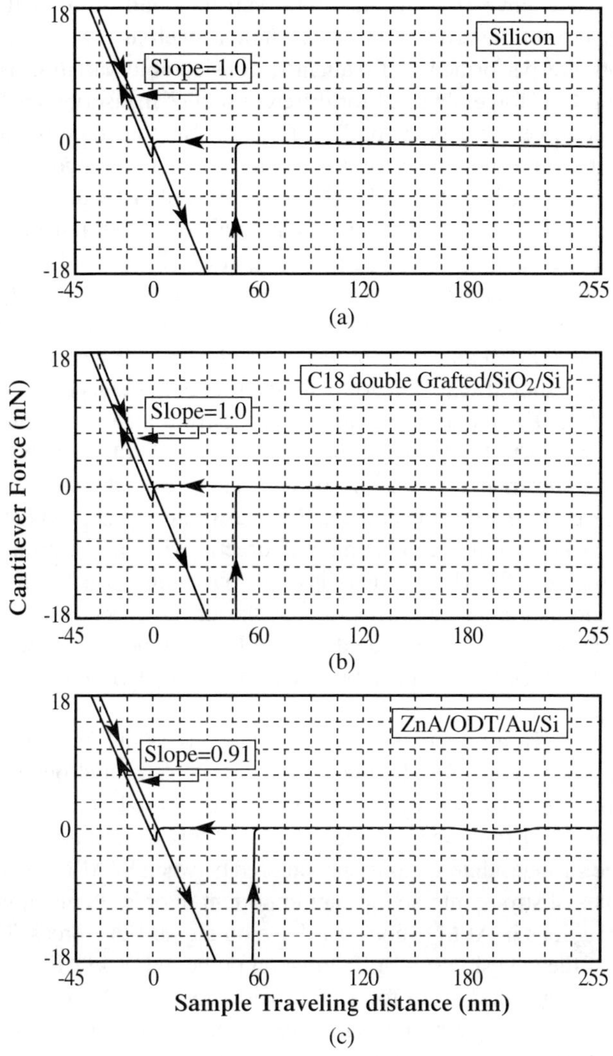

FIGURE 7.6 Force curves as a function of sample traveling distance of 18 nN for: (a) Si; (b) C18 double-grafted layer; and (c) Zn-ODT-Au-Si film. The arrows indicate the sample traveling direction, and slopes are indicated. (From Bhushan et al., *Langmuir*, 11, 3189, 1995. With permission.)

dependence of the frictional interactions on the identity of the chemical functional groups at the outermost few angstroms (Å) of microscopic contacting areas, i.e., surface free energies. Bilayer samples were prepared by immersion of gold-coated glass substrates into ethanolic solutions of 0.1 mM 16-mercaptohexadecanoic acid and 0.1 mM stearic acid for 24 h. The IR spectral analyses was used to determine the bilayer structures. AFM was used with 1 and 12 μm tube scanners. Si_3N_4 cantilevers with force constant of 0.06 N/m were used.

As mentioned above, in AFM, the probe–substrate pull-off force depends on the chemical property of the probe and the substrate as well as the solvent in which the measurements are performed.[300] Frequently, considerable variation is observed in the average pull-off force obtained for different probes measuring the same interaction.[1006] For example, the average pull-off force for Au-coated probes and substrates functionalized with a monolayer of 11-mercapto-1-undecanamide was reported in separate papers as 0.211c and 1.81a nN under ethanol. This discrepancy has been ascribed to the difference in the radius of curvature, R, between the different probes used in these experiments. A sharp probe with a small R is expected to give a smaller average pull-off force than a dull probe with large R. The common contact mechanics models used to analyze these pull-off force measurements, namely, the Derjaguin, Muller, Toporov (DMT) and Johnson, Kendall, Roberts (JKR) models, predict that the pull-off force should scale linearly with the probe tip radius of curvature.[1007] The force required in JKR model to separate a probe from adhesive contact with a flat substrate is determined by the magnitude of the work of adhesion, W_{ad}, between the probe and the substrate in a given solvent. The work of adhesion depends on interfacial energies, between the probe and the substrate. In cases where W_{ad} is known for a particular system, one can estimate indirectly AFM probe radii from pull-off force measurements.[1008] The applicability of continuum-based contact mechanics models to nanoscale contacts, in particular, the linear dependence of the pull-off force on R, has not been investigated in detail.[1009] Numerous studies have shown that $F_{pull\text{-}off}$ correlates with expectations based on the JKR theory and estimated values of W_{ad}.

In a recent study, a principal objective was to assess the functional dependence of $F_{pull\text{-}off}$ on probe radius.[1010] The interaction between two hydrophobic surfaces in water was used as a model system to correlate measured pull-off forces with tip radii. Previous AFM force measurements showed that *hydrophobic* forces in water are among the strongest interactions encountered between SAM-tailored surfaces.[1006]

The studies of strong interactions are important in order to maximize the effects that small variations in AFM probe sizes have on the pull-off forces. The functional dependence of the adhesion (pull-off) force on probe radius in AFM force measurements was investigated. Pull-off forces for hydrophobic AFM probes in contact with hydrophobic substrates were measured in water. Chemical functionalization of probes and substrates was accomplished by self-assembly of octadecanethiol on Au. Average pull-off forces for 10 SAM/Au-coated probes were measured and found to depend linearly on the probe tip radii, which were determined by SEM. Using the JKR contact mechanics model, the thermodynamic work of adhesion (W_{ad}) determined from the pull-off forces was 110 mJ/m^2, which compares favorably with estimates of W_{ad} based on reported interfacial energies. These results showed the linear dependence of the adhesive force on probe radius, as expected from contact mechanics models. They also verified that SEM can provide self-consistent and reproducible estimates of the probe radius.

A number of authors have reported measuring the radius of curvature of AFM probes,[1011] typically from SEM images.[1012] In conjunction with force measurements, the average radius of curvature for all probes used in a particular experiment is usually reported, without noting the spread encountered. Another objective of this study was to determine what the spread could be for commercial probes and whether SEM could be used reliably and reproducibly to determine *R*. Standard Si_3N_4 triangular cantilevers were used. Cleaned silicon wafers coated with Cr (5 nm) and followed by Au (100 nm) by thermal evaporation. Similarly, AFM cantilevers were coated with 3 nm of Cr and 40 nm of Au. Substrates and cantilevers were dipped immediately into a 5 mM solution of ODT in 4:1 ethanol/THF for 24 h. Force measurements were carried out after drying the specimen on a microscope equipped with a fluid cell. Cantilevers with length of 100 μm and leg width of 35 μm were used, and the force constant of each lever was determined using the well-known Cleveland procedure.[638] Resonance frequencies of coated cantilevers varied from 50 to 54.1 kHz, with the corresponding variation of force constant between 0.23 and 0.28 N/m. Approximately 200 force curves were collected with each cantilever, each with a *Z* position sweep of 500 nm at a rate of 500 nm/s. Imaging of AFM tips was performed with a field emission gun scanning electron microscope (FEG SEM) (at 200,000× magnification). Tip radii were measured by drawing a circle on the images such that an arc of the circle coincided with the tip end. The error in the measurement was estimated by drawing minimum and maximum possible arc radii to coincide with the tip. Topographic AFM images of the polycrystalline Au substrates were acquired using an oxide-sharpened Si probe. The radii of curvature of 20 Au grains were estimated by measuring their height (h) and radius (r) and using $R = (h^2 + r^2)/2$ h. For each of the 10 tips examined in this study, the average pull-off force was determined by acquiring approximately 200 force curves. The histograms showed the distribution of pull-off forces for two tips originating from the same wafer. A large (~20 nN) difference was found in the mean pull-off force, which along with the relatively small spread for both tips, resulted in completely separated distributions. Images were compared with tips of radius of curvature of 15 nm (sharp Au-coated Si_3N_4 probe) and a blunt tip (radius of curvature of 110 nm). It was found the the top view of the tip showed that the Au coating was intact, indicating that the bluntness of the tip was not due to the force measurements.

It was found that qualitatively the average pull-off force clearly depended on the probe radius.[974,1006] In these studies, data were obtained for histograms of pull-off forces with two different SAM-modified AFM tips with radii of curvature of 15 nm and 110 nm. The average pull-off force was plotted as a function of the estimated tip radius for 10 probes. A significant variation in tip radius (15 to 130 nm) was observed, with eight of the probes evenly distributed between 15 and 60 nm. It was found from these analyses that a single AFM probe can lead to significant errors (factors of two to three, typically). A good linear correlation was observed between $F_{pull\text{-}off}$ and R. On the other hand, the linear fit exhibited a nonzero intercept. Analyses of the work of adhesion from the data gave the value of Wad = 49 mJ/m^2, which is ca. half the expected value of 103 mJ/m^2. By using reported values for the interfacial energies,[3,639] the value for the thermodynamic work of adhesion between two methyl-terminated surfaces in water can be estimated. These values are 52 and 0.9 mJ/m^2 for methyl-water and methyl-methyl, respectively, yielding $W = 103$ mJ/m^2.

These data were found to be in agreement with theory. The topological shape of the polycrystalline Au surface could be modeled as a two-dimensional array of spherical caps. Thus, the tip–substrate interaction could be approximated as a contact between two half-spheres of unequal radii. From AFM images of the Au-coated substrate using oxide-sharpened Si AFM tips, the average radius of curvature was found for the Au grains to be 110 nm. These data showed that the mean pull-off force had a linear dependence on the reduced tip radius. A linear fit of the data gave a value of slope of 0.52 N/m and an intercept of 0.2 nN. These results agreed with the functional model described earlier. Furthermore, calculating the work of adhesion from the slope yields W_{ad} = 110 mJ/m^2, in agreement with values based on published interfacial energies.[629] The conclusion was, therefore, that the linear relationship between $F_{pull\text{-}off}$ and R, as predicted by models such as JKR, can be applied to AFM pull-off force measurements, which has not been shown previously. It was also concluded that the SEM method was a reliable method for estimating tip radii. Other methods for determining AFM probe radii have been developed, one of which involves scanning the tip over a step-edge[1013] or an apex having a much higher aspect ratio than the tip.[1014] Deconvolution of this image gives an estimated radius. The advantage of using SEM to determine R is the speed by which a large batch of tips can be analyzed. From these studies, it was demonstrated that SEM provides reliable measurements of R. The data also showed that there is considerable spread in tip radii for commercial AFM probes. Some of this variation in R presumably can be attributed to the microstructure of the Au coating on the tip, but the underlying shape of the Si_3N_4 is also of key importance to the overall sharpness of the probe. It becomes interesting to estimate the variation in the tip–substrate contact area for tips of different radii. According to the JKR model, the radius of the contact area at pull-off is related to the elastic modulus of the two contacting objects.[1007]

With an alkanethiol coverage of ~4 molecules/nm^2, these data corresponded to 16 and 104 molecular pairs in contact at pull-off. One may expect that a significant difference in contact area can be associated with two identically functionalized AFM probes. Assuming an average radius of curvature determined from only a few probes can result in significant over- or underestimates of tip–sample contact area and W_{ad} in pull-off measurements.

The prediction of the JKR model and that $F_{pull\text{-}off}$ scales with R was found to be not in accord with a simple model in which adhesion arises from a certain number of identical bonds between the tip and substrate. This would predict that $F_{pull\text{-}off}$ would scale with $R^{2/3}$, because the contact area is proportional to $R^{2/3}$. The dependence of the JKR model to a discrete bonding picture remains an important subject to investigate intermolecular interactions in microcontact rupture experiments. The adhesion between the hydrophobic tip and substrate in water is clearly not due to formation of discrete bonds between molecules on the tip and the substrate. The latter is the thermodynamic driving force for the adhesion that arises from the large surface energy associated with the water–monolayer interface. It was concluded that the pull-off forces in water can be measured between octadecanethiol-modified Au-coated AFM tips and substrates. For each tip, the mean pull-off force was correlated with its radius of curvature, measured from high-resolution SEM images. By taking into account the roughness of the Au substrate, the pull-off force was found to scale linearly with the reduced tip radius in accordance with the JKR model. A considerable variation was reported in probe radii, with most probes having a radius of curvature in the 20 to 60 nm range. These findings are of importance to researchers using contact mechanics models to analyze AFM pull-off force measurements.

7.13 TIME-RESOLVED ANALYSES BY STM

The STM first found wide use in the field of surface science as an atomic-scale probe of topography and electronic structure, but in applications of tunneling microscopy, now they extend from the imaging of such complex molecules as DNA to the fabrication of a structure out of single-atom building blocks. Among these powerful techniques, the STM relies on the extremely localized quantum mechanical tunneling of electrons between the sample and the tip, and therefore, it offers the advantage of an extraordinarily high three-dimensional spatial resolution. Although the possibility has long been recognized, researchers have so far met with limited success in attempting to extend that resolution to the fourth dimension, that is, to allow the STM to probe phenomena an atomic time scales as well as atomic length scales. Adding such time resolution to the capabilities of the STM requires that some important obstacles be overcome. Although the intrinsic time scale for tunneling across the junction between the sample and tip in an STM has been estimated to be of the order of f_{sec} (femtosecond) or less, stray capacitance in the wiring and the unavoidable trade-off between signal-to-noise ratio and speed in the external etectronics, limit the bandwidth in typical instruments to ~30 kHz or less. Similar difficulties arise in the field of ultrafast optics, where the speed of the electronics is no match for femtosecond optical pulses, and it is natural to adapt optical "pump and probe" methods for detecting fast transient signals to time-resolved STM experiments.

In the case of normal STM operation, one can measure the nonlinear dependence of the tunnel current on the voltage applied between the tip and the sample, without any parts moving. Among these techniques, STM depends on the extremely localized quantum mechanical tunneling of electrons between the tip and the sample. This system thus offers a three-dimensional resolution. Nonlinearity depends on the details of the electronic density of the states in the sample and tip.[1003] The I vs. V curves were related to the average barrier height and the tip–sample separation. The picosecond pulse arrangement was based upon using a laser pulse and obtaining time-resolved data.

A procedure was described to perform fast time-resolved experiments with STM.[117] As already seen, STM is the only method that provides investigations that can be carried out based upon localized quantum mechanical tunneling of electrons between the sample and the tip. This offers an observed resolution of the molecules in the three-dimensional domain. The fourth dimension has been suggested to be also possible by this method, which relates to the atomic time scale. The estimated intrinsic time scale for tunneling across the junction between the sample and the tip has been estimated to be of the order 10 f_{sec} (10^{-15} sec) or less. It was suggested that, in order for the time-resolved STM to be useful, it must be possible to relate the shape of the time-resolved current to the time dependence of the underlying processes.

References

1. Laidler, K.J., *The World of Physical Chemistry*, Oxford University Press, Oxford, 1995.
2. Adamson, A.W., *Physical Chemistry of Surfaces*, 5th ed., John Wiley & Sons, New York, 1990.
3. Birdi, K.S., Ed., *Handbook of Surface & Colloid Chemistry*, CRC Press, Boca Raton, FL, 1997; 2nd ed., 2002.
4. Christmann, K., *Introduction to Surface Physical Chemistry*, Steinkopff-Verlag, New York, 1991.
5. Binnig, G., Gerber, C.H., and Weibel, E., *Phys. Rev. Lett.*, 49, 57, 1982; Binnig, G. and Rohrer, H., *Helv. Phys. Acta*, 55, 726, 1982.
6. Binnig, G. and Rohrer, H., *Surf. Sci.*, 126, 236, 1983.
7. Binnig, G. et al., *Phys. Rev. Lett.*, 50, 120, 1983.
8. Binnig, G., Quale, C.F., and Gerber, C., *Phys. Rev. Lett.*, 56, 930, 1986.
9. Binnig, G. et al., *Europhys. Lett.*, 3, 1281, 1987.
10. Garcia, N., in *Proc. First Int. Conf. Scanning Tunneling Microsc.*, Garcia, N., Ed., *Surf. Sci.*, 181, 1987; Hansma, P.K., Ed., *Tunneling Spectroscopy*, Plenum Press, New York, 1982; 11th Int. Conf. Scanning Tunneling Microsc., July 15–20, 2001, University of British Colombia, Vancouver, Canada.
11. Feenstra, R.M., Proc. Second Int. Conf. Scanning Tunneling Microsc., *J. Vac. Sci. Technol.*, A6, 259, 1988; Sarikaya, M., *Ultramicroscopy*, 47, 1, 1992.
12. van de Leemput, L.E.C. and van Kempen, H., *Rep. Prog. Phys.*, 55, 1165, 1992.
13. Frommer, J., *Angew. Chem. Int.*, 31, 1298, 1992.
14. Parsons, E. et al., *J. Microsc.*, 174, 59, 1993.
15. Falk, R.H., *Scanning Electron Microsc.*, 80, 79, 1980.
16. Marti, O., Drake, B., and Hansma, P.K., *Appl. Phys. Lett.*, 51, 484, 1987; Drake, B. et al., *Science*, 243, 1586, 1989.
17. McNiel, I., *An Encyclopedia of the History of Technology*, Routledge, London; Oatley, C.W., The early history of the scanning electron microscope, *J. Appl. Phys.*, 53, 1, 1982; Wiesendanger, R., *Scanning Probe Microscopy and Spectroscopy*, Cambridge University Press, London; New York, 1994.
18. Joy, D.C. and Pawley, J.B., *Ultramicroscopy*, 47, 80, 1992.
19. Young, R. and Ward, J., *Electrochim. Acta*, 43, 3257, 1999.
20. Birdi, K.S., *Self-Assembly Monolayer (SAM) Structures*, Plenum Press, New York, 1999.
21. Hörber, J.K.H. et al., *Appl. Phys.*, A, 66 (entire issue), 1998.
22. Reed, M.A., *Scientific American*, 268, 118 (January), 1993.
23. Tsukada, M. et al., *Surf. Sci. Rep.*, 13, 267, 1991; Xu, J.B. et al., *Appl. Phys.*, A, 59, 155, 1994.
24. Dalidchik, F. et al., *JETP Lett.*, 65, 306, 1997.

25. Fuchs, H., *Physica Scr.*, 38, 264, 1988.
26. Wigren, R., Ph.D. thesis, Linkoping University, Sweden, 1996.
27. Mate, C.M. et al., *Surf. Sci.*, 208, 473, 1988; Tang, S.L., Bokor J., and Storz, R.H., *Appl. Phys. Lett.*, 52, 188, 1988; Kirk, M.D., Albrecht, T.R., and Quate, C.F., *Rev. Sci. Instrum.*, 59, 833, 1988.
28. Meyer, H. et al., *J. Microsc.*, 151, 269, 1988.
29. Lucas, A.A. et al., *Surf. Sci.*, 269, 27074, 1992.
30. Volmar, U.E. et al., *Appl. Phys.*, A, 66, 735, 1998.
31. Butt, H.-J. et al., *J. Microsc.*, 169, 75, 1993.
32. Siedle, P. and Butt, H.J., *Langmuir*, 11, 4, 1065, 1995.
33. Babock, K.L. and Fuchs, H., *Ultramicroscopy*, 66, 251, 1997.
34. Dammer, U. et al., *Biophys. J.*, 70, 2437, 1996.
35. Skörman, B. et al., *Langmuir*, 16, 6267, 2000.
36. Montelius, L. and Tegenfeldt, J.O., *Appl. Phys. Lett.*, 62, 2628, 1993; Montelius, L., Tegenfeldt, J.O., and van Heeren, P.J., *Vac. Sci. Technol.*, 12, 2222, B, 1994.
37. Gratter, P., Zimmermann-Edling, W., and Brodbeck, D., *Appl. Phys. Lett.*, 60, 2741, 1992.
38. Sheiko, S.S. et al., *Ultramicroscopy*, 53, 371, 1994.
39. DeRose, J.A. and Revel, J.-P., *Microsc. Microanal.*, 3, 203, 1997.
40. Heuberger, M., Dietler, G., and Schlapbach, L.J., *Vac. Sci. Technol.*, B, 14, 1250, 1996.
41. Neves, B.R.A. et al., *Ultramicroscopy*, 76, 61, 1999.
42. Atamny, F. and Baiker, A., *Surf. Sci.*, 323, 314, 1995.
43. Jensen, F., *Rev. Sci. Instrum.*, 64, 2595, 1993.
44. Griffith, J.E. et al., *J. Vac. Sci. Technol.*, B, 9, 3586, 1991.
45. Grigg, D.A. et al., *Ultramicroscopy*, 42, 1616, 1992.
46. Li, Y. and Lindsay, S.M., *Rev. Sci. Instrum.*, 62, 2630, 1991.
47. Odin, C. et al., *Surf. Sci.*, 317, 321, 1994.
48. Vesenka, J. et al., *Biophys. J.*, 65, 992, 1993; Vesenka, J., Miller, R., and Henderson, E., *Rev. Sci. Instrum.*, 65, 2249, 1994.
49. Xu, S. and Arnsdorf, M.F.J., *J. Microsc.*, 73, 199, 1994.
50. Wilson, D.L. et al., *Langmuir*, 11, 265, 1995.
51. Eppell, S.J., Zypman, F.R., and Marchant, R.E., *Langmuir*, 9, 2281, 1993.
52. Smith, D.J. et al., *Science*, 233, 872, 1986.
53. Markiewicz, P. and Goh, M.C., *Langmuir*, 10, 5, 1994.
54. Tegenfeldt, J.O. and Montelius, L., *Appl. Phys. Lett.*, 66, 1068, 1995.
55. Griffith, J.E. and Grigg, D.A., *J. Appl. Phys.*, 74, 83, 1993.
56. Villarrubia, J.S., *Surf. Sci.*, 321, 287, 1994; Villarrubia, J.S., *J. Vac. Sci. Technol.*, B, 14, 1518, 1996.
57. Keller, D., *Surf. Sci.*, 253, 353, 1991; Keller, D.J. and Chih-Chung, C., *Surf. Sci.*, 268, 333, 1992.
58. Spatz, J.P., Sheiko, S.S., and Mueller, M. *Ultramicroscopy*, 75, 1, 1998.
59. Olsson, L. et al., *J. Appl. Phys.*, 84, 4060, 1998.
60. Niedermann, Ph. and Fischer, Ï., *J. Microsc.*, 152, 93, 1988.
61. Schwarz, D. et al., *J. Microsc.*, 173, 183, 1994.
62. Keller, D. et al., *Ultramicroscopy*, 42, 1481, 1992.
63. Weihs, T.P. et al., *Appl. Phys. Lett.*, 59, 3536, 1991.
64. Horn, R.G., Israelachvili, J.N., and Pribac, F.J., *Colloid Interface Sci.*, 115, 480, 1987.
65. Landman, U., Luedtke, W.D., and Ribarsky, M.W., *J. Vac. Sci. Technol.*, A, 7, 2829, 1989.
66. Tekman, E. and Ciraci, S., *J. Phys. Condens. Matter*, 1, 3, 2613, 1991.
67. Guo, S. et al., *J. Appl. Phys.*, 77, 5369, 1995.

68. Inoue, T. et al., *Appl. Phys. Lett.*, 56, 1332, 1990.
69. Skörman, B. et al., *J. Catal.*, 181, 6, 1999.
70. Jacobsen, S.N. et al., *Surf. Sci.*, 429, 22, 1999; Jacobsen, S.N., Ph.D. thesis, Linkoping University, Sweden, 1998.
71. McCaffrey, J.P., *Microsc. Res. Technol.*, 24, 180, 1993.
72. Yang, J. and Shao, Z., *Ultramicroscopy*, 50, 157, 1993.
73. Thundat, T. et al., *J. Vac. Sci. Technol.*, 10, 630, 1992.
74. Weisenhorn, A.L. et al., *Phys. Rev.*, B, 45, 11226, 1992.
75. Goodman, F.O. and Garcia, N., *Phys. Rev.*, B, 43, 4728, 1991.
76. Pashley, M.D., Pethica, J.B., and Tabor, D., *Wear*, 100, 7, 1984.
77. Dörig, U., Ziger, O., and Pohl, D.W., *Phys. Rev. Lett.*, 65, 349, 1990.
78. Overney, R.M. et al., *Nature*, 359, 133, 1992.
79. Burnham, N.A., Colton, R.J., and Pollock, H.M., *J. Vac. Sci. Technol.*, A, 9, 2548, 1991.
80. Givargizov, E. I. et al., *Ultramicroscopy*, 82, 57, 2000.
81. Postema, A.R., de Groot, H., and Pennings, A.J., *J. Mater. Sci.*, 25, 4216, 1990.
82. Hues, S.M., Draper, C.F., and Colton, R.J., *J. Vac. Sci. Technol.*, B, 12, 2211, 1994.
83. Burnham, N.A., *Appl. Phys. Lett.*, 63, 114, 1993.
84. Wygant, J.F., *J. Am. Ceram. Soc.*, 34, 374, 1951.
85. Meyer, E. et al., *J. Vac. Sci. Technol.*, 9, 1329, B, 1991.
86. Weisenhorn, A.L. et al., *J. Vac. Sci. Technol.*, B, 9, 1333, 1991.
87. Pethica, J.B. and Oliver, W.C., *Physica Scripta,* 19, 61, 1987; Todd, J.D. and Pethica, J.B., *J. Phys.: Condens. Matter*, 1989, 1, 9823, 1989; Pethica, J.B., *Phys. Rev. Lett.*, 57, 3235, 1986; Mate, C.M. et al., *Phys. Rev. Lett.*, 59, 1942, 1987.
88. Chen, G.Y. et al., *J. Appl. Phys.*, 78, 1465, 1995; Chen, G.Y. et al., *J. Vac. Sci. Technol.*, B, 14, 1313, 1996.
89. Benfedda, M. et al., *Ultramicroscopy*, 76, 187, 1999.
90. Bachelot, R., Gleyzes, P., and Boccara, A.C., *Probe Microsc.*, 1, 89, 1997.
91. Isaacson, M, Cline, J., and Barshatzky, H., *Ultramicroscopy*, 47, 15, 1992.
92. Sandoghdar, S. et al., *J. Appl. Phys.*, 36, 6818, 1997.
93. Courjon, D., *Appl. Opt.*, 29, 3734, 1990.
94. Williamson, R.L. and Miles, M.J., *J. Appl. Phys.*, 80, 4804, 1996; Akamine, S., Kuwano, H., and Yamada, H., *Appl. Phys. Lett.*, 68, 579, 1994; Davis, R.C., Williams, C.C., and Neuzil, P., *Appl. Phys. Lett.*, 66, 2309, 1995; Mihalcea, C. et al., *Appl. Phys. Lett.*, 68, 3531, 1996; Oesterschulze, E. et al., *Appl. Phys.*, A, 66, 367, 1998; Jin, Z.,Wang, K., and Huang, W., *Appl. Phys.*, A, 66, 915, 1998; Moers, M.H.P., Gaub, H.E., and Hulst, N.F., *Langmuir*, 10, 2774, 1994.
95. Fischer, U.C., Koglin, J., and Fuchs, H., *J. Microscopy*, 176, 231, 1994.
96. Koglin, J., Fischer, U.C., and Fuchs, H., *J. Biomed. Optics,* 1, 75, 1996; Koglin, J., Fischer, U.C., and Fuchs, H., *Phys. Rev.*, B 55, 15, 1997.
97. Xu, Z. and Yoon, R.-H.J., *Colloid Interface Sci.*, 1, 34, 427, 1990.
98. Rabinovich, Y.A.I. and Yoon, R.-H., *Langmuir*, 10, 1903, 1994; Rabinovich, Y.I. and Yoon, R.-H., *Colloid Science*, A 93, 263, 1994.
99. Christenson, H.K. and Claesson, P.M., *Science*, 239 ,90, 1988.
100. Birdi, K.S., *Lipid & Biopolymer Monolayers*, Plenum Press, New York, 1989.
101. Putman, C.A.J. et al., *Langmuir*, 8, 3014, 1992.
102. Song, J.P. et al., *Surf. Sci.*, 296, 299, 1993.
103. Oberleithner, H. et al., in *Seminars in Renal Physiology*, Lang, F., Ed., S. Karger, Basel, 1994.
104. Han, W. et al., *Nature*, 386, 568, 1997.
105. Santesson, L. et al., *J. Phys. Chem.*, 99, 1038, 1995.

106. Park, S.S., *J. Phys. Chem.*, B, 102, 6020, 1998.
107. Chattoraj, D.K. and Birdi, K.S., *Adsorption and the Gibbs Surface Excess*, Plenum Press, New York, 1984; Friedbacher, G. et al., *An Introduction to Ultrathin Organic Films: From Langmuir-Blodgett to Self-Assembly*, Academic Press: New York, 1991.
108. Hansma, H.G. et al., *Science*, 256, 1180, 1992.
109. Allison, D.P. et al., *J. Vac. Sci. Technol.*, A, 11, 616, 1993.
110. Johnson, R.P.C. and Gregory, D.W., *J. Microscopy*, 171, 2, 125,1993.
111. Iri, T., Shiba, H., and Nishikawa, H., *Jpn. J. Appl. Phys.*, 31, 1441, 1992.
112. Birdi, K.S. and Vu, D.T., *Langmuir*, 10, 623, 1994; Rapaport, H. et al., *Biophys. J.*, 81, 2729, 2001; Alba, M., *Langmuir*, 17, 6498, 2001.
113. Birdi, K.S. et al., *Surf. Coat. Technol.*, 67, 183, 1994.
114. Bethell, B., Kiely, C.J., and Schiffrin, D.J., *Langmuir*, 14, 5425, 1998.
115. Lewerenz, H.J. et al., *AIDS Res. Human Retrov.*, 8, 9, 1663, 1992.
116. Shao, F. and Yang, J., *Quart. Rev. Biophys.*, 28, 195, 1995.
117. Findenegg, G.H., *J. Chem. Soc. Faraday Soc.*, 69, 1069, 1973.
118. Findenegg, G.H., *J. Chem. Soc. Faraday Soc.*, 68, 1799, 1972.
119. Groszek, A.J., *Proc. R. Soc. (London)*, 314, 473, 1970.
120. Yeo, Y.H., McGoigal, G.C., and Thomson, D.J., *Langmuir*, 9, 649, 1993.
121. Buchholz, S. and Rabe, J.P., *Angew. Chem. Int.*, 31, 189, 1992.
122. Yachoboski, K. et al., *Ultramicroscopy*, 42, 963, 1992.
123. Abe, U.R. and Buchholz, S., *Science*, 253, 424, 1991.
124. Dishner, M.H., Hemminger, J.C., and Feher, F.J., *Langmuir*, 13, 4788, 1997.
125. Ravaine, S. et al., *Langmuir*, 14, 708, 1998.
126. Delamarche, E., Michel, B., and Gerber, Ch., *Langmuir*, 10, 4103, 1994.
127. Poirier, G.E. and Tarlov, M.J., *J. Phys. Chem.*, 99, 10966, 1995.
128. Nag, K. and Keough, M.W., *Biophys. J.*, 65, 1019, 1993.
129. Smith, J.L. and Berg, J.C., *J. Phys. Chem.*, 84, 2150, 1980; Vollhardt, D., *Langmuir*, 9, 3320, 1993; Vollhardt, D., Kato, T., and Kawano, M., *J. Phys. Chem.*, 100, 4141, 1996.
130. Schwartz, D.K., Viswanathan, R., and Zasadzinski, J.A.N., *J. Phys. Chem.*, 96, 10444, 1992.
131. Mizutani, W. et al., *Jpn. J. Appl. Phys.*, 27, 1803, 1988; Horber, J.K.H. et al., *Chem. Phys. Lett.*, 145, 151, 1988.
132. Elliot, D.J.E. et al., *Langmuir*, 11, 4773, 1995.
133. McConnell, H.M., *Annu. Rev. Phys. Chem.*, 42, 171, 1991.
134. Wolthaus, L., Schaper, A., and Mobius, D., *J. Phys. Chem.*, 98, 10809, 1994.
135. Kenn, R.M. et al., *J. Phys. Chem.*, 95, 2092, 1991.
136. Kajiyama, T. et al., *Langmuir*, 8, 1563, 1992.
137. ten Grotenhuis, E. et al., *Biophys. J.*, 71, 1489, 1996.
138. Dufrune, Y.F. et al., *Langmuir*, 13, 4779, 1997.
139. Engblom, J., Engstrom, S., and Joensson, B., *J. Controlled Release*, 52, 271, 1998.
140. Wertz, P.W. et al., *J. Invest. Dermatol.*, 89, 419, 1987.
141. Nicander, I. et al., *J. Invest. Dermatol.*, 112, 72, 1999.
142. Neubauer, G. et al., *Rev. Sci. Instrum.*, 61, 2296, 1990.
143. MacRitchie, F., *Chemistry at Interfaces*, Academic Press, New York, 1990; Israelachvili, J., *Langmuir*, 10, 3774, 1994; Ipsen, J.H., Mouritsen, O.G., and Zuckermann, M.J., *J. Chem. Phys.*, 91, 1855, 1989.
144. Bibo, A.M. and Peterson, I.R., *Adv. Mater.*, 2, 309, 1990.
145. Overney, R.M. et al., *Langmuir*, 10, 1281, 1994; Overney, R.M. et al., *Langmuir*, 9, 341, 1993.
146. Hosoi, H. et al., *Jpn. J. Appl. Phys.*, 36, 6927, 1997.

147. Koleske, D.D. et al., *Mater. Res. Soc. Symp. Proc.*, 464, 377, 1997.
148. Widayati, S. and Dluhy, R.A., *Microchim. Acta Suppl.*, 14, 683, 1997.
149. Schwartz, D.K. et al., *J. Am. Chem. Soc.*, 115, 7374, 1993; Schwartz, D.K.,Visswanathan, R., and Zasadzinski, J.A., *Langmuir*, 9, 5, 1384, 1993.
150. Yang, X.M. et al., *Appl. Phys.*, A, 59, 139, 1994.
151. Harke, M. and Motschmann, H., *Langmuir*, 14, 313, 1998.
152. McConnell, H.M., Tamm, L.K., and Weis, R., *Proc. Natl. Acad. Sci., U.S.A.*, 81, 3249, 1984; Weis, R. and McConnell, H.M., *J. Phys. Chem.*, 89, 4453, 1985.
153. Benvegnu, D.J. and McConnell, H.M., *J. Phys. Chem.*, 96, 6820, 1992.
154. Klingler, J.F. and McConnell, H.M., *J. Phys. Chem.*, 97, 2962, 1993; McConnell, H.M. and Noy, B.T., *J. Phys. Chem.*, 90, 2333, 1986.
155. Yamada, R. and Uosaki, K., *Langmuir*, 13, 5218, 1997; Jenkins Bushby Boden, Yamada, R., and Uosaki, K., *Langmuir*, 855, 1998.
156. Losche, M. et al., *Phys. Chem.*, 87, 848, 1983; Peters, R. and Beck, K., *Proc. Natl. Acad. Sci., U.S.A.*, 80, 7183, 1983; Seul, M., *J. Phys. Chem*, 97, 2941, 1993; Keller, D.J., McConnell, H.M., and Moy, V.T., *J. Phys. Chem.*, 90, 2311, 1986.
157. Andelman, D., Brochard, F., and Joanny, J.F., *J. Phys. Chem.*, 86, 3673, 1987.
158. Heckl, W.M. et al., *Eur. Biophys. J.*, 84, 278, 1979.
159. Biltonen, R.L., *J. Chem. Thermodyn.*, 22, 1, 1990; Jørgensen, K. et al., *Biochim.Biophys. Acta*, 1067, 241, 1991; Tallon, J.L. and Cotterill, R.M.J., *Aust. J. Phys.*, 38, 1, 1985; Mouritsen, O.G., *Computer Studies of Phase Transition and Critical Phenomena*, Springer-Verlag, Heidelberg, 1984; Mouritsen, O.G. and Zuckermann, M.J., *Eur. Biophys. J.*, 12, 75, 1985.
160. Eklund, J.C. et al., *J. Electroanal. Chem.*, 344, 255, 1993.
161. Patrick, D.L. et al., *J. Phys. Chem.*, B, 103, 8328, 1999.
162. Sprik, M. et al., *Langmuir*, 10, 4116, 1994.
163. Sasaki, A. et al., *Jpn. J. Appl. Phys.*, 32, 2952, 1993.
164. Venkataraman, B., Breen, J.J., and Flynn, G.W., *J. Phys. Chem.*, 99, 6608, 1995.
165. Venkataraman, B. et al., *J. Phys. Chem.*, 99, 8684, 1995.
166. Kuznetsov, A.M., Sommer-Larsen, P., and Ulstrup, J., *Surf. Sci.*, 275, 52, 1992.
167. Brandow, S.L. et al., *Biophys. J. Biophys. Soc.*, 64, 898, 1993.
168. Kurnaz, M.L. and Schwartz, D.K., *J. Phys. Chem.*, 100, 11113, 1996.
169. Ho, K. and Bard, A.J., *Langmuir*, 13, 5418, 1997.
170. Terrill, R.H., Tanzer, T.A., and Bohn, P.W., *Langmuir*, 14, 845, 1998.
171. Barger, W.R., Green, J.-B.D., and Lee, G.U., *Langmuir*, 13, 4797, 1997.
172. Chernomordik, A. et al., *J. Phys. Chem.*, 99, 8690, 1995.
173. Schönenberger, C.J.A.M. et al., *Langmuir*, 8, 1133, 1992.
174. Gregoriadis, G. and Florence, A.T., *Drugs*, 45, 15, 1993.
175. Hughson, F.M., *Curr. Opin. Struct. Biol.*, 5, 507, 1995.
176. Zimmerberg, L.V., *J. Curr. Opin. Struct. Biol.*, 5, 541, 1995.
177. Smit, J.M., Bittman, R., and Wilschut, J., *J. Virol.*, 73, 8476, 1999.
178. Bernardes, C. et al., *Biochim. Biophys. Acta*, 1393, 19, 1998.
179. Kanaseki, T. et al., *J. Cell Biol.*, 137, 1041, 1997.
180. Pantazatos, D.P. and MacDonald, R.C., *J. Membrane Biol.*, 170, 27, 1999.
181. Asgharian, N. and Schelly, Z.A., *Biochim. Biophys. Acta*, 1418, 295, 1999.
182. Wong, J.Y. et al., *Biophys. J.*, 77, 1458, 1999.
183. Cescato, C., Walde, P., and Luisi, P.L., *Langmuir*, 13, 4480, 1997.
184. Parpura, V. et al., *Neuroimage*, 2, 3, 1995.
185. Garcia, R.A. et al., *Neurosci. Res.*, 52, 350, 1998.
186. Laney, D.E. et al., *Biophys. J.*, 72, 806, 1997.

187. Shibata-Seki, T. et al., *Thin Solid Films*, 273, 297, 1996; Pignataro, B. et al., *Biophys. J.*, 78, 487, 2000.
188. Egawa, H. and Furusawa, K., *Langmuir*, 15, 1660, 1999.
189. Reviakine, I. and Brisson, A., *Langmuir*, 16, 1806, 2000.
190. Kumar, S. and Hoh, J.H., *Langmuir*, 16, 9936, 2000.
191. Brian, A.A. and McConnell, H.M., *Proc. Natl. Acad. Sci., U.S.A.*, 81, 6159, 1984.
192. Fang, Y. and Yang, J., *Biochim. Biophys. Acta*, 1324, 309, 1997.
193. Butt, H.J., Jaschke, M., and Ducker, W., *Bioelectrochem. Bioenerg.*, 38, 191, 1995.
194. A-Hassan, E. et al., *Biophys. J.*, 74, 1564, 1998; Marsh, D., *Handbook of Lipid Bilayers*, CRC Press, Boca Raton, FL, 1990.
195. Lemmich, J. et al., *Eur. Biophys. J. Biophys. Lett.*, 25, 293, 1997.
196. Hansma, H.G. and Hoh, J.H., *Annu. Rev. Biophys. Biomol. Struct.*, 23, 115, 1994.
197. Heinz, W.F. and Hoh, J.H., *Trends Biotechnol.*, 17, 143, 1999.
198. Bailey, S.M. et al., *Langmuir*, 6, 1326, 1990.
199. Fang, Y. and Yang, J., *J. Phys. Chem.*, 100, 15614, 1996.
200. Needham, D. and Nunn, R.S., *Biophys. J.*, 58, 997, 1990.
201. Bailey, A.L. and Cullis, P.R., *Biochemistry*, 36, 1628, 1997.
202. Paleos, C., Kardassi, D., and Tsiourvas, D., *Langmuir*, 15, 282, 1999.
203. Iwakabe, Y. et al., *Langmuir*, 10, 3201, 1994.
204. Li, Y.Z. and Chander, M., *Science*, 253, 429, 1991.
205. Takahara, A., Ge, S., and Kajiyama, T., *Langmuir*, 11, 1341, 1995.
206. Herod, T.E. and Duran, R.S., *Langmuir*, 14, 6606, 1998.
207. Putman, C.A.J. et al., *Langmuir*, 8, 3014, 1992.
208. Sautet, J., *Surf. Sci.*, 271, 387, 1992.
209. Ducker, W.A. and Wanless, E.J., *Langmuir*, 15, 160, 1999.
210. Imae, T. and Aoki, K., *Langmuir*, 14, 1196, 1998.
211. Mirkin, C.A. and Caldwell, W.B., *Tetrahedron*, 52, 5113, 1996.
212. Dominguez, O. et al., *Langmuir*, 14, 219, 1998.
213. Kadish, K.M. and Rouff, R.S., Eds., *Recent Advances in the Chemistry and Physics of Fullerenes and Related Materials*, The Electrochemical Society: Pennington, NJ, 1994.
214. Liu, X.H. et al., *J. Am. Chem. Soc.*, 116, 5489, 1994.
215. Scholz, F. and Meyer, B., *Electroanal. Chem.*, 20, 1, 1998.
216. Chen, D., Chen, J., and Sarid, D., *Phys. Rev.*, B, 50, 109050, 1994; Yoshio, U., *Langmuir*, 15, 2788, 1999; Pedersen, H.G., *Langmuir*, 15, 3015, 1999; Rowell, R.M., *Langmuir*, 15, 2985, 1999; Neckers, D.C., *Langmuir*, 15, 2077, 1999; Zou, G., *Langmuir*, 15, 2120, 1999; Antonietti, M., *Langmuir*, 15, 1283, 1999; De Schryver, F.C., *Langmuir*, 15, 607, 1999; Sugiyama, O. et al., *J. Phys. Chem.*, B, 103, 6909, 1999; Bond, A.M., Miao, W., and Raston, C.L., *J. Phys. Chem.*, B,103, 5637, 1999; Kitamura, N., *J. Phys. Chem.*, B, 103, 4452, 1999.
217. Compton, R.G. et al., *Electroanal. Chem.*, 327, 337, 1992; Balch, A.C. et al., *Electroanal. Chem.*, 427, 137, 1997.
218. Chlistunoff, J., Cliffel, D., and Bard, A.J., *Thin Solid Films*, 257, 166, 1995.
219. Jehoulet, C., Bard, A.J., and Wudl, F., *J. Am. Chem. Soc.*, 113, 5456, 1991; Jehoulet, C. et al., *J. Am. Chem. Soc.*, 114, 4237, 1992; Szecs, A. et al., *Electroanal. Chem.*, 442, 59, 1998; Szecs, A. et al., *J. Electroanal. Chem.*, 429, 27, 1997; Szecs, A. et al., *J. Electroanal. Chem.*, 419, 39, 1996; Szecs, A. et al., *Synth. Met.*, 77, 227, 1996; Szecs, A. et al., *J. Electroanal. Chem.*, 402, 137, 1996; Szecs, A. et al., *Electrochim. Acta*, 44, 613, 1998; Compton, R.G. et al., *J. Electroanal. Chem.*, 344, 235, 1993.
220. Tomura, K. et al., *Chem. Lett.*, 199, 1365, 1994.

221. Koh, W.Y. et al., *J. Phys. Chem.*, 97, 6871, 1993.
222. Cliffel, D.E., Bard, A.J., and Shinkai, S., *Anal. Chem.*, 70, 4146, 1998; Carlisle, J.J. et al., *J. Phys. Chem.*, 100, 15532, 1996; D'Souza, F. et al., *J. Phys. Chem.*, B, 102, 212, 1998.
223. Zhou, F.M. et al., *J. Phys. Chem.*, 96, 4160, 1992.
224. Suarez, M.F. et al., *J. Phys. Chem.*, B, 103, 5637, 1999.
225. Chabre, Y. et al., *J. Am. Chem. Soc.*, 114, 764, 1992.
226. Suarez, M.F. and Compton, R.G., *J. Electroanal. Chem.*, 462, 211, 1999.
227. Atamny, F. et al., *J. Anal. Chem.*, 353, 433, 1995.
228. Krõtschmer, W., *Nanostruct. Mater.*, 6, 65, 1995.
229. Nishizawa, M., Matsue, T., and Uchida, I., *J. Electroanal. Chem.*, 353, 329, 1993.
230. Koh, W. et al., *J. Phys. Chem.*, 96, 4163, 1992.
231. Hubbard, A.T. and Anson, F.C., *Electroanal. Chem.*, 4, 129, 1970.
232. Bond, A.M. et al., *J. Chem. Soc., Faraday Trans.*, 92, 3925, 1996; Bond, A.M., Fletcher, S., and Symons, P.G., *Analyst*, 123, 1891, 1998; Chambers, J.Q., Scaboo, K., and Evans, C.D., *J. Electochem. Soc.*, 143, 3039, 1996; Scaboo, K.M. and Chambers, J.M., *Electrochim. Acta,* 8, 43, 3257, 1999; Shaw, S.J., Marken, F., and Bond, A.M. *Electroanalysis*, 8, 732, 1996; Evans, C.D. and Chambers, J.Q., *Chem. Mater.*, 6, 454, 1994.
233. Fletcher, S. et al., *J. Electroanal. Chem.*, 159, 267, 1983.
234. Li, J. et al., *Synth. Metals*, 74, 127, 1995.
235. Tatsuma, T., Kikuyama, S., and Oyama, N., *J. Phys. Chem.*, 97, 12067, 1993.
236. Weiss, P.S. et al., *Langmuir*, 14, 1284, 1998.
237. Zaera, F. and Salmeron, M., *Langmuir*, 14, 1312, 1998.
238. Giancarlo, L. et al., *Langmuir*, 14, 1465, 1998.
239. Eigler, D.M. et al., *Phys. Rev. Lett.*, 66, 1189, 1991.
240. Stensgaard, I. et al., *Surf. Sci.*, 269, 81, 1992.
241. Fasman, G.D., Ed., *Poly-Ó-Amino Acids*, vol. 1, Marcel Dekker, New York, 1967; Fasman, G.D., Ed., *Prediction of Protein Structure and the Principles of Protein Conformation*, Plenum Press, New York, 1990.
242. Lindsay, S.M. and Harrington, R.E., *Nature*, 386, 578, 1997; Seong, G.H. et al., *Analytical Chemistry,* 72, 1288, 2000.
243. Balhorn, R., *Scanning Microsc.*, 5, 625, 1991.
244. Shaiu,W.-L. et al., *J. Vac. Sci. Technol.,* A, 11, 820, 1993.
245. Thundat, T. et al., *J. Vac. Sci. Technol.,* A, 11, 824, 1993.
246. Hansma, H.G., *Science Tools from Pharmacia Biotech*, l, 3, 7, 1996.
247. Hansma, H.G., Sinsheimer, R.L., and Groppe, J., *Scanning*, 15, 296, 1993.
248. Wilson, R.J. et al., *Langmuir*, 9, 3478, 1993.
249. Henderson, E., *J. Microsc.*, 167, 77, 1992.
250. Thundat, T. et al., *J. Vac. Sci. Technol.*, A, 10, 630, 1992.
251. Mosser, G. and Brisson, A., *J. Electron Microsc. Technol.*, 18, 387, 1991.
252. Bustamante, C.J. et al., *Biochemistry,* 31, 22, 1992.
253. Hansma, H.G. et al., *Nucleic Acids*, 20, 3585, 1992.
254. Vesenka, J. et al., *Ultramicroscopy,* 42, 1243, 1992; Chen, L. et al., *Chem. Biol.*, 9, 345, 2002; Woolley, A.T. et al., *Nat. Biotechnol.*, 18, 760, 2000.
255. Thundat, T. et al., *Scanning Microsc.*, 16, 903, 1992.
256. Wilson, D.L. et al., *Langmuir,* 11, 265, 1995.
257. Birdi, K.S., *Langmuir,* 7, 3174, 1991.
258. John, S.A., *Am. J. Phys. Cell Phys.*, 267, C-1, 1995.
259. Balhorn, R., *Scanning Microsc.*, 5, 75, 1991.

260. Gibbons, A., *Science*, 253, 382, 1991.
261. Blackford, B.L., Jericho, M.H., and Mulhern, P.J., *Scanning Microsc.*, 5, 907, 1991.
262. Rees, W.A. et al., *Science*, 260, 1646, 1993.
263. Ohnishe, S. et al., *Biophys. J.*, 65, 573, 1993.
264. Breen, J.J. and Flynn, G.W., *J. Phys. Chem.*, 96, 6825, 1992.
265. de-Grooth, B.G. and Putman, C.A., *J. Microsc.*, 168, 239, 1992.
266. Radmacher, M. et al., *Science*, 257, 1900, 1992.
267. Moy, V.T., Florin, E.L., and Gaub, H.E., *Science*, 266, 257, 1994.
268. Butt, H.-J., *Biophys. J.*, 63, 578, 1992.
269. Arnoldi, M. and Fritz, M., *Appl. Phys.*, A, 66, 613, 1998.
270. Hara, M., Furuno, T., and Sasabe, H., *Biophys. J.*, 65, 583, 1993.
271. Florin, E.L., Moy, V.T., and Gaub, H.E., *Science*, 264, 415, 1994; Radmacher, M. et al., *Science*, 265, 1577, 1994; Leggett, F. et al., *Langmuir*, 9, 2356, 1993.
272. Kim, J.-J. et al., *Thin Solid Films*, 244, 700, 1994.
273. Eppell, S.J., *Langmuir*, 9, 2293, 1993.
274. Schwab, A. et al., in *Seminars in Renal Physiology*, Lang, F., Ed., S. Karger, Basel, 1994.
275. Fare, T.L et al., *Langmuir*, 8, 3116, 1992.
276. Schmiff, W.R.L. and Goddord, E.D., *Cosmetics and Toiletries Magazine*, 109, 183, 1994.
277. Weisenhorn, A.L., Rohmer, D.U., and Lorenzi, G.P., *Langmuir*, 8, 3145, 1992.
278. Stange, T.G. et al., *Langmuir*, 8, 920, 1992.
279. O'Brien, J.C. and Porter, M.D., *Chem. Rev.*, 99, 2839, 1999.
280. Balashev, K. et al., *Langmuir*, 13, 5362, 1997.
281. Li, H. et al., *Macromolecules*, 33, 465, 2000; Weng, L.-T., *Macromolecules*, 33, 665, 2000.
282. Zhao, B. et al., *Macromolecules*, 33, 8821, 2000.
283. Park, Y.-J. et al., *Langmuir*, 14, 5419, 1998.
284. Butt, H.-J. and Gerharz, B., *Langmuir*, 11, 4735, 1995.
285. Porter, T.L. et al., *J. Vac. Sci. Technol.*, A 10, 606, 1992.
286. Frazier, R.A. et al., *Langmuir*, 13, 4795, 1997.
287. Pfau, A., Schrepp, W., and Horn, D., *Langmuir*, 15, 3219, 1999.
288. Horn, D. and Linhart, F., in *Paper Chemistry*, 2nd ed., Roberts, J., Ed., Blackie & Son, Glasgow, 1996.
289. Alince, B., Vanerek, A., and van de Ven, T.G.M., *Ber. Bunsen-Ges. Phys. Chem.*, 100, 954, 1996.
290. Akesson, T., Award, C., and Jonsson, B., *J. Chem. Phys.*, 91, 2461, 1989.
291. Podgornik, R., Akesson, T., and Jonsson, B., *J. Chem. Phys.*, 102, 9423, 1995.
292. van de Ven, T.G.M. and Alince, B., *J. Colloid Interface Sci.*, 181, 73, 1996; van de Ven, T.G.M., *J. Pulp. Paper Sci.*, 23, 447, 1997.
293. Gregory, J., *Colloids Surf.*, 31, 231, 1988; Gregory, J., Polymer adsorption and flocculation, in *Industrial Water Soluble Polymers*, Finch, C.A., Ed., Royal Soc. Chem.: London, 1996.
294. Kasper, D.R., Ph.D. dissertation, California Institute of Technology, Pasadena, 1971.
295. Horn, D., Polyethylenimine-physicochemical properties and applications, in *Polymeric Amines and Ammonium Salts*, Goethals, E.J., Ed., Pergamon Press, Oxford; Elmsford, NY, 1980.
296. Akari, S., Schrepp, W., and Horn, D., *Langmuir*, 12, 857, 1996; Akari, S., Schrepp, W., and Horn, D., *Ber. Bunsen-Ges. Phys. Chem.*, 100, 1014, 1996; Akari, S. et al., *Adv. Mater.*, 7, 549, 1995.

297. Frisbie, C.D. et al., *Science*, 256, 2071, 1994.
298. Noy, A. et al., *J. Am. Chem. Soc.*, 117, 7943, 1995.
299. Wilbur, J.L. et al., *Langmuir*, 11, 825, 1995.
300. van der Vegte, E.W. and Hadziioannou, G., *Langmuir*, 13, 4357, 1997.
301. Koutsos, V. et al., *Macromolecules*, 30, 4719, 1997.
302. Furuno, T., Sasae, H., and Ikegami, A., *Ultramicroscopy*, 70, 125, 1998.
303. Spatz, J.P. et al., *Macromolecules*, 30, 3874, 1997.
304. Uchida, E. and Ikada, Y., *Macromolecules*, 30, 5464, 1997.
305. Karymov, M.A. et al., *Langmuir*, 12, 4748, 1996.
306. Siqueira, D.F., Köhler, K., and Stamm, M., *Langmuir*, 11, 3092, 1995.
307. Dziezok, P. et al., *Int. Ed. Engl.*, 36, 2812, 1997; Dziezok, P. et al., *Neue Horizonte in der Polymerforschung*, Mainz, 1998.
308. Biggs, S. and Healy, T.W., *Chem. Soc. Faraday Trans.*, 90, 3415, 1994; Biggs, S., *Langmuir*, 11, 156, 1995.
309. Stipp, S.L.S., *Langmuir*, 12, 1884, 1996.
310. Zhong, Q. et al., *Surf. Sci.*, 290, 688, 1993.
311. Claesson, P.M. et al., *Colloids Surf.*, A, 123, 341, 1997.
312. Lindquist, G.M. and Stratton, R.A.J., *Colloid Interface Sci.*, 55, 45, 1976.
313. Borkovec, M., Daicic, J., and Koper, G.J.M., *Proc. Natl. Acad. Sci., U.S.A.*, 94, 3499, 1997; Borkovec, M. and Koper, G.J.M., *Macromolecules*, 30, 2151, 1997.
314. Stone-Masui, J. and Watillon, A., *J. Colloid Interface Sci.*, 52, 479, 1975.
315. Burchard, W. and Richtering, W., in *Relaxation in Polymers*, Pietralla, M. and Pechold, W., Eds., Steinkopff-Verlag, Heidelberg; Da van der Schee, H.A. and Lyklema, J., *J. Phys. Chem.*, 88, 6661, 1984; C. Stuart, M.A., Cosgrove, T., and Vincent, B., *Adv. Colloid Interface Sci.*, 24, 143, 1986; Fleer, G.J. et al., *Polymers at Interfaces*, Chapman & Hall, London: New York, 1993; Petlicki, J., van der Ven, T.G.M., and Alince, B., *Colloids Surf.*, A83, 1989, 1994.
316. Dautzenberg, H. et al., *Polyelectrolytes–Formation, Characterization and Application*, Hanser Publishers, Munich, 1994.
317. Hu, J. et al., *Surf. Sci.*, 327, 358, 1995.
318. Bar, G. et al., *Langmuir*, 13, 3807, 1997; Magonov, S.N., Elings, V., and Whangbo, M.-H., *Surf. Sci.*, 375, 385, 1997; Bar, G. et al., *Surf. Sci.*, 444, 11, 2000; Delineau, L. et al., *Surf. Sci.*, 448, 179, 2000.
319. Maurice, P.M., *Colloids Surf.*, A107, 57, 1996.
320. Wagberg, L., Ph.D. dissertation, Royal Institute of Technology, Stockholm, Sweden, 1987; Wagberg, L. et al., *Colloids Surf.*, 27, 163, 1987; Wagberg, L. and Asell, I., *Colloids Surf.*, A104, 169, 1995.
321. Biesalski, M. and Rohe, J., Oberflöchenmodifizierung durch kovalent gebundene Polyelektrolytmonoschichten, *GDCh Vortragstagung Mainz*, 23./24.3, 1998.
322. Radeva, T. and Petkanchin, I., *J. Colloid Interface Sci.*, 196, 87, 1997; Walker, H.W. and Grant, S.B., *Colloids Surf.*, 119, 229, 1996.
323. Horn, D., *Colloid Polym. Sci.*, 65, 251, 1978; Sommer, F. et al., *Langmuir*, 11, 440, 1995.
324. Schönherr, H., Hruska, Z., and Vancso, G.J., *Macromolecules*, 33, 554, 2000.
325. Li, H. et al., *Macromolecules*, 33, 465, 2000.
326. Strobl, G.R. et al., *Macromolecules*, 19, 2683, 1986.
327. Slep, D. et al., *Langmuir*, 14, 4860, 1998.
328. Zhang, J. et al., *Langmuir*, 11, 3018, 1995.
329. Chahboun, A. et al., *J. Membrane Sci.*, 67, 295, 1992.
330. Weisenhorn, A.L. et al., *Phys. Rev.*, B, 45, 11226, 1992.

331. Gesang, T. et al., *Surf. Interface Anal.*, 23, 797, 1995.
332. Henck, S.A.J., *Vac. Sci. Technol.–Vac. Surf. Films*, 10, 934, 1992.
333. Shelley, P.H. et al., *Appl. Spectrosc.*, 50, 119, 1996.
334. Hirvi, K. et al., *Rev. Sci. Instrum.*, 65, 2735, 1994.
335. Zemek, J. et al., *Surf. Interface Anal.*, 26, 182, 1998.
336. Miyake, S., *Appl. Phys. Lett.*, 65, 980, 1994.
337. Hu, J. et al., *Surf. Sci.*, 327, 358, 1995.
338. Ton-That, C., Shard, A.G., and Bradley, R.H., *Langmuir*, 16, 2281, 2000.
339. Cleveland, J.P. et al., *Rev. Sci. Instrum.*, 64, 403, 1993.
340. Hazel, J.L. and Tsukruk, V.V.J., *Tribology-Trans. ASME*, 120, 814, 1998.
341. Zhong, Q. et al., *Surf. Sci.*, 290, 688, 1993.
342. Briggs, D. and Seah, M.P., *Practical Surface Analysis, Vol 1.: Auger and X-ray Photoelectron Spectroscopy*, John Wiley & Sons, New York, 1995.
343. Tachibana, H. et al., *Langmuir*, 16, 2975, 2000.
344. Bloor, D. and Chance, R.R., Eds., *Polydiacetylenes of NATO ASI Series E*, Nijhoff, Dordrecht, The Netherlands, 102, 1985.
345. Ogawa, T., *Polym. Sci.*, 20, 943, 1995.
346. Koshihara, S. et al., *J. Chem. Phys.*, 92, 7581, 1990.
347. Hankin, S.H.W., Downey, M.J., and Sandman, D., *J. Polymer*, 33, 5098, 1992.
348. Koshihara, S. et al., *Phys. Rev. Lett.*, 68, 1148, 1992.
349. Koshihara, S. et al., *Phys. Rev.*, B, 52, 6265, 1995.
350. Tanaka, H. et al., *Macromolecules*, 22, 1208, 1989.
351. Tieke, B., *Adv. Polym. Sci.*, 71, 79, 1985; Powers, E.T. et al., *Ang. Chem. Int. Ed.*, 41, 127, 2002.
352. Mino, N., Tamura, H., and Ogawa, K., *Langmuir*, 7, 2336, 1991.
353. Deckert, A.A. et al., *Langmuir*, 11, 643, 1995.
354. Kuriyama, K., Kikuchi, H., and Kajiyama, T., *Langmuir*, 14, 1130, 1998.
355. Sheth, S.R. and Leckband, D.E., *Langmuir*, 13, 5652, 1997.
356. Tieke, B., Lieser, G., and Wegner, G., *J. Polym. Sci., Polym. Chem. Ed.*, 17, 1631, 1979.
357. Murakami, H., Watanabe, Y., and Nakashima, N., *J. Am. Chem. Soc.*, 118, 4484, 1996.
358. Saito, A., Urai, Y., and Itoh, K., *Langmuir*, 12, 3938, 1996.
359. Matsumoto, M. et al., *J. Phys. Chem.*, B, 101, 702, 1997.
360. Seki, T., Tanaka, K., and Ichimura, K., *Macromolecules*, 30, 6401, 1997.
361. Matsumoto, M. et al., *Langmuir*, 14, 7533, 1998.
362. Tachibana, H., *J. Am. Chem. Soc.*, 120, 1479, 1998.
363. Terrettaz, S., Tachibana, H., and Matsumoto, M., *Langmuir*, 14, 7511, 1998; Tachibana, H. et al., *Thin Solid Films*, 327, 813, 1998.
364. Hinterdofer, P. et al., *Proc. Natl. Acad. Sci., U.S.A.*, 93, 3477, 1996.
365. Rief, M. et al., *Science*, 275, 1295, 1997.
366. Li, H.B. et al., *Adv. Mater.*, 10, 316, 1998.
367. Li, H. et al., *Langmuir*, 15, 2120, 1999.
368. Bond, J. and Lee, P.I., *J. Appl. Polym. Sci.*, 12, 1215, 1969.
369. Fleen, G.J. et al., *Polymers at Interfaces*, Chapman & Hall: London, New York, 1993.
370. Rief, M. et al., *Science*, 276, 1109, 1997.
371. Li, H.B. et al., *Macromol. Rapid Commun.*, 19, 609, 1998.
372. Biggs, S., *Langmuir*, 11, 156, 1995; Manne, S., Bocek, D., and Hansma, P.K., *Rev. Sci. Instrum.*, 64, 413, 1993.
373. Butt, H.J. and Jaschke, M., *Nanotechnology*, 6, 1, 1995.
374. Florin, E.L. et al., *Biosensors Bioelectron.*, 10, 895, 1995.

375. Smith, S.B., Cui, Y., and Bustamante, C., *Science*, 271, 795, 1996.
376. Kuhn, W. and Gruen, E., *Kolloid Z.*, 101, 248, 1942; Nguyen, T.Q. and Kausch, H.H., *Colloid Polym. Sci.*, 269, 1099, 1991.
377. Hild, S. et al., *J. Polym. Sci.*, B, 34, 1953, 1996.
378. Brodowsky, H.M., Boehnke, U.C., and Kremer, F., *Langmuir*, 15, 274, 1999.
379. Li, Z. et al., *Langmuir*, 11, 4785, 1995.
380. Senden, T.J., di Meglio, J.-M., and Auroy, P., *Eur. Phys. J.*, B, 3, 211, 1998.
381. Kikuchi, H., Yokoyama, N., and Kajiyama, T., *Chem. Lett.*, 11, 1107, 1997.
382. Senden, T.J., Joanny, J.-F., and di Meglio, J.-M., *Europhys. Lett.*, 41, 303, 1998.
383. Rief, M. et al., *Science*, 275, 1295, 1997.
384. Cui, Y. and Bustamante, C., *Science*, 271, 802, 1996.
385. She, H., Malotky, D., and Chaudhury, M.K., *Langmuir*, 14, 3090, 1998.
386. Courvoisier, A. et al., *Langmuir*, 14, 3727, 1998.
387. Joanny, J.-F., *Phys. Rev.*, E, 57, 6923, 1998.
388. de Gennes, P.G., *Scaling Concepts in Polymer Physics*, Cornell University Press, Ithaca, NY, 1979.
389. de Gennes, P.G., *Rep. Prog. Phys.*, 32, 187, 1969.
390. Lupine, Y. and Caille, A., *Can. J. Phys.*, 56, 403, 1978.
391. Eisenriegler, E., Kremer, K., and Binder, K., *J. Chem. Phys.*, 77, 6296, 1982.
392. Gorbunov, A.A. and Skvortsov, A.M., *J. Chem. Phys.*, 98, 5961, 1993.
393. Ingersent, K., Klein, J., and Pincus, P., *Macromolecules*, 23, 548, 1990.
394. Carslaw, H.S. and Jaeger, J.C., *Conduction of Heat in Solids*, 2nd ed., Oxford University Press, Oxford, 1959.
395. Johner, A., Bouchaud, E., and Daoud, M., *J. Phys.* (Paris), 51, 495, 1990.
396. Johner, A. and Joanny, J.-F., *Macromol. Theory Simulations*, 6, 479, 1997.
397. Gorbunov, A.A. and Skvortsov, A.M., *Adv. Colloid Interface Sci.*, 62, 31, 1995.
398. Yamamoto, S. and Kitamura, N., *J. Phys. Chem.*, B, 103, 4455, 1999.
399. Boussaad, S. and Tao, N.J., *J. Am. Chem. Soc.*, 121, 4510, 1999.
400. Pickup, J.C. and Alcock, S., *Biosensors Bioelectron.*, 6, 63, 1991.
401. Chaplin, M.F. and Bucke, C., *Enzyme Technology*, Cambridge University Press, London; New York, 1990.
402. Hall, E.A.H., *Biosensors*, Prentice Hall, Englewood Cliffs, NJ, 1991.
403. Clark, L.C., Jr. and Lyons, C., *Ann. N.Y. Acad. Sci.*, 102, 29, 1962.
404. Bowden, E.F., Hawkridge, F.M., and Blount, H.N., in *Biochemistry, Comprehensive Treatise of Electrochemistry 10*, Srinivasan, S., Ed., Plenum Press, New York, 1985.
405. Heller, A., *Acc. Chem. Res.*, 23, 128, 1990.
406. Reed, D.E. and Hawkridge, F.M., *Anal. Chem.*, 59, 2334, 1987.
407. Nassar, A.E.F., Willis, W.S., and Rusling, J.F., *Anal. Chem.*, 67, 2386, 1995.
408. Scouten, W.H., Luong, J.H.T., and Brown, R.S., *Trends Biotechnol.*, 13, 178, 1995.
409. Eddowes, M.J. and Hill, H.A.O., *J. Chem. Soc., Chem. Commun.*, 11, 771, 1977.
410. Eddowes, M.J. and Hill, H.A.O., *J. Am. Chem. Soc.*, 101, 4462, 1979.
411. Taniguchi, I. et al., *J. Chem. Soc., Chem. Commun.*, 1032, 1982.
412. Armstrong, F.A., Hill, H.A.O., and Walton, N., *J. Acc. Chem. Res.*, 21, 407, 1988.
413. Scott, D.L., Paddock, R.M., and Bowden, E.F., *J. Electroanal. Chem.*, 341, 307, 1992; Amador, S.M. et al., *Langmuir*, 9, 812, 1993.
414. Edmiston, P.L. et al., *J. Am. Chem. Soc.*, 119, 560, 1997.
415. Wood., L.L. et al., *J. Am. Chem. Soc.*, 119, 571, 1997.
416. Dubois, L.H. and Nuzzo, R.G., *Annu. Rev. Phys. Chem.*, 43, 437, 1992.
417. Cullison, J.K. et al., *Langmuir*, 10, 877, 1994.
418. Tanaka, K. and Tamamushi, R., *J. Electroanal. Chem.*, 236, 305, 1987.

419. Rojas, M.T., Han, M., and Kaifer, A.E., *Langmuir*, 8, 1627, 1992.
420. Kaufmann, J.-M., Chastel, O., and Quarin, G., *Bioelectrochem. Bioenerg.*, 23, 167, 1990.
421. Wang, J. and Lu, Z., *Anal. Chem.*, 62, 826, 1990.
422. Rusling, J. and Nassar, A.-E.F., *J. Am. Chem. Soc.*, 115, 11891, 1993.
423. Nakashima, N. et al., *J. Electroanal. Chem.*, 319, 355, 1991.
424. Salamon, Z., Vitello, L.B., and Erman, J.E., *Bioelectrochem. Bioenerg.*, 21, 213, 1989.
425. Salamon, Z., Hazzard, J.T., and Tollin, G., *Proc. Natl. Acad. Sci. U.S.A.*, 90, 6420, 1993.
426. Salamon, Z. and Tollin, G., *Biophys. J.*, 71, 848, 1996.
427. Cotton, T.M., Shultz, S.G., and van Duyne, R.P., *J. Am. Chem. Soc.*, 102, 7960, 1980.
428. Nassar, A.E.F. et al., *J. Phys. Chem.*, B, 101, 2224, 1997.
429. Barenholz, Y. et al., *Biochemistry*, 16, 2806, 1977.
430. Kendrew, J.C. et al., *Nature*, 185, 416, 1960.
431. Boussaad, S., Tao, N.J., and Arechabaleta, R., *Chem. Phys. Lett.*, 280, 397, 1997.
432. Laviron, E.J., *Electroanal. Chem.*, 101, 19, 1979.
433. Keller, C.A. and Kasemo, B., *Biophys. J.*, 75, 1397, 1998.
434. Boussaad, S. et al., *Langmuir*, 14, 6215, 1998.
435. Nicholson, R.S., *Anal. Chem.*, 37, 1351, 1965.
436. Bizzotto, D. and Lipkowski, J., *J. Electroanal. Chem.*, 409, 33, 1996.
437. Bizzotto, D., Noel, J.J., and Lipkowski, J., *Thin Solid Films*, 248, 69, 1994.
438. Bizzotto, D. and Lipkowski, J., *Prog. Surf. Sci.*, 50, 237, 1995.
439. Andersen, S.I. et al., *Carbohydr. Polym.*, 26, 299, 1995.
440. Wilkins, F. et al., *J. Microsc.*, 172, 215, 1993.
441. Gunning, A.P., McMaster, T.J., and Morris, V.J., *Carbohydr. Polym.*, 21, 47, 1993.
442. Tang, S.L. and McGhie, A.J., *Langmuir*, 12, 1088, 1996.
443. Lewerenz, H.J. et al., *AIDS Res. Hum. Retroviruses*, 8, 1663, 1992.
444. Schwartz, K. et al., *Science*, 251, 508, 1992; Zasadzinaki, J.A.N. et al., *Science*, 263, 1726, 1994.
445. Nelson, A., *Langmuir*, 13, 5644, 1997.
446. Birdi, K.S. and Gevod, V., *Colloid Polym. Sci.*, 261, 767, 1983.
447. Wilschut, J. and Hoekstra, D., *Membrane Fusion*, Marcel Dekker, New York, 1991.
448. Andersen, K.B. and Skov, H., *J. Gen. Virol.*, 70, 1921, 1989.
449. Moesby, L., Ph.D. thesis, Royal Danish School of Pharmacy, Copenhagen, 1996.
450. Kwong, P.D., *Nature*, 393, 648, 1998.
451. Burger, K.N., *Biochemistry*, 30, 1173, 1991.
452. Rafalsky, M. and Wilschut, J., *Biochemistry*, 30, 10211, 1991.
453. White, J.M., *Science*, 258, 917, 1992.
454. Pinto, L.H., *Cell*, 69, 517, 1992.
455. Phalen, T. and Kielian, M., *J. Cell Biol.*, 112, 615, 1991.
456. Chernomordik, L., Chanturiya, A.N., and Zimmerberg, J., *J. Virol.*, 68, 7115, 1994.
457. Wallach, D.F.H., *Ann. N.Y. Acad. Sci.*, 264, 124, 1975.
458. van Holde, K.E., *Physical Biochemistry*, Prentice Hall, Englewood Cliffs, NJ, 1990.
459. Umemura, K., Arakawa, H., and Ikai, A., *J. Vac. Sci. Technol.*, B(12,3), 1470, 1994.
460. Birdi, K.S. and Fasman, G.D., *J. Polym. Sci.*, A-1, 10, 2483, 1972; Birdi, K.S. and Fasman, G.D., *J. Polym. Sci.*, 42, 1099, 1973.
461. Schultz, R.D. and Asumaa, S.K., *Progress in Surface Science*, vol. 3, Academic Press, New York, 1971.
462. Lal, R. and John, S.A., *Am. J. Phys. Cell Phys.*, 266, C-1, 44, 1994.
463. Yang, J. et al., *J. Microscopy*, 171, 183, 1993.

464. Mulhern, P.J. et al., Scanning tunneling microscope '91 conference, *Ultramicroscopy*, 1992.
465. Vesenja, J., Miller, R., and Henderson, E., *Rev. Sci. Instrum.*, 65, 2249, 1994; Henderson, E., *Prog. Surf. Sci.*, 461, 39, 1994.
466. Kristensen, D., Ph.D. thesis, Royal Danish School of Pharmacy, Copenhagen, 1997.
467. Oesterhelt, D. and Stoeckenius, W., *Proc. Natl. Acad. Sci., U.S.A.*, 70, 2853, 1973.
468. Oesterhelt, D. and Stoeckenius, W., *Methods Enzymol.*, 31, 667, 1974.
469. Baldwin, J. and Henderson, R., *Ultramicroscopy*, 14, 319, 1984.
470. Henderson, R. et al., *J. Mol. Biol.*, 213, 899, 1990.
471. Kimura, Y. et al., *Nature*, 389, 206, 1997.
472. van Holde, K.E., *Physical Biochemistry*, Prentice Hall, Englewood Cliffs, NJ, 1971.
473. Hartley, P.G., Larson, I., and Scales, P.J., *Langmuir*, 13, 2207, 1997.
474. Hartley, P.G., in *Colloid and Polymer Interactions: Techniques and Applications*, Farinato, R. and Dubin, P., Eds., J. Wiley & Sons, New York, 1998.
475. Manne, S., Bocek, D., and Hansma, P.K., *Rev. Sci. Instrum.*, 64, 410, 1993.
476. Sader, J. et al., *Rev. Sci. Instrum.*, 66, 3789, 1995.
477. McCormack, D., Carnie, S.L., and Chan, D.Y.C., *J. Colloid Interface Sci.*, 169, 177, 1995.
478. Schabert, F.A., Buldt, G., and Engel, A., *Biophys. J.*, 68, 1681, 1995.
479. ller, D., Buldt, G., and Engel, A.J., *Mol. Biol.*, 249, 239, 1995; Muller, D. et al., *Biophys. J.*, 70, 1796, 1996.
480. Fisher, K.A., Yanagimoto, K., and Stoeckenius, W.J., *J. Cell Biol.*, 77, 611, 1978.
481. Hsu, K.C., Rayfield, G.W., and Needleman, R., *Biophys. J.*, 70, 2358, 1996.
482. Barabas, K. et al., *Biophys. J.*, 43, 5, 1983.
483. Rooney, E.K., Gore, M.G., and Lee, A.G., *Biochemistry*, 26, 3688, 1987.
484. Ehreberg, B. and Berezin, Y., *Biophys. J.*, 45, 663, 1984.
485. Li, Q.-G., Ni, Y.-J., and Cao, Y., *Photochem. Photobiol.*, 53, 653, 1991.
486. Riviere, M.-E. et al., *J. Arch. Biochem. Biophys.*, 284, 1, 1991.
487. Kane, V. and Mulvaney, P., *Langmuir*, 14, 3303, 1998.
488. Green, J.-B.D., Novoradovsky, A., and Lee, G.U., *Langmuir*, 15, 238, 1999.
489. Dorn, I.T. et al., *Langmuir*, 14, 4836, 1998.
490. Antognozzi, M. et al., *Polymer*, 41, 4232, 2000.
491. McMaster, T.J. et al., *Probe Microscopy*, 1, 43, 1997.
492. Hobbs, J.K. et al., *Polymer*, 39, 2437, 1998.
493. Ratner, B. and Tsukruk, V.V., *Scanning Probe Microscopy in Polymers*, ACS Symposium Series, American Chemical Society, Washington, DC, 1998.
494. James, P.J. et al., *Polymer*, 41, 4223, 2000.
495. Leclere, Ph. et al., *Langmuir*, 12, 4317, 1996.
496. Cleveland, J.P. et al., *Appl. Phys. Lett.*, 72, 2613, 1998.
497. Tamayo, J. and Garcia, R., *Appl. Phys. Lett.*, 73, 2926, 1998.
498. Betzig, E., Finn, P.L., and Weiner, J.S., *Appl. Phys. Lett.*, 60, 2485, 1992.
499. Yang, P.C., Chen, Y., and Vaeziravani, M., *J. Appl. Phys.*, 71, 2499, 1992.
500. Davy, S., Spajer, M., and Courjon, D., *Appl. Phys. Lett.*, 73, 18, 2594, 1998.
501. Brunner, R., Marti, O., and Hollricher, O., *J. Appl. Phys.*, 86, 7100, 1999.
502. Yeager, H.L. and Steck, A., *J.Electrochem. Soc.*, 128, 1880, 1986; Hu, H.-W., Carson, G.A., and Granick, S., *Phys. Rev. Lett.*, 66, 2758, 1991.
503. Dreyfus, B. et al., *J. Phys.*, 51, 1341, 1990.
504. Hoh, J.H., Heinz, W.F., and Hassan, E.A., Force volume support note 240, Digital Instruments, Plainview, NY, 1997.
505. Cappella, B. and Dietler, G., *Surf. Sci. Rep.*, 34, 1, 1999.

506. Reynaud, C. et al., *Surf. Interface Anal.*, 30, 185, 2000.
507. Raiteri, R. et al., *Phys. Chem. Chem. Phys.*, 1, 4881, 1999.
508. Beake, B.D., Leggett, G.J., and Shipway, P.H., *Surf. Interface Anal.*, 27, 1084, 1999.
509. Chen, X. et al., *Ultramicroscopy*, 75, 171, 1998.
510. Nam, A.J. et al., *J. Vac. Sci. Technol.*, B13, 1556, 1995.
511. McLean, S.R. and Sauer, B.B., *J. Polym. Sci., Part B, Polym. Phys.*, 37, 859, 1999.
512. Kuptsov, A.H. and Zhizhin, G.N., *Handbook of Fourier Transform Raman and Infrared Spectra of Polymers*, Elsevier, Amsterdam; New York, 1998.
513. Eisenberg, A. and King, M., *Ion-Containing Polymers, Physical Properties and Structure*, vol. 2, Academic Press, New York, 1977.
514. Kuhle, A. et al., *Appl. Phys. A Solids Surf.*, 66, 329, 1998.
515. Behrend, O.P. et al., *Appl. Phys. Lett.*, 75, 2551, 1999.
516. SanPaulo, A. and Garcia, R., *Biophys. J.*, 78, 1599, 2000.
517. Garcia, R. and SanPaulo, A., *Ultramicroscopy*, 82, 79, 2000.
518. Whangbo, M.H., Magonov, S.N., and Bengel, H., *Probe Microscopy*, 1, 23, 1997.
519. McLean, S.R. and Sauer, B.B., Polyurethanes and other block copolymers, *Macromolecules*, 30, 8314, 1997.
520. Sauer, B.B., McLean, R.S., and Thomas, R.R., *Langmuir*, 14, 3045, 1998.
521. Hsu, W.Y. and Gierke, T.D., *J. Membr. Sci.*, 13, 307, 1983.
522. Drummond-Roby, M.A. and Wetsel, G.C., *Appl. Phys. Lett.*, 69, 3689, 1996.
523. Littke, W., Haber, M., and Guntherodt, H.-J., *J. Crystal Growth*, 122, 80, 1992; Parbhu, A.N., Bryson, W.G., and Lal, R., *Biochemistry*, 38, 11755, 1999.
524. Pieczko, M.E. and Breen, J.J., *Langmuir*, 1412, 11, 1995.
525. Lin, J.-S. et al., *Langmuir*, 14, 4843, 1998.
526. Stabel, A. et al., *Langmuir*, 11, 1427, 1995.
527. Eggleston, C.M. and Hochella, M.F., *Science*, 254, 983, 1991.
528. Dove, P.M. and Hochella, M.F., *Geochim. Cosmochem. Acta*, 57, 705, 1993.
529. Niu, C. and Lieber, C.M., *J. Phys. Chem.*, 96, 3419, 1992.
530. Birdi, K.S. and Vu, D.T., *Probe Microscopy*, 1, 99, 1997.
531. Tanford, C., *The Hydrophobic Effect*, John Wiley & Sons, New York, 1980.
532. Tao, N.J. and Shi, Z., *J. Phys. Chem.*, 98, 1464, 1994.
534. Benard, J., Ed., *Adsorption on Metal Surfaces*, Elsevier, Amsterdam; New York, 1983.
535. Widring, C.A. and Porter, M.D., *J. Am. Chem. Soc.*, 9, 2805, 1911.
536. Zubimendi, J.L., Salvarezza, R.C., and Arvia, A.J., *Langmuir*, 12, 2, 1996.
537. Vicente, J.L. et al., *Langmuir*, 12, 19, 1996.
538. Jelley, E.E., *Nature*, 138, 1009, 1936; 139, 631, 1937.
539. Scheibe, G., *Angew. Chem.*, 49, 563, 1936; 50, 212, 1937.
540. De Boer, S., Vink, K.J., and Wiersma, D.A., *Chem. Phys. Lett.*, 137, 99, 1987.
541. Spano, F.C. and Mukamel, S., *Phys. Rev.*, A, 40, 5783, 1989.
542. Scherer, P.O. and Fischer, S.F., *Chem. Phys.*, 86, 269, 1984.
543. Daltrozzo, E. et al., *Photo. Sci. Eng.*, 18, 441, 1974.
544. Scheibe, G., *Angew. Chem.*, 52, 631, 1939.
545. Czikkely, V., Foersterling, H.D., and Kuhn, H., *Chem. Phys. Lett.*, 6, 11, 1970.
546. Yao, H., Ikeda, H., and Kitamura, N., *J. Phys. Chem.*, 102, 7691, 1998.
547. Kobayashi, T., Ed., *J Aggregates*, World Scientific, Singapore, 1996.
548. Wolthaus, L., Schaper, A., and Moebius, D., *Chem. Phys. Lett.*, 225, 322, 1994.
549. Saijo, H. and Shiojiri, M., *J. Imaging Sci. Technol.*, 40, 111, 1996.
550. Saijo, H. and Shiojiri, M., *J. Cryst. Growth*, 1996, 166, 930.
551. Higgins, D.A. et al., *J. Am. Chem. Soc.*, 118, 4049, 1996.
552. Daehne, L., Tao, J., and Mao, G., *Langmuir*, 14, 565, 1998.

553. Sugiyama, S. et al., *Chem. Lett.*, 19, 37, 1999.
554. Yao, H. et al., *J. Phys. Chem.*, B, 103, 4452, 1999.
555. Ono, S.S. et al., *J. Phys. Chem.*, B, 103, 33, 6909, 1999.
556. Schmidt, H., *J. Vac. Sci. Technol.*, A, 8, 388, 1990.
557. Drake, B. et al., *Science*, 1989, 243, 1586.
558. Hartman, H. et al., *Clays, Clay Mineral*, 38, 337, 1990.
559. Marchetti, A.P., Salzberg, C.D., and Walker, E.I.P., *J. Chem. Phys.*, 64, 4693, 1976.
560. Misawa, K. et al., *Chem. Phys. Lett.*, 220, 251, 1994.
561. Kobayashi, T. and Misawa, K., *J. Lumin.*, 72, 38, 1997.
562. Kobayashi, T., *Supramol. Sci.*, 5, 343, 1998.
563. Dammeier, Von B. and Hoppe, W., *Acta Crystallogr.*, B27, 2365, 1971.
564. Daehne, L., Horvath, A., and Weiser, G., *Chem. Phys.*, 196, 307, 1995.
565. Feyter, S. et al., *9th Int. Conf. STM/Spectr. and Related Techniques*, July, 1997, Hamburg.
566. McKendry, R. et al., *Nature*, 391, 566, 1998.
567. Yablon, D.G., Giancarlo, L.C., and Flynn, G.W., *J. Phys. Chem.*, 104, 7627, B, 2000.
568. Blake, C.C.F. et al., *Nature*, 196, 1173, 1962.
569. Nisman, R., Smith, P., and Vansco, J.G., *Langmuir*, 10, 1667, 1995.
570. Eng, L.M. et al., *Appl. Phys.*, A59, 145, 1994; Yoshimura, H., Ebina, S., and Nagayama, K., *Langmuir*, 11, 1711, 1995.
571. Schonhen, H., Sneuvy, D., and Vancso, G.J., *Polym. Bull.*, 30, 567, 1993.
572. McLean, R.S., Doyle, M., and Sauer, B.B., *Macromolecules*, 33, 6541, 2000.
573. Schonherr, H., Zdenek H., and Vancso, G.J., *Macromolecules*, 33, 4532, 2000.
574. Lacapere, J.-J., Stokes, D.L., and Chatenay, D., *Biophys. J.*, 63, 303, 1992.
575. Keith, H.D. et al., *J. Appl. Phys.*, 30, 1485, 1959.
576. Meille, S.V. et al., *Macromolecules*, 27, 2615, 1994.
577. Lotz, B., Kopp, S., and Dorset, D.C.R., *Acad. Sci.* (Paris), series IIb, 319, 187, 1994.
578. Brockner, S. et al., *Prog. Polym. Sci.*, 16, 361, 1991.
579. Lotz, B., Wittmann, J.C., and Lovinger, A., *Polymer*, 37, 4979, 1996.
580. Toulouse, G., *Comm. Phys.*, 2, 115, 1977.
581. Puterman, M. et al., *J. Polym. Sci.*, 15, 805, 1977.
582. Cartier, L., Spassky, N., and Lotz, B., *C.R. Acad. Sci.* (Paris), IIb, 322, 429, 1996.
583. Stocker, W. et al., *Macromolecules*, 31, 807, 1998.
584. Leugering, H.J., *Makromol. Chem.*, 109, 204, 1967.
585. Garbarczyk, J. and Paukszta, D., *Colloid Polym. Sci.*, 263, 985, 1985; New Japan Chemical Co., Ltd., European Patent EP 93101000.3, Japanese Patents JP 34088/92, JP 135892/92, JP 283689/92, and JP 324807/92, 1992.
585. Varga, J., in *Poly(propylene): Structure, Blends and Composites*, vol. 1, Karger-Kocsis, J., Ed., Chapman & Hall, London; New York, 1995.
586. Lotz, B. et al., *Polym. Bull.*, 25, 101, 1991.
587. Garbarczyk, J., *Acta Crystallogr.*, C41, 1062, 1985.
588. Guanyi, S. et al., German Patent P 36 10 644, 1986.
589. Zhou, G. et al., *Makromol. Chem.*, 187, 633, 1986.
590. Tjong, S.C., Shen, J.S., and Li, R.K.Y., *Polym. Eng. Sci.*, 36, 100, 1996.
591. Wolfschwenger, J. and Bernreitner, K., PCD-Polymere Gesellschaft m.b. H., European Patent EP 682066 A1 951115, 1995.
592. Wittmann, J.C. and Lotz, B., *Prog. Polym. Sci.*, 15, 909, 1990.
593. Turner-Jones, A., Aizlewood, J.M., and Beckett, D.R., *Makromol. Chem.*, 75, 134, 1964.
594. Stocker, W. et al., *Macromolecules*, 27, 6677, 1994.

595. Kopp, S., Wittmann, J.C., and Lotz, B., *Polymer*, 35, 916, 1994.
596. Rybnikar, F., *J. Macromol. Sci.–Phys.*, B30, 201, 1991.
597. Barrat, A. et al., *Europhys. Lett.*, 20, 633, 1992.
598. Ohrer, R.S., Henrich, V.E., and Bonnell, D.A., *Science*, 250, 12391, 1990.
599. Israelachvili, J.N. and Michelle, L.G., *Langmuir*, 5, 288, 1989.
600. Laslowski, J. and Kitchener, J.A., *J. Colloid Interface Sci.*, 29, 670, 1969.
601. Tsutumi, K. and Takahashi, H., *Colloid Polym. Sci.*, 263, 506, 1985.
602. Barthel, H., *Colloids Surf.*, 101, 217, 1995.
603. Wei, W. et al., *J. Colloid Interface Sci.*, 157, 154, 1993.
604. Schonherr, H. and Vancso, G.J., *Langmuir*, 13, 1567, 1997.
605. O'Shea, S.J., Welland, M.E., and Rayment, T., *Langmuir*, 9, 1826, 1993.
606. Tsao, Y.-H. et al., *Langmuir*, 7, 3154, 1991.
607. Banga, R. and Yarwood, J., *Langmuir*, 11, 4393, 1995.
608. Fujii, M. et al., *Langmuir*, 10, 984, 1994.
609. Flinn, D.H., Guzonas, D.A., and Yoon, R.-H., *Colloids Surf.*, 87, 163, 1994.
610. Mulvaney, P. and Ciersig, M., *J. Chem. Soc., Farady Trans.*, 92, 3137, 1996.
611. Kardassi, D., Tsiourvas, D., and Paleos, C.M., *J. Colloid Interface Sci.*, 186, 203, 1997.
612. Junno, T. et al., *Appl. Phys. Lett.*, 66, 3295, 1995.
613. Butt, H.-J., Kuropka, R., and Christensen, B., *Colloid Polym. Sci.*, 272, 1218, 1994; Utsugi, H., *Chem. Phys. Lett.*, 11, 593, 1973.
614. Fuji, M. et al., *Adv. Powder Technol.*, 8, 210, 1997; Fuji, M. et al., *Langmuir*, 16, 3281, 2000.
615. Kiselev, A.V., Kuznetsov, B.V., and Lanin, S.N., *J. Colloid Interface Sci.*, 69, 148, 1979.
616. Snyder, L.R. and Ward, J.W., *J. Phys. Chem.*, 70, 3941, 1966.
617. Carrott, P.J.M., Roberts, R.A., and Sing, K.S.W., *Langmuir*, 10, 984, 1994.
618. Fuji, M. et al., *Adv. Powder Technol.*, 10, 187, 1999.
619. Fuji, M. et al., *J. Soc. Powder Technol., Jpn.*, 33, 740, 1996.
620. Markiewicz, P. and Goh, C.M., *Langmuir*, 10, 5, 1994.
621. Utsugi, H. et al., *Colloids Surf.,* 21, 528, 1972.
622. Kiseleve, V.A. et al., *Kolloid. Zh.*, 22, 671, 1960.
623. Gobet, J. and Kovats, E., *Adsorption Sci. Technol.*, 1, 285, 1984.
624. Koberstein, E. and Voll, M., *Z. Phys. Chem.*, Neue Folge, 71, 275, 1970.
625. Barthel, H., in *Chemically Modified Surfaces, Proc. 4th Symp. Chem. Modif. Surf.*, Mottola, H.A. and Steinmetz, J.R., Eds., Elsevier, Amsterdam; New York, 1992.
626. Chikazawa, M., Kanazawa, T., and Shikizai, H., *Rev. Mod. Phys.,* 57, 456, 1984.
627. Kessaissia, Z., Papirer, E., and Donnet, J.-B., *J. Colloid Interface Sci.*, 79, 257, 1981.
628. Takahashi, H. and Hyomen, A., 57, 495, 1984.
629. Israelachvili, J.N. and Parshley, R.M., *Nature*, 300, 341, 1982; Parshley, R.M. et al., *Science*, 229, 1088, 1985; Christenson, H.K. et al., *J. Phys. Chem.*, 94, 8004, 1990.
630. Fujii, M. et al., *Langmuir*, 11, 405, 1995.
631. Drucker, W.A., *Langmuir*, 10, 3279, 1994; Lin, X.-Y., Creuzet, F., and Arribart, H., *J. Phys. Chem.*, 97, 7272, 1993.
632. Fowkes, F.M., *J. Adhesion Sci. Technol.*, 1, 7, 1987.
633. Sprycha, K., *J. Colloid Interface Sci.*, 127, 1, 1989.
634. Bousse, L. et al., *J. Colloid Interface Sci.*, 147, 22, 1991; Healey, T.W. et al., *Electroanal. Chem.*, 80, 57, 1997.
635. Butt, H-J., *Biophys. J.*, 60, 1438, 1991; Butt, H.-J. and Guckenberger, R., *Ultramicroscopy*, 46, 375, 1992.

636. Ducker, W.A., Senden, T.J., and Pashley, R.M., *Langmuir*, 8, 1831, 1992.
637. Hoh, J.H. and Engel, A., *Langmuir*, 9, 3310, 1993.
638. Manne, S. et al., *Rev. Sci. Instrum.*, 64, 403, 1993.
639. Derjaguin, B.V. and Kusakov, M., *Acta Physicochim. URSS*, 10, 25, 1939.
640. Israelachvili, J.N., *Intermolecular and Surface Forces*, Academic Press, New York, 1985; Israelachvili, J.N. and Pashley, R.M., *Nature*, 300, 341, 1982.
641. Pashley, R.M. et al., *Science*, 229, 1088, 1985.
642. Hough, D.B. and White, L.R., *Adv. Colloid Interface Sci.*, 14, 3, 1980.
643. Bain, C.D. et al., *J. Am. Chem. Soc.*, 111, 321, 1989.
644. Hu, K. and Bard, A.J., *Langmuir*, 14, 4790, 1998.
645. Lee, M.-T. et al., *Langmuir*, 14, 6419, 1998; Lee, B.-W. and Clark, N.A., *Langmuir*, 14, 5495, 1998.
646. Rutland, M.W. and Senden, T.J., *Langmuir*, 9, 412, 1993.
647. Stuart, J.K. and Hlady, V., *Langmuir*, 11, 1368, 1995.
648. Schnur, J.M., *Science*, 262, 512, 1993.
649. Friggeri, A., van Veggel, F.C.J.M., and Reinhoudt, D.N., *Langmuir*, 14, 5457, 1998.
650. Schwarz, A. et al., *Langmuir*, 14, 5526, 1998.
651. Kamat, P.V. and Chandrasekharan, N., *J. Phys. Chem.*, 104, 10855, 2000.
652. Steigerwald, M.L. and Brus, L.E., *Acc. Chem. Res.*, 23, 183, 1990.
653. Wang, Y. and Herron, N., *Science*, 273, 632, 1996.
654. Henglein, A., *J. Phys. Chem.*, 97, 5457, 1993.
655. Kamat, P.V., Composite semiconductor nanoclusters, in *Semiconductor Nanoclusters Physical, Chemical and Catalytic Aspects*, Kamat, P.V. and Meisel, D., Eds., Elsevier Science, Amsterdam; New York, 1997.
656. Mulvaney, P., Giersig, M., and Henglein, A., *J. Phys. Chem.*, 97, 7061, 1993.
657. Pileni, M.P., *New J. Chem.*, 8, 693, 1998.
658. Henglein, A., *Langmuir*, 14, 6738, 1998.
659. Link, S. and El-Sayed, M.A., *J. Phys. Chem.*, B, 103, 4212, 1999.
660. Trau, M. et al., *Nature*, 374, 437, 1995; Nakato, Y. et al., *Electrochemistry*, 68, 556, 2000.
661. Gilbert, S.E., Cavalleri, O., and Kern, K., *J. Phys. Chem.*, B, 100, 12, 1996.
662. Cavalleri, O. et al., *Langmuir*, 14, 7292, 1998.
663. Hickman, J.J. et al., *Science*, 252, 688, 1991.
664. Chen, S. et al., *Science*, 280, 2098, 1998.
665. Elghanian, R. et al., *Science*, 277, 1078, 1997.
666. Baba, K., Takakuwa, M., and Miyagi, M., *Optical Rev.*, 4, 411, 1997.
667. Amihood, D., Katz, E., and Willner, I., *Langmuir*, 11, 1313, 1995.
668. Colvin, V.L., Goldstein, A.N., and Alivisatos, A.P., *J. Am. Chem. Soc.*, 114, 5221, 1992.
669. Fan, H. and Lopez, G.P., *Langmuir*, 13, 119, 1997.
670. Sarathy, K.V. et al., *J. Phys. Chem.*, B, 101, 9876, 1997.
671. Chen, S., *J. Phys. Chem.*, B, 104, 663, 2000.
672. Brust, M. et al., *Adv. Mater.*, 7, 795, 1995.
673. Giersig, M. and Mulvaney, P., *Langmuir*, 9, 3408, 1993.
674. Giersig, M. and Mulvaney, P., *J. Phys. Chem.*, B, 97, 6334, 1993.
675. Hayward, R.C., Saville, D.A., and Aksay, I.A., *Nature*, 404, 56, 2000.
676. Kiely, C.J. et al., *Nature*, 396, 444, 1998.
677. Brust, M. et al., *J. Chem. Soc., Chem. Commun.*, 11, 801, 1994.
678. Fink, J. et al., *Chem. Mater.*, 10, 922, 1998.
679. Thomas, G. and Kamat, P.V., *J. Am. Chem. Soc.*, 122, 2655, 2000.

680. Brust, M. et al., *Langmuir*, 14, 5425, 1998.
681. Fujiwara, H., Yanagida, S., and Kamat, P.V., *J. Phys. Chem.*, B, 103, 2589, 1999.
682. Kim, S.-H. et al., *J. Phys. Chem.*, B, 103, 10, 1999.
683. Linnert, T., Mulvaney, P., and Henglein, A., *J. Phys. Chem.*, 97, 679, 1993.
684. Kreibig, U. and Vollmer, M., *Optical Properties of Metal Clusters*, Springer-Verlag, Heidelberg, 1995.
685. Khazraji, A.C. et al., *J. Phys. Chem.*, B, 103, 693, 1999.
686. Chandrasekharan, N. and Kamat, P.V., *J. Phys. Chem.*, 104, 10851, 2000.
687. Imae, P. et al., *Langmuir*, 14, 5631, 1998.
688. Wold, D.J. and Frisbie, C.D., *J. Am. Chem. Soc.*, 122, 29701, 2000.
689. Yaliraki, S.N., Kemp, M., and Ratner, M.A., *J. Am. Chem. Soc.*, 121, 3428, 1999; Ratner, M.A. et al., *Ann. N.Y. Acad. Sci.*, 852, 1998; Jortner, J. and Ratner, M., *Molecular Electronics*, Blackwell Scientific, Oxford, 1997.
690. Bumm, L.A. et al., *J. Phys. Chem.*, B, 103, 8122, 1999.
691. Xue, Y. et al., *Phys. Rev.*, 59, 7852, 1999.
692. Tian, W. et al., *J. Chem. Phys.*, 109, 2874, 1998.
693. Cygan, M.T. et al., *J. Am. Chem. Soc.*, 120, 2721, 1998.
694. Zhou, S. et al., *Chem. Phys. Lett.*, 297, 77, 1998.
695. Datta, S. et al., *Phys. Rev. Lett.*, 79, 2530, 1997.
696. Magoga, M. and Joachim, C., *Phys. Rev.*, B, 56, 4722, 1997.
697. Dhirani, A. et al., *J. Chem. Phys.*, 106, 5249, 1997.
698. Bumm, L.A. et al., *Science*, 271, 1705, 1996.
699. Joachim, C. et al., *Phys. Rev. Lett.*, 74, 2102, 1995.
700. Dorig, U. et al., *Phys. Rev.*, B, 48, 1711, 1993.
701. Slowinski, K., Slowinska, K., and Majda, M., *J. Phys. Chem.*, B, 103, 8544, 1999.
702. Delville, M., Tsionsky, M., and Bard, A., *Langmuir*, 14, 2774, 1998.
703. Sachs, S.B. et al., *J. Am. Chem. Soc.*, 119, 10563, 1997.
704. Slowinski, K. et al., *J. Am. Chem. Soc.*, 119, 11910, 1997.
705. Smalley, J.F. et al., *J. Phys. Chem.*, 99, 13141, 1995.
706. Forster, R.J. and Faulkner, L.R., *J. Am. Chem. Soc.*, 116, 5453, 1994.
707. Chidsey, C.E.D., *Science*, 251, 919, 1991.
708. Chen, J. et al., *Science*, 286, 1550, 1999.
709. Zhou, C. et al., *Appl. Phys. Lett.*, 71, 611, 1997.
710. Reed, M.A. et al., *Science*, 278, 252, 1997.
711. Slowinski, K., Fong, H.K.Y., and Majda, M., *J. Am. Chem. Soc.*, 121, 7257, 1999.
712. Haag, R. et al., *J. Am. Chem. Soc.*, 121, 7895, 1999.
713. Rampi, M.A., Schueller, O.J.A., and Whitesides, G.M., *Appl. Phys. Lett.*, 72, 1781, 1998.
714. Tans, S.J., Verschueren, A.R.M., and Dekker, C., *Nature*, 393, 49, 1998.
715. Metzger, R.M., *Acc. Chem. Res.*, 32, 950, 1999.
716. Collier, C.P. et al., *Science*, 285, 391, 1999.
717. Stoddard, J.F. and Heath, J.R., *Science*, 289, 1172, 2000.
718. Metzger, R.M. et al., *J. Am. Chem. Soc.*, 119, 10455, 1997.
719. Fischer, C.M. et al., *Surf. Sci.*, 362, 905, 1996.
720. Dorogi, M. et al., *Phys. Rev.*, B, 52, 9071, 1995.
721. Kelley, T.W., Granstrom, E.L., and Frisbie, C.D., *Adv. Mater.*, 11, 261, 1999.
722. Salmeron, M. et al., *Langmuir*, 9, 3600, 1993; Loiacono, M.J., Granstrom, E.L., and Frisbie, C.D., *J. Phys. Chem.*, B, 102, 1679, 1998.
723. Dai, H., Wong, E.W., and Lieber, C.M., *Science*, 272, 523, 1996.
724. Yano, K. et al., *Appl. Phys. Lett.*, 68, 188, 1996.

725. Klein, D. and McEuen, P., *Appl. Phys. Lett.*, 66, 2478, 1995.
726. De Wolf, P. et al., *J. Vac. Sci. Technol.*, A, 13, 1699, 1995.
727. Alperson, B. et al., *Phys. Rev.*, B, 52, 17017, 1995.
728. Heddleson, J. et al., *J. Vac. Sci. Technol.*, B, 12, 317, 1994; Ouyang, M., Huang, J.L., and Lieber, C.M., *Annu. Rev. Phys. Chem.*, 53, 201, 2002; Odom, T.W., Huang, J., and Lieber, C.M., *Ann. N.Y. Acad. Sci.*, 960, 203, 2002.
729. Bain, C.D. et al., *J. Am. Chem. Soc.*, 111, 321, 1989.
730. Schimmel, T.H., Fuchs, H., and Lux-Steiner, M., *Phys. Stat. Sol.*, A, 131, 47, 1992.
731. Barrett, R.C. and Quate, C.F., *J. Appl. Phys.*, 70, 2725, 1991; Duan, X. et al., *Nano Letters*, 2, 487, 2002; Gudiksen, M.S., Wang, J., and Lieber, C.M., *J. Phys. Chem*, B, 106, 4036, 2002; Hadley, P., Nakanishi, T., and Dekker, C., *Science*, 294, 1317, 2001.
732. Polewska, W. et al., *J. Phys. Chem.*, B, 103, 10440, 1999; Newman, R.C. and Sierdazki, K., *Science*, 263, 1708, 1994.
733. Eigler, D.M., Lutz, C.P., and Rudge, W.E., *Nature*, 352, 600, 1991; Demir, U. and Shannon, C., *Langmuir*, 10, 2794, 1994.
734. Wignall, G.D., Farrar, N.R., and Morris, S., *J. Mater. Sci.*, 25, 69, 1990.
735. Breuer, O. et al., *J. Appl. Polym. Sci.*, 64, 1097, 1997.
736. Salome, L. and Carmona, F., *Carbon*, 29, 599, 1991; Hjelm, R.P., Seeger, P.A., and Wampler, W.A., *Polym. Polym. Compos.*, 1, 53, 1993.
737. Carmona, F., *Physica Acta,* 157, 461, 1989.
738. Beaucage, G. et al., *J. Polym. Sci.*, B, 37, 1105, 1999.
739. Heaney, M.B., *Phys. Rev.*, B, 52, 12477, 1995; Viswanathan, R. and Heaney, M.B., *Phys. Rev. Lett.*, 75, 4433, 1995.
740. Oakes, J.A. and Sandberg, C.L., *IEEE Trans. Ind. Appl.*, IA-9, 462, 1973.
741. Doljack, F.A., *IEEE Trans. Comput., Hybrids, Manuf. Technol.*, CHMT-4, 372, 1981.
742. Marr, D.W.M. et al., *Macromolecules*, 30, 2120, 1997.
743. Oakey, J. et al., *Macromolecules*, 32, 5399, 1999; Pfau, A., Janke, A., and Heckman, W., *Surf. Interface Anal.*, 27, 410, 1999; Oakey, J. et al., *Macromolecules*, 33, 5198, 2000.
744. Keita, B. et al., *New J. Chem.*, 21, 851, 1997.
745. Maas, S. and Gronski, W., *Kautschuk Gummi Kunststoffe*, 47, 409, 1994.
746. Vansco, G.J. et al., *J. Colloid Polym. Sci.*, 275, 181, 1997.
747. Lee, J.-C. et al., *J. Appl. Polym. Sci.*, 64, 797, 1996.
748. Maas, S. and Gronski, W., *Rubber Chem. Tech.*, 68, 652, 1995.
749. Batiashvili, M.S., Lomtadze, T.T., and Georkhelidze, N.N., *Int. Polym. Sci. Technol.*, 17, 76, 1990.
750. Hawkins, W.L., *Polymer Stabilization*, Wiley Interscience, New York, 1972.
751. Holmstrom, A., *Durability of Macromolecular Materials*, American Chemical Society, Washington, DC, 45, 1979.
752. Gilroy, H.M., *Durability of Macromolecular Materials*, American Chemical Society, Washington, DC, 63, 1979.
753. Wignall, G.D. and Bates, F.S., *J. Appl. Crystallogr.*, 20, 445, 1987.
754. Dubner, W.S., Schultz, J.M., and Wignall, G.D., *J. Appl. Crystallogr.*, 23, 469, 1990.
755. Magonov, S.N. and Whangbo, M.-H., *Adv. Mater.*, 6, 355–371, 1994.
756. Howald, L. et al., *Phys. Rev.*, B, 49, 5651, 1994.
757. Zhou, H. and Wilkes, G.L., *Polymer*, 38, 5735, 1997.
758. Stanjek, H., Niederbudde, E.A., and Hausler, W., *Clay Miner.*, 27, 3, 1992.
759. Klug, H.P. and Alexander, L.E., *X-ray Diffraction Procedures for Polycrystalline and Amorphous Materials*, John Wiley & Sons, New York, 1954.

760. Spells, S.J. and Hill, M.J., *Polymer*, 32, 2716, 1991.
761. Kida, N., Ito, M., and Kaido, H., *J. Appl. Polym. Sci.*, 61, 1345, 1996.
762. Leblanc, J.L. and Stragliati, B., *J. Appl. Polym. Sci.*, 63, 959, 1997.
763. Avasthi, D.K. et al., *9th Int. Conf. STM/Spectroscopy and Related Techniques*, Hamburg, July, 1997.
764. Cai, W.-B. et al., *Langmuir*, 14, 6992, 1998; Poirier, G.E., Hance, B.K., and White, J.M., *J. Phys. Chem.*, 97, 5965, 1993.
765. Thompson, D., *J. Microsc.*, 152, 627, 1988; Whitman, L.J. et al., *Science*, 251, 1206, 1991.
766. Itaya, K., *J. Physical Chem.*, B, 102, 49, 10034, 1998.
767. Correa-Duarte, M.A. et al., *Langmuir*, 14, 6430, 1998.
768. Henglein, A., *Chem. Rev.*, 89, 1861, 1989.
769. Galembeck, F. et al., in *Fine Particles Science and Technology: From Micro to Nanoparticles*, Pelizzetti, E., Ed., Kluwer, Dordrecht, 1996.
770. Matijevic, E., *Chem. Mater.*, 5, 426, 1993.
771. Goodwin, J.W. et al., *Brit. Polym. J.*, 10, 173, 1978.
772. Dimitrov, A.S., Tiwa, T., and Nagayama, K., *Langmuir*, 15, 5257, 1999.
773. Brown, K. and Natan, M.J., *Langmuir*, 14, 726, 1998; Stöber, W., Fink, A., and Bohn, E., *J. Colloid Interface Sci.*, 26, 62, 1968.
774. Van Helden, A.K. and Vrij, A., *J. Colloid Interface Sci.*, 78, 312, 1980.
775. Kirkland, J.J., *J. Chromatogr.*, 185, 273, 1979.
776. Kops-Werkhoven, M.M. and Fijnaut, H.M., *J. Chem. Phys.*, 74, 1618, 1981.
777. Cardoso, A.H., Leite, C.A.P., and Galembeck, F., *J. Braz. Chem. Soc.*, 10, 497, 1999.
778. Cardoso, A.H., Leite, C.A.P., and Galembeck, F., *Langmuir*, 15, 4447, 1999.
779. Terris, B.D. et al., *J. Vac. Sci. Technol.*, 8, 374, A, 1990.
780. Shaw, D.J., *Introduction to Colloid and Surface Chemistry*, Butterworth & Co., London, 1996.
781. Goldstein, J.I. et al., *Scanning Electron Microscopy and X-ray Microanalysis*, Plenum Press, New York, 1990.
782. Iler, R.K., *The Chemistry of Silica*, Wiley Interscience, New York, 1979.
783. Yamane, M., Inoue, S., and Yasumori, J., *J. Non-Cryst. Solids*, 63, 45, 1984.
784. Wood, D.L. et al., *J. Am. Ceram. Soc.*, 66, 693, 1983.
785. Harris, M.T., Brunson, R.R., and Byers, C.H., *J. Non-Cryst. Solids*, 121, 397, 1990.
786. Bogush, G.H. and Zukoski, C.F., in *Ultrastructure Processing of Advanced Ceramic*, Mackenzie, J.D., Ed., Wiley Interscience, New York, 1988.
787. Bailey, J.K. and Mecartney, M.L., *Colloids Surf.*, A, 63, 151, 1992.
788. Birdi, K.S., *Fractals in Chemistry, Geochemistry and Biophysics*, Plenum Press, New York, 1994.
789. Boukari, H., Lin, J.S., and Harris, M.T., *Chem. Mater.*, 9, 2376, 1997.
790. Dimonie, V.L., El-Aasser, M.S., and Vanderhoff, J.W., *Polym. Mater. Sci. Eng.*, 58, 821, 1988.
791. McCarley, R.L., Hedricks, S.A., and Bard, A.J., *J. Phys. Chem.*, 96, 10089, 1992; Sugimura, H. and Nakagri, N., *Langmuir*, 11, 3623, 1995.
792. Bruckner-Lea, C.J. et al., *Langmuir*, 9, 3612, 1993.
793. Yang, M.X. et al., *Langmuir*, 14, 1458, 1998.
794. Sommerfeld, D.A., Cambron, R.T., and Beebe, T.O., *J. Phys. Chem.*, 94, 8926, 1990.
795. Robin, L.M., Hendricks, S.A., and Bard, A.J., *J. Phys. Chem.*, 96, 10089, 1992.
796. Scneir, P.J. and Hansma, P.K., *Langmuir*, 3, 1025, 1987.
797. Tachibana, H. et al., *Appl. Phys. Lett.*, 61, 2420, 1992; Tachibana, H. et al., *Langmuir*, 16, 2975, 2000.

798. Gao, X. and Weaver, M.J., *J. Phys. Chem.*, 97, 8685, 1993.
799. van Hove, M.A. et al., *Adv. Quantum Chem.*, 20, 1, 1989.
800. Gao, X. and Weaver, M.J., *J. Phys. Chem.*, 97, 8685, 1993; Tao, N.J. and Lindsay, S.M., *J. Appl. Phys.*, 70, 5141, 1991.
801. Nejoh, H., *Nature*, 353, 640, 1991.
802. Fan, Fu.-R.F. and Bard, A.J., *Science*, 267, 871, 1995.
803. Beebe, T.P., Jr., *J. Phys. Chem.*, B, 102, 10799, 1998.
804. Struber, U. and Kuppers, J., *Surface Science Letters*, 294, 924, 1993.
805. Balakrishnan,V. and Bottani, C.E., *J. Aggregates,* World Scientific, Singapore, 1986.
806. Andreasen, G. et al., *Langmuir*, 15, 1, 1999.
807. Bachand, G.D. et al., *Nano Letters*, 1, 42, 2001.
808. Sone, J. et al., *Nanotechnology*, 10, 135, 1999.
809. Block, S.M., *Cell*, 93, 1998.
810. Schnitzer, M.J. and Block, S.M., *Nature*, 388, 386, 1997.
811. Mazumdar, M. et al., *Proc. Natl. Acad. Sci., U.S.A.*, 93, 6522–6556.
812. Kitamura, K. et al., *Nature*, 397, 129, 1999.
813. Lohman, T., Thorn, K., and Vale, R.D., *Cell*, 93, 9, 1998.
814. Wang, M. et al., *Science*, 282, 902, 1998.
815. Berry, R. and Berg, H., *Proc. Natl. Acad. Sci., U.S.A.*, 94, 14433, 1997.
816. DeRosier, D., *Cell*, 93, 17, 1998.
817. Noji, H. et al., *Nature*, 386, 299, 1997.
818. Yashuorida, R. et al., *Cell*, 93, 1117, 1998.
819. Montemagno, C.D. and Bachand, G.D., *Nanotechnology*, 10, 225, 1999.
820. Bachand, G.D. and Montemagno, C.D., *Biomedical Microdevices*, 2, 179, 2000.
821. Soong, R.K. et al., *Science*, 290, 1555, 2000.
822. Kinosita, K., Jr., et al., *Philos. Trans. R. Soc. London Ser. B-Biol. Sci.*, 355, 473, 2000.
823. Kinosita, K., Jr., *Faseb J.*, 14, 1567, 2000.
824. Omote, H. et al., *Nanotechnology,* 10, 506, 1999.
825. Wada, Y. and Futai, M., *Proc. Natl. Acad. Sci., U.S.A.*, 96, 7780, 1999.
826. Hisabori, T., Kondoh, A., and Yoshida, M., *FEBS Lett.*, 463, 35, 1999.
827. Sabbert, D., Engelbrecht, S., and Junge, W., *Nature*, 381, 623, 1996.
828. Barklis, E. et al., *J. Biol. Chem.*, 273, 7177, 1998.
829. Lanfermeijer, F.C., Venema, K., and Palmgreen, M.G., *Protein Express. Purif.*, 12, 29, 1998.
830. Soong, R.K. et al., *Tech. Proc. 2nd Int. Conf. Modeling and Simulation of Microsyst.*, San Juan, Puerto Rico, April 19–21, 1999.
831. Chou, S.Y., Krauss, P.R., and Renstrom, P.J., *J. Vac. Sci. Technol.*, B, 14, 4129, 1996.
832. Chou, S.Y. et al., *J. Vac. Sci. Technol.*, B, 15, 2897, 1997.
833. Li, M.T. et al., *Appl. Phys. Lett.*, 76, 673, 2000.
834. Marcus, R.B., Ravi, T.S., and Gmitter, T., *Appl. Phys. Lett.*, 56, 236, 1990.
835 Herrasti, P. et al., *J. Phys. Rev.*, A, 45, 7440, 1990.
836. Albers, W.M., Ph.D. thesis, VTT Technical Research Center of Finland, 1999.
837. Nicoline, C.A. et al., *Langmuir*, 13, 5507, 1997.
838. Dziri, L.D. et al., *Langmuir*, 14, 4853, 1998.
839. Bergman, A.A. et al., *Langmuir*, 14, 6785, 1998.
840. Xu, F. et al., *Langmuir*, 14, 6505, 1998.
841. Taguchi,Y. et al., *Langmuir*, 14, 6550, 1998.
842. Jain, R.K. and Lind, R.C., *J. Opt. Soc. Am.*, 73, 647, 1983.
843. Hache, F., Ricard, D., and Flytzanis, C., *J. Opt. Soc. Am.*, B., 3, 1647, 1986.
844. Yumoto, J., Fukushima, S., and Kubodera, K., *Opt. Lett.*, 12, 832, 1987.

845. Haus, J.W. et al., *J. Opt. Soc. Am.*, B, 6, 797, 1989.
846. Fukumi, K. et al., *J. Appl. Phys.*, 75, 3075, 1994.
847. Grabert, H. and Devoret, M.H., Eds., *Single-Charge Tunneling*, Plenum Press: New York, 1992.
848. Amman, M. et al., *Phys. Rev.*, 43, 1146, 1991.
849. Andres, R.P. et al., *Science*, 272, 1323, 1996.
850. Perez, A. et al., *J. Mater. Res.*, 2, 910, 1987.
851. Nasu, H. et al., *Jpn. J. Appl. Phys.*, 28, 862, 1989.
852. Kay, E., *Z. Phys. D.*, 3, 251, 1986.
853. Yanagi, H. et al., *Chem. Mater.*, 10, 1258, 1998.
854. Dislich, H., *Angew. Chem., Int. Ed. Engl.*, 6, 1879, 1996.
855. Matsuoka, J. et al., *Ceram. Soc. Jpn.*, 100, 599, 1992.
856. Kozuka, H. and Sakka, S., *Chem. Mater.*, 5, 222, 1993.
857. Weaver, S. et al., *Langmuir*, 12, 4618, 1996.
858. Akbarian, F., Dunn, B.S., and Zink, J.I., *J. Phys. Chem.*, 99, 3892, 1995.
859. Akbarian, F., Dunn, B.S., and Zink, J.I., *J. Raman Spectrosc.*, 27, 775, 1996.
860. Becker, R.S., Golovchenko, J.A., and Swartzentruber, B.S., *Nature*, 325, 419, 1987; Eigler, D.M. and Schweizer, E.K., *Nature*, 524, 419, 1990.
861. Mamin, H.J. and Rugar, D., *Appl. Phys. Lett.*, 61, 1003, 1992.
862. Day, H.C. and Allee, D.R., *Appl. Phys. Lett.*, 62, 2691, 1993.
863. Campbell, P.M. and Snow, E.S., *Appl. Phys. Lett.*, 66, 1388, 1995.
864. Hosaka, S. et al., *J. Vac. Sci. Technol.*, B, 13, 1307, 1995.
865. Betzig, E., Finn, P.L., and Weiner, J.S., *Appl. Phys. Lett.*, 60, 2484, 1992.
866. Toda, T. et al., *Appl. Phys. Lett.*, 73, 517, 1998.
867. Quinn, M. and Mills, G., *J. Phys. Chem.*, 98, 9840, 1994.
868. Halperin, W.P., *Rev. Mod. Phys.*, 58, 533, 1986.
869. Kreibig, U., *Z. Phys. D.*, 3, 239, 1986.
870. Keefer, K.D., *Better Ceramics through Chemistry*, Brinker, C.J., Clark, D.E., and Ulrich, D.R., Eds., North-Holland, Amsterdam, 1984.
871. Franke, K. et al., *Surf. Sci. Lett.*, 302, 283, 1994.
872. Gruverman, A., Auciello, O., and Tokumoto, H., *Integrated Ferroelectrics*, 19, 49, 1998.
873. Gruverman, A., Auciello, O., and Tokumoto, H., *Annu. Rev. Mater. Sci.*, 28, 101, 1998.
874. Hidaka, T. et al., *J. Integrated Ferroelectrics*, 17, 319, 1997.
875. Yoo, I.K. et al., *J. Mater. Res. Soc. Symp. Proc.*, 493, 299, 1998.
876. Eng, L., *Nanotechnology*, 10, 405, 1999.
877. Ganpule, C.S. et al., *Appl. Phys. Lett.*, 75, 409, 1999.
878. Colla, E.L. et al., *Appl. Phys. Lett.*, B, 72, 2763, 1998.
879. Hong, S. et al., *J. Appl. Phys.*, 86, 607, 1999.
880. Harnagea, C. et al., *Jpn. J. Appl. Phys.*, 38, 2, 1255, 1999.
881. Hong, J.W. et al., *Appl. Phys. Lett.*, 75, 3183, 1999.
882. Jo, W., Kim, D.C., and Hong, J.W., *Appl. Phys. Lett.*, 76, 390, 2000.
883. Hong, J.W. et al., *Phys. Rev.*, 58, 5078, 1998.
884. Likodimos, V. et al., *Appl. Phys.*, B, 87, 443, 2000.
885. Luo, E.Z. et al., *Phys. Rev.*, 61, 203, 2000.
886. Chen, X.Q. et al., *J. Vac. Sci. Technol.*, 17, 1930, 1999.
887. Pompe, W. et al., *J. Appl. Phys.*, 74, 6012, 1993.
888. Hamazaki, S.-I. et al., *J. Phys. Soc. Jpn.*, B, 64, 3660, 1995.
889. Terris, B.D. et al., *J. Phys. Rev. Lett.*, 63, 2669, 1989.
890. Foster, C.M. et al., *J. Appl. Phys.*, B, 81, 2349, 1997.
891. Chen, Q. and Swain, G., *Langmuir*, 14, 7017, 1998.

892. Vincent, C. and Scrosati, B., *Modern Batteries: An Introduction to Electrochemical Power Sources*, 2nd ed., Arnold, London, 1997.
893. Ebert, L.B., *Annu. Rev. Mater. Sci.*, 6, 181, 1976.
894. Miras, M.C. et al., *J. Electroanal. Chem.*, 338, 279, 1992.
895. Zhou, P. and Fischer, J.E., *Phys. Rev.*, B53, 12643, 1996.
896. Carr, K.E., *Carbon*, 8, 155, 1970.
897. Biberacher, W. et al., *Mater. Res. Bull.*, 17, 1385, 1982.
898. Quate, C.F. and Gerber, C., *Phys. Rev. Lett.*, 56, 933, 1986; Siegenthaler, H., in *Scanning Tunneling Microscopy II*, Wiesendanger, R. and Guentherodt, H.J., Eds., Springer-Verlag, Heidelberg, 1993; Gewirth, A.A. and Siegenthaler, H., *Nanoscale Probe of the Solid/Liquid Interface; NATO ASI Series E: Applied Science*, vol. 288, Kluwer, Dordrecht, 1995.
899. Inaba, M. et al., *Langmuir*, 12, 1535, 1996.
900. Chu, A.C., Josefowicz, J.Y., and Farrington, G.C., *J. Electrochem. Soc.*, 144, 4161, 1997.
901. Kinoshita, K., *Carbon: Electrochemical and Physicochemical Properties*, John Wiley & Sons, New York, 1988.
902. Beck, F., Junge, H., and Krohn, H., *Electrochim. Acta*, 26, 799, 1981.
903. Kang, F., Zhang, T.-Y., and Leng, Y., *J. Phys. Chem. Solids*, 57, 883, 1996.
904. Alsmeyer, D.C. and McCreery, R.L., *Anal. Chem.*, 64, 1528, 1992.
905. Goss, C.A. et al., *Anal. Chem.*, 65, 1378, 1993.
906. Zhang, B.L. et al., *Electrochim. Acta*, 40, 733, 1995.
907. Zhang, J. and Wang, E., *J. Electroanal. Chem.*, 399, 83, 1995.
908. Staub, R., Alliata, D., and Nicolini, C., *Rev. Sci. Instrum.*, 66, 2513, 1995.
909. Brett, C.M.A. and Brett, A.M.O., *Electrochemistry Principles, Methods and Applications*, Oxford University Press, Oxford, 1993.
910. Pierson, H.O., *Handbook of Carbon, Graphite, Diamond and Fullerenes*, Noyes Publishing, Park Ridge, NJ, 1993.
911. Aronson, S., Lemont, S., and Weiner, J., *Inorg. Chem.*, 10, 1296, 1971.
912. Hõring, P. et al., *Appl. Phys.*, A , 66, 481, 1998.
913. Beck, F. and Krohn, H., *Synth. Met.*, 7, 193, 1983.
914. Alliata, D. et al., *5th Int. Conf. Nanometer-Scale Sci. Technol.* (NANO 5), Birmingham, UK, 1998.
915. Alliata, D. et al., *Electrochem. Commun.*, 1, 5, 1999.
916. Winter, M., Novak, P., and Monnier, A.J., *Electrochem. Soc.*, 145, 428, 1998.
917. Hooley, J.G., *Mater. Sci. Eng.*, 31, 17, 1977.
918. Shi, H. et al., *J. Electrochem. Soc.*, 143, 3466, 1996; Moret, R., *Intercalation of Layered Materials; NATO ASI Series*, Dresselhaus, M.S., Ed., Plenum Press, New York, 1986.
919. Besenhard, J.O. et al., *Synth. Met.*, 7, 185, 1983.
920. Rhodes, E., Proc. of the Fifth Int. Power Symp., Brighton, 1966, in *Power Sources 1966*, Collins, D.A., Ed., Pergamon Press, Oxford; Elmsford, NY, 1966.
921. Hathcock, K.W. et al., *Anal. Chem.*, 67, 2201, 1995.
922. Schonherr, H. et al., *J. Am. Chem. Soc.*, 122, 3679, 2000.
923. Bishop, A.R. and Nuzzo, R.G., *Curr. Opinion Colloid Interface Sci.*, 1, 127, 1996.
924. Dubois, L.H. and Nuzzo, R.G., *Annu. Rev. Phys. Chem.*, 43, 437, 1992.
925. Carpick, R.W. and Salmeron, M., *Chem. Rev.*, 97, 1163, 1997.
926. Lio, A., Charych, D.H., and Salmeron, M., *J. Phys. Chem.*, B, 101, 3800, 1997.
927. Rubinstein, I. et al., *Nature*, 332, 426, 1988.
928. Rubinstein, I. et al., *Nature*, 337, 217, 1989.

929. Turyan, I. and Mandler, D., *Anal. Chem.*, 66, 58, 1994.
930. Gafni, Y. et al., *Eur. J. Chem.*, 2, 759, 1996.
931. Kepley, L.J., Crooks, R.M., and Ricco, A.J., *Anal. Chem.*, 64, 3191, 1992.
932. Schierbaum, K.-D. et al., *Science*, 265, 1413, 1994.
933. Mrksich, M., Grunwell, J.R., and Whitesides, G.M., *J. Am. Chem. Soc.*, 117, 12009, 1995.
934. Mrksich, M., Sigal, G.B., and Whitesides, G.M., *Langmuir*, 11, 4383, 1995.
935. Huisman, B.-H. et al., *Adv. Mater.*, 8, 561, 1996.
936. Kumar, A. et al., *Acc. Chem. Res.*, 28, 219, 1995.
937. Kumar, A., Biebuyck, H.A., and Whitesides, G.M., *Langmuir*, 10, 1498, 1994.
938. Wilbur, J.L. et al., *Adv. Mater.*, 6, 600, 1994.
939. Kumar, A. and Whitesides, G.M., *Science*, 263, 60, 1994.
940. Jackman, J.R., Wilbur, J.L., and Whitesides, G.M., *Science*, 269, 664, 1995.
941. Xia, Y. and Whitesides, G.M., *Adv. Mater.*, 7, 471, 1995.
942. Briggs, D. and Seah, M.P., Eds., *Practical Surface Analysis*, Wiley Interscience, New York, 1992.
943. Popovitz-Biro, R., Lahav, M., and Milstein, D., *Science*, 278, 2100, 1997.
944. Neogi, P., Neogi, S., and Stirling, C.J.M., *J. Chem. Soc., Chem. Commun.*, 1134, 1993.
945. van Ryswyk, H. et al., *Langmuir*, 12, 6143, 1996.
946. Somorjai, G.A., *Chem. Rev.*, 96, 1223, 1996.
947. Touzov, I. and Gorman, C.B., *J. Phys. Chem.*, 101, 5263, 1997.
948. Werts, M.P.L., van der Vegte, E.W., and Hadziioannou, G., *Langmuir*, 13, 4939, 1997.
949. Schonherr, H., Hruska, Z., and Vancso, G.J., *Macromolecules*, 31, 3679, 1998.
950. Schonherr, H. and Vancso, G.J., *J. Polym. Sci. B, Polym. Phys.*, 36, 2486, 1998.
951. Gidalewitz, D. et al., *Angew. Chem., Int. Ed. Engl.*, 36, 955, 1997.
952. Pearce, R. and Vancso, G.J., *Macromolecules*, 30, 5843, 1997.
953. Kaupp, G. and Haak, M., *Angew. Chem., Int. Ed. Engl.*, 35, 2774, 1996.
954. Kautek, W., Dieluweit, S., and Sahre, M., *J. Phys. Chem.*, B, 101, 2709, 1997.
955. Wall, J.T., Grieser, F., and Zukoski, C.F., *J. Chem. Soc., Faraday Trans.*, 93, 4017, 1997.
956. Nelles, G. et al., *Appl. Phys.*, A, 66, 1261, 1998; Nelles, G. et al., *Langmuir*, 14, 546, 1998.
957. Bamberg, E., Ringsdorf, H., and Butt, H.-J., *Langmuir*, 14, 808, 1998.
958. Giessibl, F.J., *Science*, 267, 68, 1995.
959. Guthner, P., *J. Vac. Sci. Technol.*, B, 14, 2428, 1996.
960. Kitamura, S. and Iwatsuki, M., *Jpn. J. Appl. Phys.*, 34, 145, 1995.
961. Kitamura, S. and Iwatsuki, M., *Jpn. J. Appl. Phys.*, 35, 668, 1996.
962. Sugawara, Y. et al., *Science*, 270, 1646, 1995.
963. Bammerlin, M. et al., *Phys. Rev. Lett.*, 56, 838, 1986.
964. Guntherodt, H.-J., *Probe Microsc.*, 1, 3, 1997.
965. Alves, C.A., Smith, E.L., and Porter, M.D., *J. Am. Chem. Soc.*, 114, 1222, 1992.
966. Pan, J., Tao, N., and Lindsay, S.M., *Langmuir*, 9, 1556, 1993.
967. Alves, C.A. and Porter, M.D., *Langmuir*, 9, 3507, 1993.
968. Liu, G.-Y. and Salmeron, M., *Langmuir*, 10, 367, 1994.
969. Liu, G.-Y. et al., *J. Chem. Phys.*, 101, 4301, 1994.
970. Wolf, H. et al., *Phys. Chem.*, 99, 7102, 1995.
971. Jaschke, M. et al., *Phys. Chem.*, 100, 2290, 1996.
972. Schonherr, H. et al., *Langmuir*, 13, 1576, 1997.
973. Frisbie, C.D. et al., *Science*, 265, 2071, 1994.
974. Noy, A. et al., *J. Am. Chem. Soc.*, 117, 7943, 1995.

975. Vezenov, D.V. et al., *J. Am. Chem. Soc.*, 119, 2006, 1997.
976. Noy, A., Vezenov, D.V., and Lieber, C.M., *Annu. Rev. Mater. Sci.*, 27, 381, 1997.
977. Wong, S.S. et al., *Nature*, 394, 52, 1998; Wong, S.S. et al., *J. Am. Chem. Soc.*, 120, 8557, 1998; Cheung, C.L. et al., *J. Phys. Chem.*, B, 106, 2429, 2002; Lieber, C.M., *Nano Letters*, 2, 81, 2002.
978. Israelachvili, J.N., *Intermolecular and Surface Forces*, 2nd ed., Academic Press, New York, 1991; Israelachvili, J., *Intermolecular and Surface Forces*, Academic Press, New York, 1992.
979. Thomas, R.C. et al., *J. Phys. Chem.*, 98, 4493, 1994.
980. Green, J.-B.D. et al., *J. Phys. Chem.*, 99, 10960, 1995.
981. Akari, S. et al., *Adv. Mater.*, 7, 549, 1995.
982. Sinniah, S.K. et al., *J. Am. Chem. Soc.*, 118, 89251, 1996.
983. Schonherr, H. and Vancso, G.J., *Macromolecules*, 30, 6391, 1997; van der Vegte, E.W. and Hadziioannou, G., *Langmuir*, 13, 4357, 1997; McKendry, R. et al., *Nature*, 391, 566, 1998.
984. Duevel, R.V. and Corn, R.M., *Anal. Chem.*, 64, 337, 1992.
985. Zong, K. et al., *Synth. Commun.*, 27, 157, 1997.
986. Chechik, V. et al., *Langmuir*, 14, 3003, 1998.
987. Schonherr, H. et al., *Langmuir*, 15, 5541, 1999; Hansen, W.N., *J. Opt. Soc. Am.*, 58, 380, 1968.
988. Tortonese, M. and Kirk, M., *Proc. SPIE*, 3009, 53, 1997.
989. Nuzzo, R.G., Dubois, L.H., and Allara, D.L. *J. Am. Chem. Soc.*, 112, 558, 1990.
990. Engquist, I., Lestelius, M., and Liedberg, B., *Langmuir*, 13, 4003, 1997.
991. Porter, M.D. et al., *J. Am. Chem. Soc.*, 109, 3559, 1987.
992. Sondag, A.H.M., Tol, A.J.W., and Touwslager, F.J., *Langmuir*, 8, 1127, 1992.
993. Takami, T. et al., *Langmuir*, 11, 3876, 1995.
994. Schonherr, H. et al., *Langmuir*, 12, 3898, 1996.
995. Tsao, M.-W. et al., *Langmuir*, 16, 1734, 2000.
996. Harrison, L.G., in *Comprehensive Chemical Kinetics, Vol. 2, The Theory of Kinetics*, Bamford, C.H. and Tipper, C.F.H., Eds., Elsevier, Amsterdam; New York, 1969.
997. Wang, J. et al., *J. Am. Chem. Soc.*, 119, 12796, 1997.
998. Delamarche, E., *Langmuir*, 10, 2869, 1994.
999. Anguiano, E., Pancorbo, M., and Aguilar, M.J., *Microscopy*, 172, 223, 1993.
1000. Zhou, Y. et al., *Langmuir*, 14, 660, 1998.
1001. Omoike, A. et al., *Langmuir*, 14, 4731, 1998.
1002. Tian, F., Xiao, X., and Loy, M.M., *Langmuir*, 15, 244, 1999.
1003. Nunes, G. and Freeman, M.R., *Science*, 262, 1029, 1993.
1004. Lehmann, T. et al., *J. Microscopy*, 151, 277, 1988.
1005. Jones, L.A. and Thomas, D.F., *J. Vac. Sci. Technol.*, A, 10, 636, 1992; Bao, G.W., Troemel, M., and Li, S.F.Y., *Appl. Phys.*, A, 66, 1283, 1998.
1006. Takano, H. et al., *Chem. Rev.*, 99, 2845, 1999.
1007. Derjaguin, B.V., Muller, V.M., and Toporov, P.J., *J. Colloid Interface Sci.*, 53, 2, 1975; Johnson, K.L., Kendall, K., and Roberts, A.D., *Proc. R. Soc. London*, A, 324, 301, 1971.
1008. Clear, S.C. and Nealey, P.F., *J. Colloid Interface Sci.*, 213, 238, 1999.
1009. Sugawara, Y. et al., *Wear*, 168, 13, 1993.
1010. Skulason, H. and Frisbie, C.D., *Langmuir*, 16, 6294, 2000.
1011. Tsukruk, V.V. and Bliznyuk, V.N., *Langmuir*, 14, 446, 1998.
1012. Kidoaki, S. and Matsuda, T., *Langmuir*, 15, 7639, 1999.
1013. Ogletree, D.F., Carpick, R.W., and Salmeron, M., *Rev. Sci. Instrum.*, 67, 3298, 1996.
1014. Atamny, F., Baiker, A., and Atamny, F., *Surf. Sci.*, 323, 414, 1995.

Index

A

B

C

D

E

F

G

H

I

J

K

L

M

N

O

P

Q

R

S

W

X